AF454464

Crop Adaptation to Elevated CO$_2$ and Temperature

Crop Adaptation to Elevated CO$_2$ and Temperature

Editor

James Bunce

MDPI • Basel • Beijing • Wuhan • Barcelona • Belgrade • Manchester • Tokyo • Cluj • Tianjin

Editor
James Bunce
Adaptive Cropping Systems
Laboratory
USDA-ARS, Beltsville
Agricultural Research Center
Beltsville, MD
United States

Editorial Office
MDPI
St. Alban-Anlage 66
4052 Basel, Switzerland

This is a reprint of articles from the Special Issue published online in the open access journal *Plants* (ISSN 2223-7747) (available at: www.mdpi.com/journal/plants/special_issues/Crop_Temperature).

For citation purposes, cite each article independently as indicated on the article page online and as indicated below:

LastName, A.A.; LastName, B.B.; LastName, C.C. Article Title. *Journal Name* **Year**, *Volume Number*, Page Range.

ISBN 978-3-0365-3390-2 (Hbk)
ISBN 978-3-0365-3389-6 (PDF)

Contents

About the Editor

James Bunce

The editor served for 40 years in the USDA-Agricultural Research Service as a Plant Physiologist, retiring in 2016. He studied plant gas exchange, plant–water relationships, crop responses to water stress, elevated CO_2, high temperature stress, and crop adaptation to climate change.

plants

Editorial

Crop Adaptation to Elevated CO$_2$ and Temperature

James Bunce

Adaptive Cropping Systems Laboratory, USDA-ARS, Beltsville, MD 20705-2350, USA; buncejames49@gmail.com

Citation: Bunce, J. Crop Adaptation to Elevated CO$_2$ and Temperature. *Plants* **2022**, *11*, 453. https://doi.org/10.3390/plants11030453

Received: 21 January 2022
Accepted: 4 February 2022
Published: 7 February 2022

Publisher's Note: MDPI stays neutral with regard to jurisdictional claims in published maps and institutional affiliations.

There is no ambiguity about the fact that both atmospheric CO$_2$ levels and air temperatures are continuing to increase. It has only recently been recognized that the combination of these changes is likely to have a net negative impact on the production of many of our most important food crops. Therefore, concurrent with efforts to reduce emissions of CO$_2$ and other gases which warm the atmosphere, efforts should be made to adapt crops to the conditions of elevated CO$_2$ levels and temperature.

In response to the solicitation of articles on this topic, twelve articles have been published in this Special Issue of *Plants*, reflecting strong current research interest in this topic, as well as the diversity of relevant approaches.

Gavelienė et al. [1] tested the effects of warming on the root morphology of two species of lupine, one invasive and one noninvasive, and found that the two species had contrasting responses, which might affect their adaptation to climate warming.

Gardi et al. [2] examined growth and water use efficiency responses to elevated CO$_2$ levels among 15 landrace and 15 released lines of barley from Ethiopia and found a large diversity of responses, suggesting that genetic improvement should be feasible in this important crop species.

Marcos-Barbero et al. [3] screened sixty bread wheat genotypes for grain yield at elevated CO$_2$ and high-temperature conditions and found a large range of yields under those conditions, identifying genotypes that displayed promise of adaptation to climate change.

Jurkoniene et al. [4] examined the effects of warming on the IAA content and ethylene production of two lupine species with contrasting invasiveness and found more flexible responses in the invasive species.

Barickman et al. [5] compared the growth of basil at low, moderate, and high temperatures at ambient and elevated CO$_2$ levels and found that elevated CO$_2$ levels reduced photosynthesis at high temperatures but increased it at moderate and lower growth temperatures.

Ben Marium et al. [6] conducted a meta-analysis concerning the impacts of elevated CO$_2$ levels, elevated temperature, and drought on the yield and grain quality of cereals and found that the beneficial yield responses to elevated CO$_2$ levels were offset by both high temperatures and drought stress, with a general negative impact of elevated CO$_2$ levels on nutritional quality.

Chen and Setter [7] examined the responses of tuber formation in potato to elevated temperature and CO$_2$ levels and found that elevated CO$_2$ levels partly compensated for the inhibition of tuber growth caused by elevated temperatures and that high temperatures at tuber initiation were especially important in this species.

Jayawardena et al. [8] examined the responses of nitrogen uptake and metabolism to elevated CO$_2$ levels and temperature in tomato in great detail and found that the decreased nitrogen uptake and assimilation in response to the combined treatments probably resulted from decreased plant demand for nitrogen.

Bourgault et al. [9] tested the hypothesis that an elevated CO$_2$ level only increases root growth in topsoil, not at depth. They conducted a detailed root-growth analysis in a FACE experiment with lentil, and found that in some cases, root growth at depth also increased at elevated CO$_2$ levels.

Ma et al. [10] analyzed the response of sugar-metabolism-related genes to elevated CO_2 level treatment in the goji berry to provide a molecular explanation of the reduced sugar content of these fruits when plants are grown at elevated CO_2 levels.

Wang and Liu [11] provided a review of the effects of heat and elevated CO_2 levels on the yield and grain quality of wheat, one of the crops in which negative effects of climate change on grain quality were first noticed.

Ziska [12] reviewed data concerning whether newer crop varieties are better adapted than older ones to high CO_2 levels and suggested that examining the genetic responses of weedy relatives of crops to the changes in atmospheric CO_2 that have recently occurred may provide a useful source of genetic traits, which could improve the responses of crops to future CO_2 levels.

I hope that this compilation of research papers and reviews illustrates the broad range of relevant research on the topic of Crop Adaptation to Elevated CO_2 and Temperature and stimulates additional research on this critical topic.

Funding: This research received no external funding.

Institutional Review Board Statement: Not applicable.

Informed Consent Statement: Not applicable.

Data Availability Statement: Not applicable.

Conflicts of Interest: The authors declare no conflict of interest.

References

1. Gavelienė, V.; Jurkonienė, S.; Jankovska-Bortkevič, E.; Švegždienė, D. Effects of Elevated Temperature on Root System Development of Two Lupine Species. *Plants* **2022**, *11*, 192. [CrossRef] [PubMed]

2. Gardi, M.W.; Malik, W.A.; Haussmann, B.I.G. Impacts of Carbon Dioxide Enrichment on Landrace and Released Ethiopian Barley (*Hordeum vulgare* L.) Cultivars. *Plants* **2021**, *10*, 2691. [CrossRef] [PubMed]

3. Marcos-Barbero, E.L.; Pérez, P.; Martínez-Carrasco, R.; Arellano, J.B.; Morcuende, R. Screening for Higher Grain Yield and Biomass among Sixty Bread Wheat Genotypes Grown under Elevated CO_2 and High-Temperature Conditions. *Plants* **2021**, *10*, 1596. [CrossRef] [PubMed]

4. Jurkonienė, S.; Jankauskienė, J.; Mockevičiūtė, R.; Gavelienė, V.; Jankovska-Bortkevič, E.; Sergiev, I.; Todorova, D.; Anisimovienė, N. Elevated Temperature Induced Adaptive Responses of Two Lupine Species at Early Seedling Phase. *Plants* **2021**, *10*, 1091. [CrossRef] [PubMed]

5. Barickman, T.C.; Olorunwa, O.J.; Sehgal, A.; Walne, C.H.; Reddy, K.R.; Gao, W. Yield, Physiological Performance, and Phytochemistry of Basil (*Ocimum basilicum* L.) under Temperature Stress and Elevated CO_2 Concentrations. *Plants* **2021**, *10*, 1072. [CrossRef] [PubMed]

6. Ben Mariem, S.; Soba, D.; Zhou, B.; Loladze, I.; Morales, F.; Aranjuelo, I. Climate Change, Crop Yields, and Grain Quality of C_3 Cereals: A Meta-Analysis of [CO_2], Temperature, and Drought Effects. *Plants* **2021**, *10*, 1052. [CrossRef] [PubMed]

7. Chen, C.-T.; Setter, T.L. Role of Tuber Developmental Processes in Response of Potato to High Temperature and Elevated CO_2. *Plants* **2021**, *10*, 871. [CrossRef] [PubMed]

8. Jayawardena, D.M.; Heckathorn, S.A.; Rajanayake, K.K.; Boldt, J.K.; Isailovic, D. Elevated Carbon Dioxide and Chronic Warming Together Decrease Nitrogen Uptake Rate, Net Translocation, and Assimilation in Tomato. *Plants* **2021**, *10*, 722. [CrossRef] [PubMed]

9. Bourgault, M.; Tausz-Posch, S.; Greenwood, M.; Löw, M.; Henty, S.; Armstrong, R.D.; O'Leary, G.L.; Fitzgerald, G.J.; Tausz, M. Does Elevated [CO_2] Only Increase Root Growth in the Topsoil? A FACE Study with Lentil in a Semi-Arid Environment. *Plants* **2021**, *10*, 612. [CrossRef] [PubMed]

10. Ma, Y.; Devi, M.J.; Reddy, V.R.; Song, L.; Gao, H.; Cao, B. Cloning and Characterization of Three Sugar Metabolism Genes (*LBGAE*, *LBGALA*, and *LBMS*) Regulated in Response to Elevated CO_2 in Goji Berry (*Lycium barbarum* L.). *Plants* **2021**, *10*, 321. [CrossRef] [PubMed]

11. Wang, X.; Liu, F. Effects of Elevated CO_2 and Heat on Wheat Grain Quality. *Plants* **2021**, *10*, 1027. [CrossRef]

12. Ziska, L.H. Crop Adaptation: Weedy and Crop Wild Relatives as an Untapped Resource to Utilize Recent Increases in Atmospheric CO_2. *Plants* **2021**, *10*, 88. [CrossRef]

Article

Effects of Elevated Temperature on Root System Development of Two Lupine Species

Virgilija Gavelienė *, Sigita Jurkonienė *, Elžbieta Jankovska-Bortkevič and Danguolė Švegždienė

Nature Research Centre, Institute of Botany, Akademijos Str. 2, 08412 Vilnius, Lithuania;
elzbieta.jankovska@gamtc.lt (E.J.-B.); dangasv@gmail.com (D.Š.)
* Correspondence: virgilija.gaveliene@gmail.com (V.G.); sigita.jurkoniene@gamtc.lt (S.J.)

Abstract: The aim of this study was to assess the effect of elevated temperature on the growth, morphology and spatial orientation of lupine roots at the initial stages of development and on the formation of lupine root architecture at later stages. Two lupine species were studied—the invasive *Lupinus polyphyllus* Lindl. and the non-invasive *L. luteus* L. The plants were grown in climate chambers under 25 °C and simulated warming at 30 °C conditions. The angle of root curvature towards the vector of gravity was measured at the 48th hour of growth, and during a 4-h period after 90° reorientation. Root biometrical, histological measurements were carried out on 7-day-old and 30-day-old plants. The elevation of 5 °C affected root formation of the two lupine species differently. The initial roots of *L. polyphyllus* were characterized by worse spatial orientation, reduced growth and reduced mitotic index of root apical meristem at 30 °C compared with 25 °C. The length of primary roots of 30-day-old lupines and the number of lateral roots decreased by 14% and 16%, respectively. More intense root development and formation were observed in non-invasive *L. luteus* at 30 °C. Our results provide important information on the effect of elevated temperature on the formation of root architecture in two lupine species and suggest that global warming may impact the invasiveness of these species.

Keywords: gravitropic angle of curvature; initial root; invasiveness; lateral root number; primary root; root system architecture; simulated warming

check for
updates

Citation: Gavelienė, V.; Jurkonienė, S.; Jankovska-Bortkevič, E.; Švegždienė, D. Effects of Elevated Temperature on Root System Development of Two Lupine Species. *Plants* **2022**, *11*, 192. https://doi.org/10.3390/plants11020192

Academic Editor: James Bunce

Received: 22 December 2021
Accepted: 8 January 2022
Published: 12 January 2022

Publisher's Note: MDPI stays neutral with regard to jurisdictional claims in published maps and institutional affiliations.

1. Introduction

The world is experiencing ongoing global climate change, which can have serious consequences on plants, including changes in the availability of certain nutrients. For understanding the effects of climate warming on plant root systems, particularly their spatial distribution, it is essential to predict plant performance and community recovery in a warming climate. Compared with shoots, much less is known about how roots, especially root system architecture (RSA), may respond to elevated temperature. In addition, limited information is available on the specificities of the effects of elevated temperatures on the development of the root system in invasive plants. How does an increase in temperature change the intensity and the direction of root formation? To answer this question, researchers have compared the responses of plants with different RSAs in their studies [1,2]. The ability of a plant to take up nutrients is closely associated with the size and morphology of its root system [1,3]. Any changes in the growth or morphological modifications of root systems may provoke undesirable consequences in nutrient uptake [4]. It is recognized that many aspects of plant metabolism are accelerated by elevated temperatures [5,6]. Other environmental factors such as water, nutrients and temperature also have a strong influence on root structure [7]. Roots need an optimal temperature range to have a proper growth rate and function. In general, the optimal root temperature tends to be lower than the optimal shoot temperature [8,9]. It is evident that the optimum root temperature of plants varies depending on the species. Within this range, higher temperatures are generally associated with modified root-to-shoot ratios, while further increases in temperature would reduce root development and cause a change in RSA, thus reducing the root-to-shoot ratio [10]. For instance, some plants tend to produce more extensive root systems in elevated temperatures. An increase in temperature slows down lateral root growth in adult maize plants and promotes the development of long axial roots to reach deeper soil layers for water [11,12]. However, in potatoes, the initiation and elongation of adventitious and lateral roots were inhibited by increasing temperature. Another effect of warmer soil on potatoes is the swelling of the root cap meristem and the bending of the root tip. The alteration of root growth in these plants appeared due to a reduced rate of cell division [13,14]. Similarly, in sorghum, the high root zone temperature reduced the rate of root elongation and cell production rate [15]. The response of RSA to elevated temperature can be species-specific, as different species often have different optimum temperatures for root growth [16,17]. Literature data show that the effect of increasing temperature on root growth of plant seedlings can be promotive, inhibitive or first promotive then inhibitive after an optimum temperature is reached [18,19]. Even for species sharing the same habitat, their RSA can have species-specific responses to increased temperature [20]. Differences in the RSA of plant species may determine the intensity and direction of root formation in response to elevated temperatures. At high temperatures, the negative root response may have been intensified, with a competitive advantage going to species with larger and more rapidly forming roots.

Literature data indicate that greater root resilience plays a key role in plants adapting to high temperatures [21–23] in all stages of root development, including tropisms and the formation of new organs [24,25]. Furthermore, the oriented plant growth, which is collectively referred to as tropism [26,27], is influenced by various environmental factors, such as light, temperature, water and gravity. Gravitropism is an important tropic response that triggers asymmetric cell elongation in plant organs in response to gravity. It proceeds through three sequential steps: gravity perception, signal transduction and asymmetric cell elongation in the responding plant organs [28,29]. The roots grow downward, and the shoots grow upward, showing positive and negative gravitropic responses, respectively [30,31]. The well-known Cholodny-Went hypothesis illustrates that gravitropic stimuli result in differential cell elongation in the responding organs [32–35]. It has been shown that gravitropic perception occurs in the columella cells in the roots upon gravity stimulation [36,37]. The gravitropic response of plant organs is influenced by a variety of environmental signals. The best understood are the effects of light and temperature. Many scientists agree that climate change will alter habitat biodiversity and increase vulnerability to invasion. However, there is little information on the impact of potentially increasing global temperatures on the growth and development of alien plant species at the early stages of development. Moreover, one of the selected lupine species is invasive in Lithuania, *L. polyphyllus*, and there is very limited information on the specificities of the effect of elevated temperatures on the root system development of invasive plants. Therefore, in this research, RSA traits of seedlings of two lupine species (*L. polyphyllus* and *L. luteus*) with different spreading performances for understanding their responses to temperature change were studied. We hypothesized that increased temperature may differentially affect root growth, spatial orientation and root architecture of non-native plant species, thereby influencing them to become invasive. Studies on plant root system adaptive responses to altered temperature can provide the knowledge needed for the efficient management of invasive species. Thus, the goal of the current study was to investigate root growth, morphology and spatial orientation of two alien lupine species during the early growth stage at the elevated temperature.

2. Results

2.1. The Initial Root Growth at 25 °C and 30 °C

2.1.1. Angle of Curvature of Initial Roots at 25 °C and 30 °C

After 48 h of seedling growth, the spatial orientation and growth direction of the roots of both lupines depended on the temperature: the angle of curvature of the primary roots of the invasive *L. polyphyllus* with respect to the gravitational vector was 6.2° at 25 °C, and 20.8° at 30 °C. The initial roots of the non-invasive *L. luteus* showed a better orientation towards the gravity vector at 30 °C (Table 1, Figure 1).

Table 1. Influence of 25 °C and 30 °C temperatures on the angle of curvature of the initial roots of *L. polyphyllus* and *L. luteus* seedlings grown vertically for 48 h.

Plant Species	Lupinus polyphyllus		Lupinus luteus	
Temperature	25 °C	30 °C	25 °C	30 °C
Angle of curvature, degrees	6.2 ± 0.53 a	20.8 ± 0.95 b	14.2 ± 1.21 c	6.8 ± 0.43 a

Values presented are the mean values of four replications with standard deviation. Different lowercase letters indicate significant differences between test variants at $p < 0.05$.

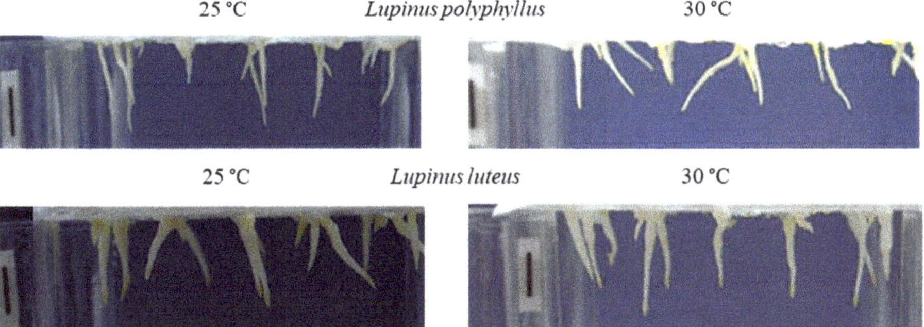

Figure 1. Spatial orientation of the initial roots of *L. polyphyllus* and *L. luteus* seedlings at 25 °C and 30 °C after 48 h. Scale bar, 10 mm.

2.1.2. Gravitropic Response of Initial Roots to 90° Reorientation

The strongest root response to gravitropic irritation in both lupine species was found to occur within the first hour. The gravitropic bending of *L. polyphyllus* roots after 1 h was 16° greater at 25 °C than at 30 °C. The gravitropic bending of *L. luteus* roots was more intensive at 30 °C. The gravitropic response of the roots of both lupine species to a 90° reorientation was closer to the direction of gravity after 4 h (Figure 2).

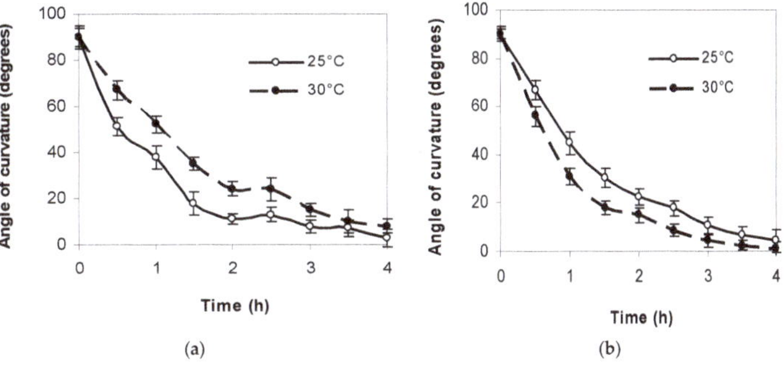

Figure 2. The dynamics of gravitropic response of *L. polyphyllus* (**a**) and *L. luteus* (**b**) roots to 90° reorientation at 25 °C and 30 °C.

2.1.3. Growth of Primary Roots of 7-Day-Old Seedlings at 25 °C and 30 °C

Morphometric studies showed that the length of roots of the invasive lupine grown at 30 °C for seven days was approximately 12% lower than that of the plants grown at 25 °C (Figure 3), while the roots of the non-invasive lupine grew up to 13% longer at 30 °C.

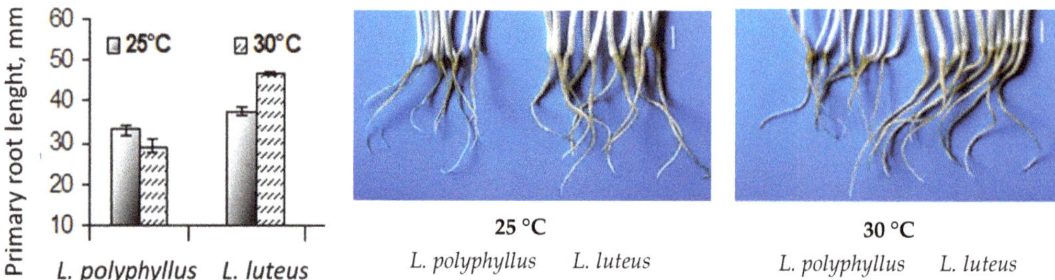

Figure 3. Effect of 25 °C and 30 °C temperature on root growth parameters of seven-day-old seedlings of two lupine species. Scale bar, 10 mm.

We found that the root-to-shoot ratio of both species decreased at 30 °C (Table 2). This index, in the case of *L. polyphyllus*, decreased crucially by 65% and in the case of *L. luteus* by 22%.

Table 2. Influence of 25 °C and 30 °C temperature on root-to-shoot ratio of the seven-day-old seedlings of *L. polyphyllus* and *L. luteus*.

Plant Species	*Lupinus polyphyllus*		*Lupinus luteus*	
Temperature	25 °C	30 °C	25 °C	30 °C
Root-to-shoot ratio	0.182 ± 0.03 a	0.063 ± 0.01 b	0.217 ± 0.03 a	0.169 ± 0.01 a

Values presented are the mean values of four replications with standard deviation. Different lowercase letters indicate significant differences between test variants, at $p < 0.05$.

2.1.4. Root Apex Development at 25 °C and 30 °C

Cytomorphological analysis of the root cap columella of *L. polyphyllus* showed that the length of the cells in the individual rows of the columella varied with temperature (Figure 4). From the seventh row of the columella onwards, cell length increased more at 25 °C than at 30 °C. The changes in cell length in the *L. luteus* columella were substantially different from that of *L. polyphyllus*. The cell length of the columella at 30 °C was greater starting from the second row onwards. This trend was observed in all the following rows.

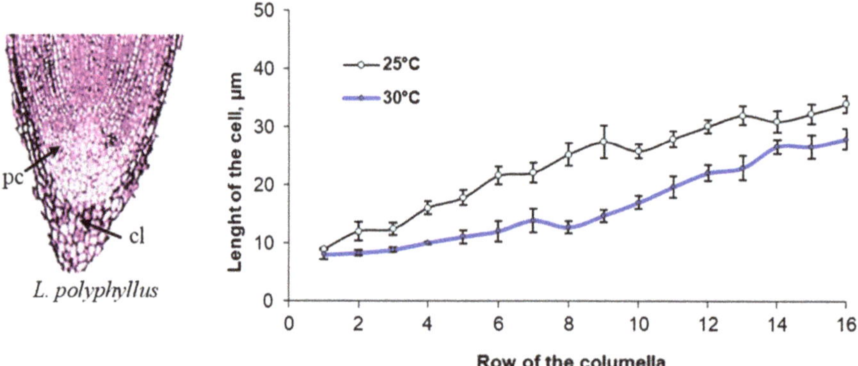

Figure 4. *Cont.*

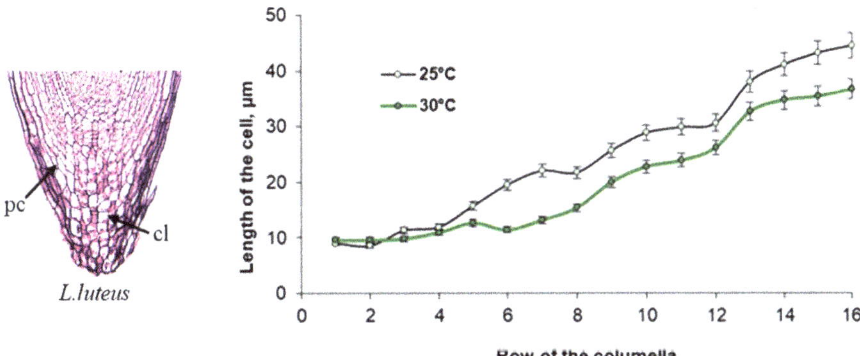

Figure 4. Impact of 25 °C and 30 °C temperature on the length of cells in the columella (cl) rows of the primary root cap (pc) (from the initial cells) of *L. polyphyllus* and *L. luteus*.

Determination of the cell division mitotic index (MI) in *L. polyphyllus* root apical meristem preparations showed that the cell MI value decreased by 12% in the test variant at 30 °C as compared with 25 °C (Figure 5a). Contrary, the calculation of MI in the non-invasive *L. luteus* root apical meristem indicated a significant increase at 30 °C. By observing the cross-sections of the invasive lupine root apex, we determined that meristem cells occurred in the prophase, metaphase and some even in the anaphase in the test variant at 25 °C, whereas most cells of plants grown at 30 °C were found in the prophase (Figure 5b).

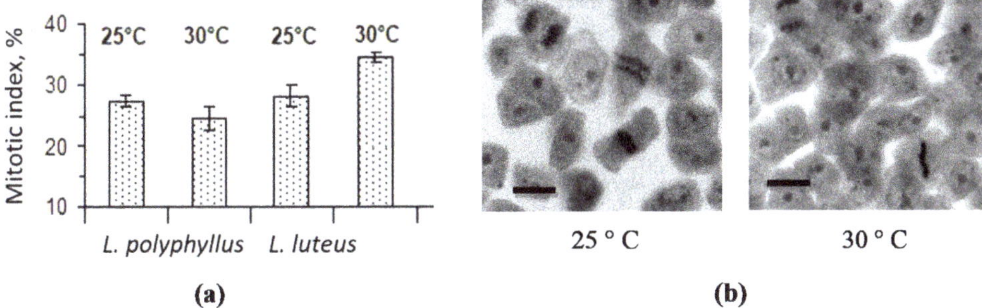

Figure 5. Effect of 25 °C and 30 °C temperatures on *L. polyphyllus* and *L. luteus* root apex meristem cells mitotic activity (**a**), fragments of primary root apical meristem pressed preparations from *L. polyphyllus* seedlings (**b**). Scale bar, 20 μm.

2.2. The Development of 30-Day-Old Lupine Roots at 25 °C and 30 °C

The data of the morphometric measurements showed that simulated 5 °C warming affected invasive *Lupinus polyphyllus* root formation:primary root length decreased by 14% and the number of lateral roots by 16%. The length of the primary root and the number of lateral roots of non-invasive *L. luteus* were higher at 30 °C (Figures 6 and 7).

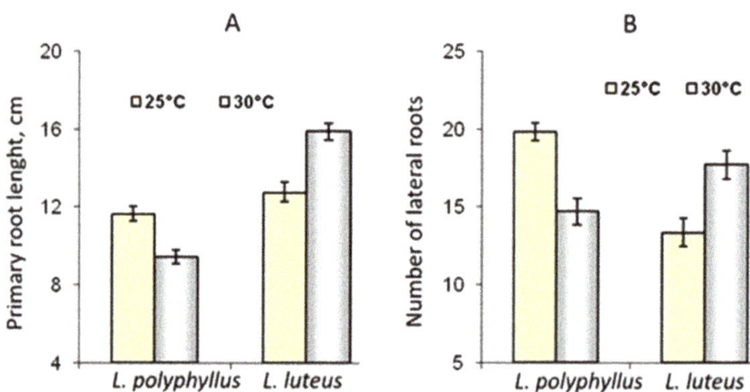

Figure 6. Effect of 25 °C and 30 °C temperature on the length of primary roots (**A**) and the number of lateral roots (**B**) of invasive *L. polyphyllus* and non-invasive *L. luteus* plants grown in soil for 30 days.

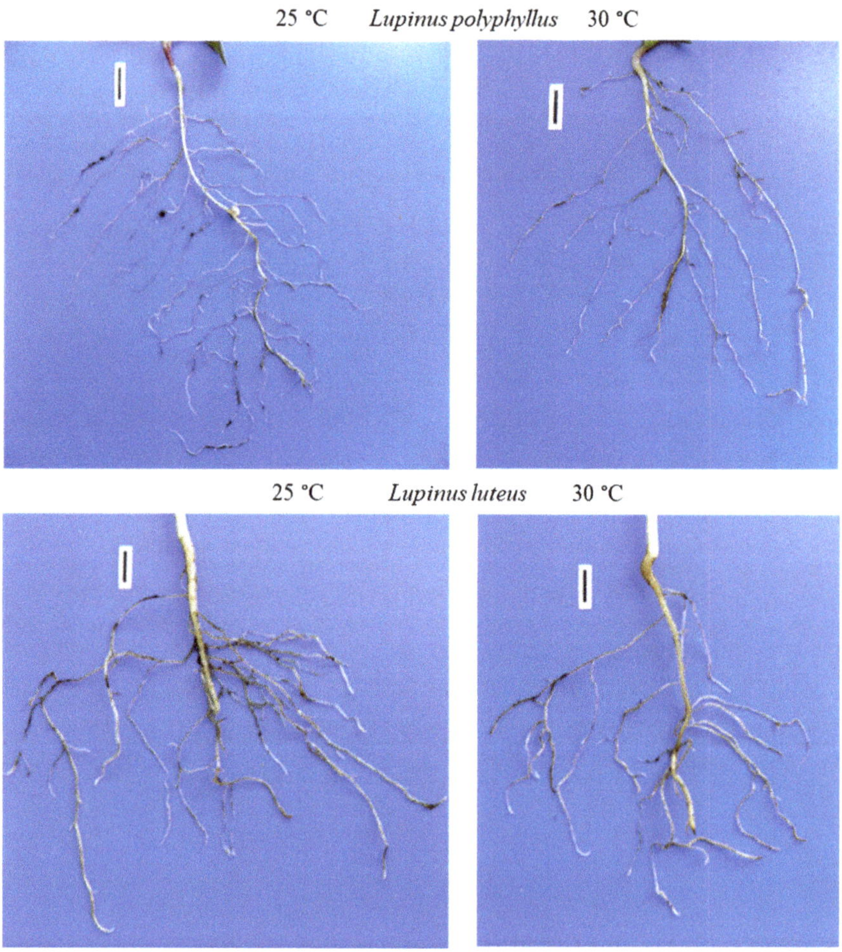

Figure 7. Roots of invasive *L. polyphyllus* and non-invasive *L. luteus* grown in the soil for 30 days. Scale bar, 10 mm.

3. Discussion

Temperature is one of the most important variables affecting plant growth. The effect of elevated temperature on aboveground plant parts has been well studied, while the effect on roots is less understood [2,17]. Roots need an optimal temperature range to grow and function properly. In general, the optimum root temperature is usually lower than the optimum shoot temperature. Literature shows that the effect of increasing temperature on root growth of plant seedlings can be either stimulatory, inhibitory or, once the optimum temperature is reached, initially stimulatory and then inhibitory [9,38,39]. In particular, it is important to study plants with different RSAs to understand the response of root development to temperature changes [1,2]. During seed germination, the growth of the seedling's primary root and its ability to orient itself in space (gravitropism) are critical characteristics for seedling establishment and survival [35]. To answer the question of whether there are differences in the ability of the primary root of the two lupine species to respond to gravity and whether elevated temperature influences this process, an analysis of the direction of root growth in relation to gravity was conducted. This study showed that after 48 h of growth, the spatial orientation and subsequent growth direction of both lupine roots depended on the ambient temperature:angle of curvature of the initial roots of the invasive *L. polyphyllus*, with respect to the gravitational vector, was 6.2° at 25 °C, and 20.8° at 30 °C. Thus, the initial roots of the non-invasive *L. luteus* were better oriented towards the Earth's gravity vector at 30 °C.

The vertical orientation of emerging roots is typically the first response of plants to gravity [40,41]. Sensing of the gravity stimulus ultimately triggers a signaling network orchestrated by the phytohormone auxin, which is key to the coordination of directional root growth in response to gravity [42–44]. Although root gravitropism has been studied extensively, no conclusive data on the onset of gravisensing is established. The inception of gravisensitivity in flowering plant roots after various periods of static orientation (gravistimulation) of imbibed seeds was studied [7,37,43]. Their results indicate that after gravistimulation (90° reorientation), gravitropic bending of flowering plant roots was established in 6 h along the gravity vector. These results well coincide with ours. Our data showed that after 1 h of gravistimulation (90° reorientation), the gravitropic bending of *L. polyphyllus* roots was 16° greater at 25 °C than at 30 °C; differently, the *L. luteus* roots response was more intensive at 30 °C. The gravitropic response of the initial roots of both lupine species to a 90° reorientation was closer to the direction of gravity after 4 h, both at 25 °C and at 30 °C. These data suggest that the initial roots of invasive lupines are less able to grow in the gravitational direction in a 5 °C warmer environment. Thus, dependence between the increase of environmental temperature and the inception of root gravitropic competence was determined. However, these parameters are not applicable to the description of RSA with complex geometry.

It has been shown that the effect of elevated temperature on the root growth of plant seedlings can be either activating or inhibiting in plants with a higher proportion of roots [2]. The morphometric tests carried out in this study showed that after seven days, the primary root growth of invasive lupines slowed down by 12% at 30 °C as compared to plants grown at 25 °C, while the root growth of non-invasive lupines accelerated by 13% at 30 °C. Elevated temperature is associated with a reduced root-to-shoot ratio, and a further increase in temperature limits root development and alters RSA [10]. We determined that the root-to-shoot ratio was reduced in both species at 30 °C; however, the roots of *L. luteus* were less sensitive to warming temperatures. It was obvious that this index, in the case of *L. polyphyllus*, decreased crucially.

Literature data indicate that the size of the root cap, the proportion of the columella in a root cap and meristem cell division were related to the growth of the roots [44,45]. It is known that the apical root growth correlated with the size of the columella and the number of cap cells in the plant root apex [46]. An increase in temperature promotes the initial growth of the roots of *Arabidopsis* seedlings and, at the same time, affects the elongation of columella cells [40]. In the current study, cytomorphological analysis of the root cap columella of *L. polyphyllus* showed that the length of the cells in the individual rows of the columella varied with temperature (Figure 4)—the increase of columella cell length was more intensive at 25 °C than at 30 °C. Nevertheless, *L. polyphyllus* cell length was greater at 30 °C already from the second row of the columella onwards. The apical meristem of roots provides cell regeneration, and the transition zone between the meristem and the cell extension zone enables the apex, directly or indirectly through the secondary signal, to sense changing environmental parameters and respond to changes in cell division [47,48]. Furthermore, anatomical-cytological analysis of apical meristem cells in the invasive lupine root apex showed that cell division was intense at a lower temperature. By observing the cross-sections of root apex, we determined that in the test variant at 25 °C, meristem cells occurred in the prophase, metaphase and some even in the anaphase, whereas most cells of plants grown at 30 °C were found in the prophase.

The data on differences in root size of *L. polyphyllus* and *L. luteus* resulting from temperature change suggests that the elevated temperature may be more difficult for invasive lupines to adapt to. The architecture of the root system is determined by the development of both primary and lateral roots [49,50]. The plant root system takes up water and dissolved nutrients from the soil; therefore, the size and extent of the root system have important implications for plant development [7,51]. Our results show that the two species of lupine seedlings grown in the soil for 30 days responded differently to changes in temperature. The most significant changes were observed in root length and lateral root formation. Plants of the invasive lupine had a larger root system at 25 °C, and the root size of non-invasive lupine generally increased at 30 °C. Under the elevated temperature, non-invasive plants produced more extensive root systems.

Our results provide key information concerning the elevated temperature on the formation of root architecture of two lupine species and suggest that the elevated temperature affects species invasiveness. In the early stages of growth (after 48 h), the spatial orientation of the initial roots of both lupines depended on the temperature—the angle of curvature of the initial roots of *L. polyphyllus* was closer to the gravity vector than *L. luteus* at 25 °C. The initial roots of the non-invasive *L. luteus* showed a better orientation towards the gravity vector at 30 °C. These processes were important for the subsequent formation of root architecture—the dynamics of gravitropic response of *L. polyphyllus* and *L. luteus* initial roots to 90° reorientation showed that the gravitropic bending of *L. luteus* roots was more intensive at 30 °C. Simulated warming (5 °C) affected *L. polyphyllus* root formation as the initial roots were characterized by disrupted gravitropic orientation to the gravity vector; the cell division mitotic index (MI) of root apical meristem decreased by 12% at 30 °C as compared with 25 °C. The temperature of 30 °C triggered the non-invasive *L. luteus* root development, formation and spatial orientation, both in the initial and later stages of development. After 30 days of growth, seedlings of the two lupine species responded differently to elevated temperature—the invasive lupine formed a larger root system at 25 °C, and the non-invasive lupine root size increased at 30 °C. Bearing in mind that global warming tends to enhance species invasiveness and the northward spread, among other issues, these findings provide important information on the effect of increased temperature on the formation of plant root architecture and suggest that elevated temperature alters the invasiveness of alien species due to changes in root architecture.

4. Materials and Methods

4.1. The Initial Root Growth at 25 °C and 30 °C

Two different lupine species—invasive *L. polyphyllus* and non-invasive *L. luteus* [52]—seeds were harvested in a natural environment in Lithuania and used as plant material. Seeds were soaked for 5 h in tap water at room temperature and then germinated in climate chambers (Climacell, Czech Republic) at 90% relative humidity in the dark at two different temperatures: at 25 °C (optimal temperature for lupine) and 30 °C (simulated climate warming temperature) [53]. For root system architecture exploration, seeds were sown in 7 cm diameter pots containing a mixture of vegetable compost 90%, peat 9%, ash of deciduous trees 1% and fertilizer NPK and grown 30 days in growth chambers with 12 h light/dark photoperiod, at 25 °C and 30 °C.

4.2. The Measurement of the Angle of Root Curvature

The assessment of root-growth patterns is based on the measurement of angular deviation of the root tip from the vertical axis. For assay of roots gravitropic response, 30 soaked seeds of both lupines were planted in gaps in transparent plexiglass boxes filled with distilled water so that protruding roots could grow freely downwards, i.e., towards the action of the gravitational force. The seedlings grew in the germinators at 25° and 30 °C, and relative humidity of 90% in the dark. The angle of root curvature towards the vector of gravity was measured at the 48th hour of growth.

4.3. Determination of Gravitropic Response of Roots to 90° Reorientation

Seeds were germinated on wet filter paper for 21 h and then planted on a sterile control medium (1% agar [w/v]) in square Petri dishes. The seeds with initial roots were fastened by agar and oriented so that roots could orient freely along the agar surface for 24 h in a vertical orientation. The dynamics of root curvature as an angle towards the gravitropic vector were measured at the 4-h period of the reorientation in a 90-degree position.

4.4. Morphometrical Tests

Measurements of root length and root-to-shoot ratio were performed on 7-day-old seedlings grown in tap water in the dark at 25 °C and 30 °C. The length of the primary root and the number of lateral roots of the two species of lupine were measured after 30 days of growth at 25 °C and 30 °C in soil.

4.5. Anatomical-Cytometrical Analysis of Primary Root Development
4.5.1. Cytometrical Investigations

Primary roots were excised from roots of 10 seedlings (7-day-old). The prepared samples were fixed in a formalin:acetic acid:ethanol (1:1:20) (FAA) mixture, dehydrated in a graded ethanol series, embedded in paraffin and cut with a rotary microtome Leica RM2125 into 10–15 μm sections. Serial longitudinal sections were stained with periodic acid-Schiff's reagent, and the length of statocytes in the columella rows of the root cap were measured with a light microscope and a digital video camera (Olympus) (DP-11). The images were analyzed using the SigmaScan Pro (Jandel Scientific Software) program.

4.5.2. Determination of Mitotic Index

For estimation of primary root apical meristem cells' mitotic activity, the roots were fixed in acetic acid:ethanol mixture (1:3). After 4 days of fixation, roots were washed from the mixture; the apical meristem zone was excised and dyed with acetocarmine, whereas cell walls were macerated with chloral hydrate [54]. In temporary squash preparations by a light microscope (Nikon Eclipse 80i), 6 cell mitoses phases were counted and mitotic index (MI) calculated. MI—cell number in mitosis per 1000 cells of the analyzed object (expressed in per mille $^o/_{oo}$). MI = (M/N) 1000, where M—number of mitoses, N—cell number. For each variant, 20 primary root apical meristems were analyzed.

4.6. Statistical Analysis

Tests were provided with three biological replicates. For morphometrical measurements, roots of 40 seedlings were analyzed for each variant. The data presented are mean values ± standard deviation of three experiments with four replicates in each. The data were statistically examined using analysis of variance (ANOVA) and tested for significant mean differences ($p < 0.05$) using Tukey's test.

5. Conclusions

Elevated temperature impacted the formation of root architecture of two lupine species while influencing their invasiveness.

During the early stages of growth, the spatial orientation of the initial roots was temperature-dependent: at 25 °C, the angle of curvature of the initial roots of *L. polyphyllus* was closer to the gravity vector than that at 30 °C, while *L. luteus* were better oriented towards the gravity vector at 30 °C.

The dynamics of the gravitropic response of initial roots to 90° reorientation confirmed that the gravitropic bending of *L. luteus* roots was more intense at 30 °C; meanwhile, *L. polyphyllus* was at 25 °C.

The simulated warming (5 °C) had an effect on *L. polyphyllus* root formation: the mitotic index of cell division in the root apical meristem was reduced by 12% at 30 °C compared to 25 °C.

After 30 days of cultivation at different temperatures, the root system of the invasive lupine was better developed at 25 °C, whereas the root size of the non-invasive lupine increased at 30 °C.

The current study provides important information on the effect of elevated temperature on the formation of plant root architecture and suggests that global warming is altering the invasiveness of alien species through changes in root architecture.

Author Contributions: Conceptualization, S.J. and V.G.; methodology, D.Š. and V.G.; formal analysis, S.J. and E.J.-B.; investigation, V.G. and D.Š.; data curation, S.J. and V.G.; writing—original draft preparation, V.G.; visualization, D.Š., writing—review and editing, E.J.-B. and S.J.; supervision, S.J. All authors have read and agreed to the published version of the manuscript.

Funding: This research was supported by the Nature Research Centre R&D III programme.

Data Availability Statement: The data supporting reported results can be found in scientific reports of the Laboratory of Plant Physiology of Institute of Botany of Nature Research Centre, where archived datasets generated during the study are included.

Acknowledgments: We thank the staff of the Laboratory of Plant Physiology of the Nature Research Centre for support and help provided.

Conflicts of Interest: The authors declare no conflict of interest.

References

1. Alexander, J.M.; Diez, J.M.; Levine, J.M. Novel competitors shape species' responses to climate change. *Nature* **2015**, *525*, 515–518. [CrossRef]
2. Luo, H.; Xu, H.; Chu, C.; He, F.; Fang, S. High temperature can change root system architecture and intensify root interactions of plant seedlings. *Front. Plant Sci.* **2020**, *11*, 160. [CrossRef] [PubMed]
3. Lynch, J. Root architecture and plant productivity. *Plant Phys.* **1995**, *109*, 7–13. [CrossRef] [PubMed]
4. Rubinigg, M.; Wenisch, J.; Elzenga, J.T.M.; Stulen, I. NaCI salinity affects lateral root development in *Plantago maritima*. *Func. Plant Biol.* **2004**, *3*, 775–780. [CrossRef] [PubMed]
5. Radville, L.; McCormack, M.L.; Post, E.; Eissenstat, D.M. Root phenology in a changing climate. *J. Exp. Bot.* **2016**, *67*, 3617–3628. [CrossRef]
6. Hayes, S.; Schachtschabel, J.; Mishkind, M.; Munnik, T.; Arisz, S.A. Hot topic: Thermosensing in plants. *Plant Cell Environ.* **2020**, *44*, 2018–2033. [CrossRef]
7. Ma, Z.; Chang, S.X.; Bork, E.W.; Steinaker, D.F.; Wilson, S.D.; White, S.R. Climate change and defoliation interact to affect root length across northern temperate grasslands. *Funct. Ecol.* **2020**, *34*, 2611–2621. [CrossRef]
8. Koevoets, I.T.; Venema, J.H.; Elzenga, J.T.M.; Testerink, C. Roots withstanding their environment: Exploiting root system architecture responses to abiotic stress to improve crop tolerance. *Front. Plant Sci.* **2016**, *7*, 1335. [CrossRef]
9. De Lima, C.F.F.; Kleine-Vehn, J.; de Smet, I.; Feraru, E. Getting to the root of belowground high temperature responses in plants. *J. Exp. Bot.* **2021**, *72*, 7404–7413.
10. Ribeiro, P.R.; Fernandez, L.G.; de Castro, R.D.; Ligterink, W.; Hilhorst, H.W. Physiological and biochemical responses of *Ricinus communis* seedlings to different temperatures: A metabolomics approach. *BMC Plant Biol.* **2014**, *14*, 223. [CrossRef]
11. Hund, A.; Fracheboud, Y.; Soldati, A.; Stamp, P. Cold tolerance of maize seedlings as determined by root morphology and photosynthetic traits. *Eur. J. Agron.* **2008**, *28*, 178–185. [CrossRef]

12. Calleja-Cabrera, J.; Boter, M.; Oñate-Sánchez, L.; Pernas, M. Root growth adaptation to climate change in crops. *Front. Plant Sci.* **2020**, *11*, 544. [CrossRef] [PubMed]

13. Sattelmacher, B.; Marschner, H.; Kühne, R. Effects of the temperature of the rooting zone on the growth and development of roots of potato (*Solanum tuberosum*). *Ann. Bot.* **1990**, *65*, 27–36. [CrossRef]

14. Joshi, M.; Fogelman, E.; Belausov, E.; Ginzberg, I. Potato root system development and factors that determine its architecture. *J. Plant Physiol.* **2016**, *205*, 113–123. [CrossRef]

15. Pardales, J.R.; Kono, Y.; Yamauchi, A. Epidermal cell elongation in sorghum seminal roots exposed to high root-zone temperature. *Plant Sci.* **1992**, *81*, 143–146. [CrossRef]

16. Rogers, E.D.; Benfey, P.N. Regulation of plant root system architecture: Implications for crop advancement. *Curr. Opin. Biotechnol.* **2015**, *32*, 93–98. [CrossRef]

17. Gray, S.B.; Brady, S.M. Plant developmental responses to climate change. *Dev. Biol.* **2016**, *419*, 64–77. [CrossRef]

18. Forbes, P.; Black, K.; Hooker, J. Temperature-induced alteration to root longevity in *Lolium perenne*. *Plant Soil.* **1997**, *190*, 87–90. [CrossRef]

19. Seiler, G.J. Influence of temperature on primary and lateral root growth of sunflower seedlings. *Environ. Exp. Bot.* **1998**, *40*, 135–146. [CrossRef]

20. Bardgett, R.D.; Mommer, L.; de Vries, F.T. Going underground: Root traits as drivers of ecosystem processes. *Trends Ecol. Evol.* **2014**, *29*, 692–699. [CrossRef]

21. Nagel, K.A.; Kastenholz, B.; Jahnke, S.; van Dusschoten, D.; Aach, T.; Mühlich, M.; Truhn, D.; Scharr, H.; Terjung, S.; Walter, A.; et al. Temperature responses of roots: Impact on growth, root system architecture and implications for phenotyping. *Funct. Plant Biol.* **2009**, *36*, 947. [CrossRef]

22. Aidoo, M.K.; Bdolach, E.; Fait, A.; Lazarovitch, N.; Rachmilevitch, S. Tolerance to high soil temperature in foxtail millet (*Setaria italica* L.) is related to shoot and root growth and metabolism. *Plant Physiol. Biochem.* **2016**, *106*, 73–81. [CrossRef]

23. Wani, S.H.; Kumar, V.; Saroj, V.S.; Sah, K. Phytohormones and their metabolic engineering for abiotic stress tolerance in crop plants. *Crop J.* **2016**, *4*, 162–176. [CrossRef]

24. Lucas, M.; Godin, C.; Jay-Allemand, C.; Laplaze, L. Auxin fluxes in the root apex co-regulate gravitropism and lateral root initiation. *J. Exp. Bot.* **2008**, *59*, 55–66. [CrossRef]

25. Miyazawa, Y.; Takahashi, H. Molecular mechanisms mediating root hydrotropism: What we have observed since the rediscovery of hydrotropism. *J. Plant Res.* **2020**, *133*, 3–14. [CrossRef]

26. Morita, M.T. Directional gravity sensing in gravitropism. *Annu. Rev. Plant Biol.* **2010**, *61*, 705–720. [CrossRef] [PubMed]

27. Muthert, L.W.F.; Izzo, L.G.; van Zanten, M.; Aronne, G. Root tropisms: Investigations on Earth and in space to unravel plant growth direction. *Front. Plant Sci.* **2020**, *10*, 1807. [CrossRef] [PubMed]

28. Moulia, B.; Fournier, M. The power and control of gravitropic movements in plants: A biomechanical and systems biology view. *J. Exp. Bot.* **2009**, *60*, 461–486. [CrossRef] [PubMed]

29. Taniguchi, M.; Furutani, M.; Nishimura, T.; Nakamura, M.; Fushita, T.; Iijima, K.; Baba, K.; Tanaka, H.; Toyota, M.; Tasaka, M.; et al. The Arabidopsis LAZY1 family plays a key role in gravity signalling within statocytes and in branch angle control of roots and shoots. *Plant Cell* **2017**, *29*, 1984–1999. [CrossRef] [PubMed]

30. Morita, M.T.; Tasaka, M. Gravity sensing and signaling. *Curr. Opin. Plant Biol.* **2004**, *7*, 712–718. [CrossRef]

31. Leitz, G.; Kang, B.H.; Schoenwaelder, M.E.; Staehelin, L.A. Statolith sedimentation kinetics and force transduction to the cortical endoplasmic reticulum in gravity-sensing *Arabidopsis* columella cells. *Plant Cell* **2009**, *21*, 843–860. [CrossRef]

32. Strohm, A.K.; Baldwin, K.L.; Masson, P.H. Molecular mechanisms of root gravity sensing and signal transduction. *Wiley Interdiscip. Rev. Dev. Biol.* **2012**, *1*, 276–285. [CrossRef]

33. Sato, E.M.; Hijazi, H.; Bennett, M.J.; Vissenberg, K.; Swarup, R. New insights into root gravitropic signalling. *J. Exp. Bot.* **2015**, *66*, 2155–2165. [CrossRef] [PubMed]

34. Masson, P.H.; Tasaka, M.; Morita, M.T.; Guan, C.; Chen, R.; Boonsirichai, K. *Arabidopsis thaliana*: A model for the study of root and shoot gravitropism. *Arab. Book* **2002**, *1*, e0043. [CrossRef]

35. Nakamura, M.; Nishimura, T.; Morita, M.T. Gravity sensing and signal conversion in plant gravitropism. *J. Exp. Bot.* **2019**, *70*, 3495–3506. [CrossRef]

36. Friml, J.; Wisniewska, J.; Benkova, E.; Mendgen, K.; Palme, K. Lateral relocation of auxin efflux regulator PIN3 mediates tropism in *Arabidopsis*. *Nature* **2002**, *415*, 806–809. [CrossRef]

37. Ma, Z.; Hasenstein, K.H. The onset of gravisensitivity in the embryonic root of flax. *Plant Physiol.* **2006**, *140*, 159–166. [CrossRef] [PubMed]

38. Domisch, T.; Finér, L.; Lehto, T.; Smolander, A. Effect of soil temperature on nutrient allocation and mycorrhizas in Scots pine seedlings. *Plant Soil* **2002**, *239*, 173–185. [CrossRef]

39. Martins, S.; Montiel-Jorda, A.; Cayrel, A.; Huguet, S.; Roux, C.P.; Ljung, K.; Vert, G. Brassinosteroid 720 signaling-dependent root responses to prolonged elevated ambient temperature. *Nat. Commun.* **2017**, *8*, 309. [CrossRef] [PubMed]

40. Merkys, A.I. *Geotropic Reaction of Plants*, 1st ed.; Mintis: Vilnius, Lithuania, 1973; p. 264. (In Russian)

41. Su, S.H.; Gibbs, N.M.; Jancewicz, A.L.; Masson, P.H. Molecular mechanisms of root gravitropism. *Curr. Biol.* **2017**, *27*, R964–R972. [CrossRef] [PubMed]

42. Leyser, O. Auxin signaling. *Plant Physiol.* **2018**, *176*, 465–479. [CrossRef] [PubMed]

43. Küpers, J.J.; Oskam, L.; Pierik, R. Photoreceptors regulate plant developmental plasticity through auxin. *Plants* **2020**, *9*, 940. [CrossRef] [PubMed]
44. Zhang, Y.; Xiao, G.; Wang, X.; Zhang, X.; Friml, J. Evolution of fast root gravitropism in seed plants. *Nat. Commun.* **2019**, *10*, 3480. [CrossRef] [PubMed]
45. Pilet, P.E. Importance of the cap cells in maize root gravireaction. *Planta* **1982**, *156*, 95–96. [CrossRef]
46. Ransom, J.S.; Moore, R. Geoperception in primary and lateral roots of *Phaseolus vulgaris* (*Fabaceae*). I. Structure of columella cells. *Am. J. Bot.* **1983**, *70*, 1048–1056. [CrossRef] [PubMed]
47. Kawata, S.; Katano, M.; Yamazaki, K. The growing directions and the geotropic responses of rice crown roots. *Jpn. J. Crop Sci.* **1980**, *49*, 301–310. [CrossRef]
48. Verbelen, J.P.; De Cnodder, T.; Le, J.; Vissenberg, K.; Baluška, F. The root apex of *Arabidopsis thaliana* consists of four distinct zones of cellular activities: Meristematic zone, transition zone, fast elongation zone, and growth terminating zone. *Plant Signal. Behav.* **2006**, *1*, 296–304. [CrossRef]
49. Casimiro, I.; Marohant, A.; Bhalero, R.P.; Beeckmann, T.; Dhoope, S.; Swarup, R.; Graham, N.; Inze, D.; Sandberg, G.; Casero, P.J.; et al. Auxin transport promotes *Arabidopsis* lateral root initiation. *Plant Cell* **2001**, *13*, 843–852. [CrossRef]
50. Overvoorde, P.; Fukaki, H.; Beeckman, T. Auxin control of root development. *Cold Spring Harb. Perspect. Biol.* **2010**, *2*, a001537. [CrossRef]
51. Wu, Q.; Pagès, L.; Wu, J. Relationships between root diameter, root length and root branching along lateral roots in adult, field-grown maize. *Ann. Bot.* **2016**, *117*, 379–390. [CrossRef] [PubMed]
52. CABI's Invasive Species Compendium. Available online: https://www.cabi.org/isc/datasheet/31710 (accessed on 5 January 2022).
53. Wang, R.L.; Zeng, R.S.; Peng, S.L.; Chen, B.M.; Liang, X.T.; Xin, X.W. Elevated temperature may accelerate invasive expansion of the liana plant *Ipomoea cairica*. *Weed Res.* **2011**, *51*, 574–580. [CrossRef]
54. Paulauskas, A.; Slopšytė, G.; Morkūnas, V. *Methods and Practicals of General Genetic Investigations*, 1st ed.; Infrorastras: Vilnius, Lithuaniai, 2003; pp. 23–30. (In Lithuania)

Article

Impacts of Carbon Dioxide Enrichment on Landrace and Released Ethiopian Barley (*Hordeum vulgare* L.) Cultivars

Mekides Woldegiorgis Gardi [1,*], Waqas Ahmed Malik [2] and Bettina I. G. Haussmann [3]

[1] Institute of Cropping Systems and Modelling, University of Hohenheim, Fruwirthstrasse 14, 70599 Stuttgart, Germany

[2] Institute of Biostatistics, University of Hohenheim, 70599 Stuttgart, Germany; W.Malik@uni-hohenheim.de

[3] Institute of Plant Breeding, Seed Science and Population Genetics, University of Hohenheim, 70599 Stuttgart, Germany; bettina.haussmann@uni-hohenheim.de

[*] Correspondence: mekides.gardi@uni-hohenheim.de; Tel.: +49-17-7773-4262

Abstract: Barley (*Hordeum vulgare* L.) is an important food security crop due to its high-stress tolerance. This study explored the effects of CO_2 enrichment (eCO_2) on the growth, yield, and water-use efficiency of Ethiopian barley cultivars (15 landraces, 15 released). Cultivars were grown under two levels of CO_2 concentration (400 and 550 ppm) in climate chambers, and each level was replicated three times. A significant positive effect of eCO_2 enrichment was observed on plant height by 9.5 and 6.7%, vegetative biomass by 7.6 and 9.4%, and grain yield by 34.1 and 40.6% in landraces and released cultivars, respectively. The observed increment of grain yield mainly resulted from the significant positive effect of eCO_2 on grain number per plant. The water-use efficiency of vegetative biomass and grain yield significantly increased by 7.9 and 33.3% in landraces, with 9.5 and 42.9% improvement in released cultivars, respectively. Pearson's correlation analysis revealed positive relationships between grain yield and grain number ($r = 0.95$), harvest index ($r = 0.86$), and ear biomass ($r = 0.85$). The response of barley to eCO_2 was cultivar dependent, i.e., the highest grain yield response to eCO_2 was observed for *Lan_15* (122.3%) and *Rel_10* (140.2%). However, *Lan_13*, *Land_14*, and *Rel_3* showed reduced grain yield by 16, 25, and 42%, respectively, in response to eCO_2 enrichment. While the released cultivars benefited more from higher levels of CO_2 in relative terms, some landraces displayed better actual values. Under future climate conditions, i.e., future CO_2 concentrations, grain yield production could benefit from the promotion of landrace and released cultivars with higher grain numbers and higher levels of water-use efficiency of the grain. The superior cultivars that were identified in the present study represent valuable genetic resources for future barley breeding.

Keywords: barley; biomass; CO_2 enrichment; *Hordeum vulgare* L.; water-use efficiency; yield

Citation: Gardi, M.W.; Malik, W.A.; Haussmann, B.I.G. Impacts of Carbon Dioxide Enrichment on Landrace and Released Ethiopian Barley (*Hordeum vulgare* L.) Cultivars. *Plants* **2021**, *10*, 2691. https://doi.org/10.3390/plants10122691

Academic Editor: James Bunce

Received: 20 November 2021
Accepted: 4 December 2021
Published: 7 December 2021

Publisher's Note: MDPI stays neutral with regard to jurisdictional claims in published maps and institutional affiliations.

check for
updates

1. Introduction

The global demand for food crops is increasing and may continue to do so for decades. A 70−100% increase in the cereal food supply by 2050 is required to feed the predicted world population of over nine billion people [1]. In terms of production and consumption, barley (*Hordeum vulgare* L.) is one of the most important cereal crops in the world following wheat, maize, and rice. It is cultivated both in highly productive agricultural systems and at the subsistence level in marginal environments [2]. Ethiopia is the second-largest barley producer in Africa, accounting for nearly 25% of the total production [3]. It has been cultivated in Ethiopia for the last 5000 years and accounts for 8% of the total cereal production in the country [4]. In the 2017/18 growing season, the national area coverage was 975,300 ha, with the production and productivity values of barley being approximately 2.1 million tons and 2.17 tons ha^{-1}, respectively [3]. It is grown at elevations from 1500 to over 3500 m above sea level (m.a.s.l) and is predominantly cultivated between 2000 and 3000 m.a.s.l. [5,6].

Ethiopian barley germplasm has been used internationally as a source of useful genes due to its improved traits, including improved protein quality and disease and drought tolerance [5,7]. Long-term geographic isolation and adaptation to diverse climatic conditions and soil types resulted in a high level of variation between cultivars [8]. The crop is primarily used as a type of food and beverage in more than 20 different ways, which reflects its cultural and nutritional importance [9]. Despite its importance and morphological variations, one key challenge in barley breeding is the issue of developing cultivars that can face the challenges of changing climatic conditions [10]. Changes in the global atmospheric CO_2 concentration constitute one of the most important and well-known examples of global climate change. The current increase in CO_2 will likely continue into future decades and may bring the concentration close to 550 ppm by 2050 [11,12]. Elevated CO_2 (eCO_2) levels are known to have positive effects on photosynthetic processes, and consequently, on plant growth in C3 plant species, mainly through the modification of water and nutrient turnover [13–15]. Thus, as CO_2 is fundamental for plant production, understanding cultivar behavior and the targeted exploitation of this resource via plant breeding could optimize yields and contribute to future food security [16–18].

Several CO_2-enrichment studies regarding major cereal species, i.e., barley [19–21], wheat [22,23], and rice [24], reported substantial intraspecific variation between cultivars regarding plant growth and yield in response to eCO_2 enrichment. In contrast, another study regarding different cultivars of wheat reported non-significant intraspecific variation in yield responses [25]. To the best of our knowledge, no information is currently available regarding the response of Ethiopian barley cultivars to eCO_2. Therefore, the present study aimed to evaluate the growth, yield formation, and water-use efficiency response of Ethiopian barley cultivars under current and future CO_2 concentrations.

2. Results

2.1. Plant Height and Biomass Allocation Pattern

Significant impacts caused by CO_2 enrichment were observed for several yield variables in both the landrace and released cultivars, except in the variables of leaf biomass fraction, the number of ears per plant, and thousand-grain weight. The interaction between CO_2 and the cultivars also had a significant effect on most of the yield variables (Table 1). The average plant height of the landrace and released cultivars in the ambient CO_2 (aCO_2) condition were 101.9 and 94.5 cm, respectively (Table 1). The effect of CO_2 enrichment was observed in the variable of plant height, with an increase of 7.6% in landraces and 6.7% in released cultivars (Figure 1). The average vegetative biomass of the landrace was 35.6 g dry weight per plant in the aCO_2 condition (Table 1), while the released cultivars had 39.4 g dry weight per plant (Table 1). Significant increases in vegetative biomass, by 7.6 and 9.4%, respectively, were recorded across the landrace and released cultivars in the eCO_2 condition (Figure 1). The increase observed in vegetative biomass was mainly due to the significant effect of eCO_2 on the stem biomass in both the landrace and released cultivars (Table 1). As shown in Figure 2, a negative correlation between vegetative biomass and grain yield ($r = -0.51$, $p < 0.05$) as well as harvest index ($r = -0.85$, $p < 0.001$) was observed.

2.2. Grain Yield Parameters

Grain yield and its parameters were significantly affected by genotype/cultivars, CO_2 treatment, and their interaction in both the landrace and released cultivars (Table 1). The average grain yield of the landrace was 8.1 g dry weight per plant, resulting from 13.8 ears and 146 grains per plant. On the other hand, the released cultivars had a grain yield of 6.7 g dry weight per plant from 12.8 ears and 134 grains per plant, on average, under the aCO_2 conditions. Increases in the grain yield of the landrace and released cultivars, by 34.1 and 40.6%, respectively, were recorded under the eCO_2 condition (Table 1 and Figure 1). All yield components contributed significantly to the increase in grain yield, except for the number of ears. The number of grains per plant showed the largest increase of 32.2% in the landrace and 31.3% in the released cultivars (Table 1 and Figure 1). In accordance with this, the harvest index increased by 14.3% (landraces) and 23.3% (released cultivars) in the eCO_2 condition. The eCO_2 condition was recorded to have a significant effect on thousand-grain weight for the released cultivars; the thousand-grain weight increased by 10.4% on average, while the change was not significant in the landrace (Figure 1).

Table 1. Analysis of variance results. Mean and standard error (S.E.) of phenological parameters of landrace (Gen) and released cultivars (Cul) under ambient and elevated CO_2 conditions, as well as their interactions.

Variables	Cultivar	aCO_2	eCO_2	S.E.	Δ %	CO_2	Gen/Cul	$CO_2 \times$ Gen/Cul
Plant height (cm)	Landrace	101.9	109.6	3.8	7.6	***	*	*
	Released	94.5	100.8	3.8	6.7	***	***	ns
Vegetative biomass (g plant^{-1})	Landrace	35.6	38.3	2.0	7.6	***	***	***
	Released	39.4	43.1	2.0	9.4	***	***	***
Stem biomass (g plant^{-1})	Landrace	19.3	21.4	1.4	10.9	***	***	***
	Released	21.5	23.7	1.3	10.2	***	***	***
Leaf biomass (g plant^{-1})	Landrace	11.2	11.4	0.7	1.8	ns	***	***
	Released	12.6	13.3	0.7	5.6	***	**	*
Ear biomass (g plant^{-1})	Landrace	13.2	16.5	1.8	25.0	***	***	ns
	Released	11.9	15.6	1.8	31.1	***	**	ns
Chaff (awn) biomass (g plant^{-1})	Landrace	5.1	5.5	1.6	9.0	**	***	***
	Released	5.2	6.1	1.2	17.6	***	***	***
Number of ears (plant^{-1})	Landrace	13.8	15.2	1.8	10.2	ns	**	ns
	Released	12.8	13.4	2.2	4.7	ns	*	ns
Number of grain (plant^{-1})	Landrace	146.0	193.0	30.9	32.2	***	***	***
	Released	134.0	176.0	34.2	31.3	***	***	ns
Grain yield (g plant^{-1})	Landrace	8.1	10.9	1.7	34.1	***	***	***
	Released	6.7	9.42	1.7	40.6	***	***	**
Thousand-grain weight (g)	Landrace	54.5	56.2	3.1	3.1	ns	***	ns
	Released	49.2	54.3	5.6	10.4	*	**	ns
Harvest index	Landrace	0.21	0.24	0.03	14.3	**	***	ns
	Released	0.16	0.20	0.03	23.3	***	***	***
Total water use (WU_T, L plant^{-1})	Landrace	9.2	9.1	0.1	−1.1	*	***	***
	Released	9.3	9.3	0.1	0.0	ns	***	***
Water-use efficiency of vegetative biomass (WUE_B, g L^{-1})	Landrace	3.8	4.1	0.1	7.9	***	***	***
	Released	4.2	4.6	0.1	9.5	***	***	***
Water-use efficiency of grains (WUE_G, g L^{-1})	Landrace	0.9	1.2	0.1	33.3	***	***	ns
	Released	0.7	1.0	0.1	42.9	***	***	ns

Significance level: $p < 0.001$ (***); $p < 0.01$ (**); $p < 0.05$ (*); and non-significant (ns).

In Figure 2, the correlation analysis revealed that grain yield had a positive and strong association with the number of grains ($r = 0.95$, $p < 0.001$), ear biomass ($r = 0.91$, $p < 0.001$) and harvest index ($r = 0.86$, $p < 0.001$). In addition, the performance of the genotypes/cultivars regarding the response of grain yield under the aCO_2 condition versus the eCO_2 condition had a significant and positive correlation in the landrace ($r = 0.64$, $p = 0.01$) and released cultivars ($r = 0.93$, $p < 0.001$), as shown in Figure 3. Among the landrace cultivars, *Lan_15* displayed the highest yield, while *Lan_7* displayed the lowest yield under both the ambient and elevated CO_2 conditions. Comparing the released cultivars, the highest grain yield was recorded for *Rel_4*, and *Rel_10* had the lowest yield. Moreover, a strong and positive correlation of cultivars was recorded for grain number per plant under the aCO_2 condition versus the eCO_2 condition (Figure 3); however, the best genotypes under aCO_2 were not always the best genotypes under eCO_2 in terms of both number of grains and grain yield.

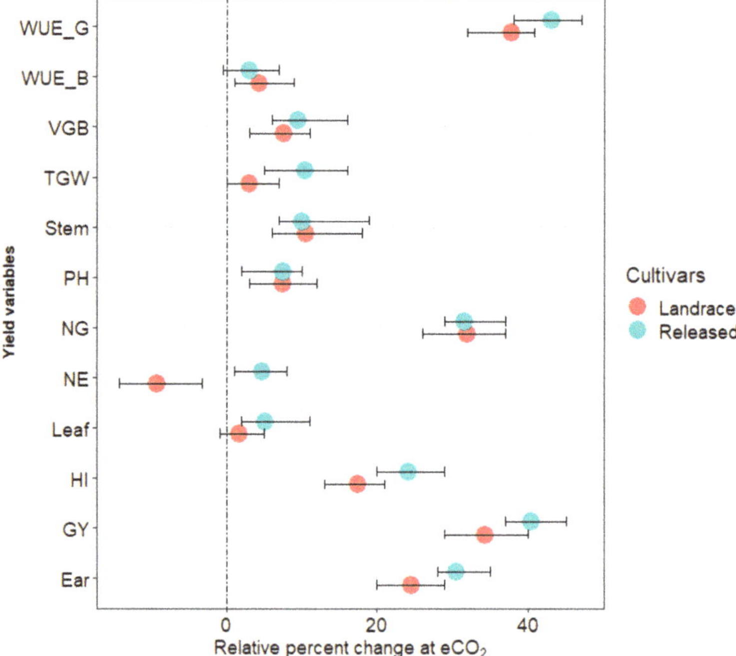

Figure 1. Relative effect of eCO$_2$ condition on plant height, biomass fractions, yield components, and water-use efficiency of barley. Average relative changes due to CO$_2$ enrichment against aCO$_2$ are presented, with error bars representing their standard errors. Ear: ear biomass; GY: grain weight; HI: harvest index; Leaf: leaf biomass; NE: number of ears; NG: grain number; PH: plant height; Stem: stem biomass; TGW: thousand-grain weight; VGB: vegetative biomass; WUE_B: water-use efficiency of vegetative biomass; and WUE_G: water-use efficiency of grain.

2.3. Water-Use Efficiency

The variables of water-use efficiency of vegetative biomass (WUE_B) and grain (WUE_G) were significantly affected by the CO$_2$ condition and type of cultivar ($p < 0.001$), as shown in Table 1. However, their interaction did not affect the response of total water use in both the landrace and released cultivars. In the aCO$_2$ condition, the landrace cultivar used 9.2 L plant^{-1} of WU_T, and had 4.7 g L^{-1}, WUE_B, and 0.9 g L^{-1} WUE_G (Table 1). On the other hand, the released cultivars used 9.3 L plant^{-1} of WU_T and had 4.9 g L^{-1} WUE_B, and 0.7 g L^{-1} WUE_G (Table 1). The levels of total water consumption of water by the landrace and released cultivars were not significantly different under the different CO$_2$ levels. The effect of CO$_2$ enrichment was higher in the response of WUE_G than WUE_B. WUE_G was increased by 33.3% in landraces and 42.9% in the released cultivars (Table 1 and Figure 1). In comparison, *Lan_15* and *Rel_4* showed the highest WUE_G among the landrace and released cultivars, respectively, while the lowest WUE_G was observed in *Lan_6*, *Lan_7*, and *Rel_10* (Figure 4a,b).

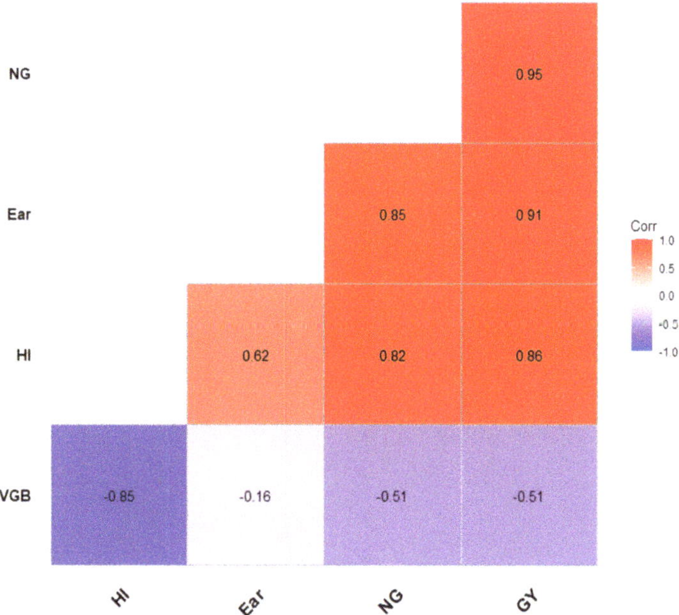

Figure 2. Correlation between grain yield, yield parameters, and water-use efficiency. VGB: vegetative biomass; Ear: ear biomass; NG: number of grains; GY: grain yield; WUE_G: water-use efficiency of grains; HI: harvest index. The value shows Pearson's correlation coefficient. The minus sign indicates a negative correlation between the variables.

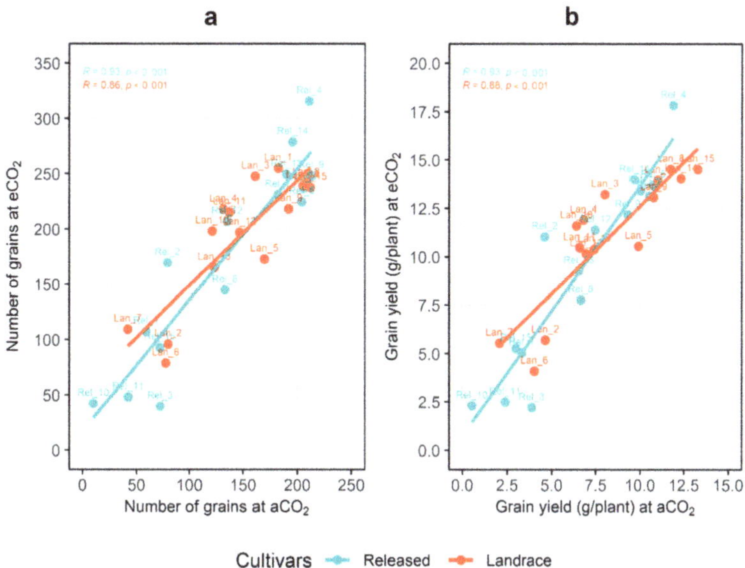

Figure 3. Mean response of landrace and released cultivars under elevated (500 ppm) CO_2 plotted against mean response under ambient (400 ppm) CO_2, where responses refer to (**a**) number of grains per plant and (**b**) grain yield (in grams) per plant.

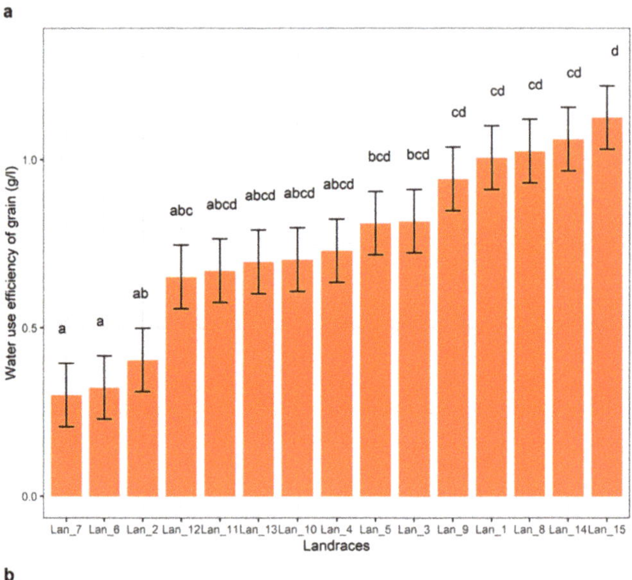

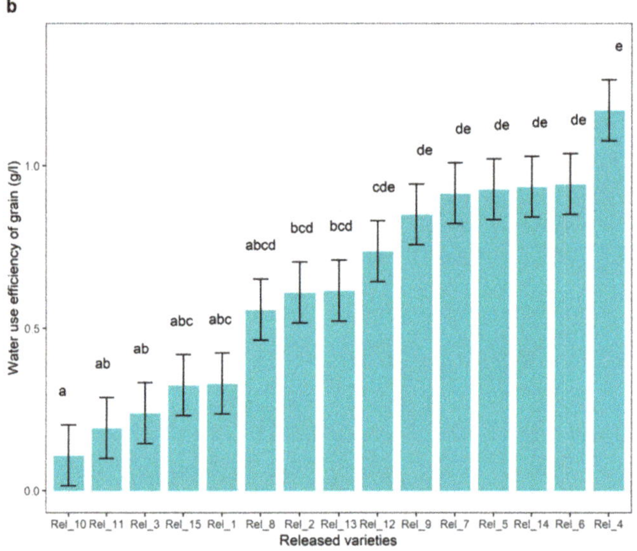

Figure 4. Mean response of landrace (**a**) and released cultivars (**b**) regarding the water-use efficiency of grains (WUE_G, g L^{-1}). The letters indicate the significant level between genotypes/cultivars. Mean values sharing a letter are not significantly different.

3. Discussion

3.1. The Overall Effect of eCO_2 on Vegetative Biomass, Grain Yield, and Water-Use Efficency

Atmospheric CO_2 enrichment is expected to contribute to the required increase in grain yield production in the future [15,26,27]. Our findings from the climate chamber experiment, where the eCO_2 condition was applied as a single factor, correspond well with findings in previously published data. In the present study, on average, vegetative biomass was increased by 7.6% in landraces and 9.4% in the released cultivars, respectively. The enhancement was predominantly due to higher biomass allocation towards ear and stem biomass. The eCO_2 condition was observed to have a significant effect on the response of leaf biomass in the released cultivars alone. In line with the present results, findings from CO_2 enrichment studies regarding barley reported the significant enhancement of vegetative biomass due to higher CO_2 concentrations [28–30]. A previous study [15] summarized the biomass response of the C3 species and reported an average enhancement of vegetative biomass by 16% under eCO_2 conditions. Comparable results were also reported regarding other C3 crops, such as wheat [31] and rice [26].

In the present study, the released cultivars had a higher relative grain yield increase (40.6%) under the eCO_2 condition as compared to the landraces (34.1%). This supports the hypothesis that enhanced net-photosynthesis in eCO_2 conditions was unconsciously targeted through breeding. However, surprisingly, the landrace group had higher actual grain yield production levels under both the aCO_2 and eCO_2 conditions. In support of this finding [28], grain yield was determined via grain number per plant and ear biomass, which indicates that CO_2 enrichment and the acquisition of extra carbon were carried forward to the grains rather than the biomass yield. Previous studies regarding barley [19,20] and wheat [32,33] reported the positive correlation of grain yield with grain number. In the current study, an average enhancement of thousand-grain weight by 10.4% due to eCO_2 conditions was recorded in the released cultivars, whereas the response was not significantly affected in the landraces. In line with our findings, a study regarding wheat reported an enhancement of thousand-grain weight by 3.8–7.0% [34]; on the other hand, a non-significant effect of eCO_2 conditions on the thousand-grain weight of barley and wheat was reported in other studies [20,35]. The effects of eCO_2 conditions on the harvest index have been reviewed in rice, wheat, and soybean, with contradictory results. In the present study, the harvest index was increased by 23.3 and 14.3% in the released and landrace cultivars, respectively, under the eCO_2 condition. Similarly, in [27], a significant increase in the harvest index was also displayed in rice under eCO_2 conditions, which was contrary to a decrease in harvest indexes related to soybean and wheat [26,36]. The actual grain yield of landrace observed in the present study was higher compared to that of the released cultivars; however, the positive effect of eCO_2 was greater in the released cultivars. Accordingly, the relative percentage change of the harvest index was observed to be higher for the released cultivars compared to the landraces. Our finding supports the effort of breeding to reduce the percentage of vegetative biomass to increase the harvest index of crops, which is in line with the findings of [17].

As CO_2 levels rise above the current ambient level, photosynthesis is commonly enhanced and transpiration is frequently reduced, resulting in greater water efficiency and increased plant growth and productivity [37]. In the present study, a significant improvement regarding WUE_G was displayed. Average enhancements in the values of WUE_G by 33.3 and 42.9% were observed in the landrace and released cultivars, respectively, under the eCO_2 condition. In agreement with these findings, previous studies reported that eCO_2 conditions had a significant effect on the WUE_G and WUE_B values of barley and other crops. For instance, a study regarding two barley cultivars reported a significant enhancement of water-use efficiency of vegetative biomass and grain under well-watered conditions [17]. Furthermore, increases in WUE values by 20% under well-watered and by 42% under drought conditions, due to the presence of eCO_2, were reported [29]. Regarding wheat, the authors of [38,39] reported a significant enhancement of WUE_B and WUE_G values due to high eCO_2 conditions. On the other hand, the author of [40] revealed a clear reduction in the water consumption of barley under eCO_2 conditions. The current study, as well as several previous studies, revealed that eCO_2 conditions cause increases in water-use efficiency values by increasing growth and yield more so than by increasing water consumption. This would be beneficial for use in future food production, especially in water-limited areas.

3.2. Cultivar Specific Responses to eCO_2 on Barley Production

In this study, a wide range of intraspecific variation was observed in the responses of the measured yield parameters to the eCO_2 condition, from negative to large increments.

The response of grain yield to the eCO_2 condition ranged from -25% (*Lan_14*) to $+122.3\%$ (*Lan_15)* in the landraces, while the released cultivars showed a 42% reduction in grain yield (*Rel_3*) to an increment of 140.2% (*Rel_10*) under the eCO_2 condition. High grain yield and stability were found among landraces and the released cultivars. The landraces originated and were grown in different altitudes, indicating that suitable resources for climate resilience are available from different areas. The highest yielding landraces were *Lan_15, Lan_8, Lan_1, Lan_9*, and *Lan_6* under both the aCO_2 and eCO_2 conditions. The highest yielding landraces were grown in various parts of Ethiopia between 1642 and 3570 m.a.s.l, indicating the diversity and potential of choosing cultivars for future climate conditions. On the other hand, the highest yielding released cultivars were *Rel_4, Rel_5, Rel_6, Rel_7*, and *Rel_10*, which were characterized by early maturation, high yields, and resistance to lodging and leaf diseases (*Pyrenophora teres* and *Rhynchosporium secalis*). As shown in our findings, CO_2 enrichment studies regarding different barley cultivars reported a significant variation among cultivars in the response of grain yield and its parameters [20,21,41]. The greater enhancement of ear biomass per plant and improvement regarding WUE_G values significantly contributed to the observed grain yield gain in the highest yielding cultivars. In line with these findings, several studies have reported that barley yield responses to eCO_2 conditions are mostly cultivar dependent [19,23,42]. Studies involving other C3 crops have also reported significant differences between cultivars tested in future climate change scenarios. Variations in the responses to eCO_2 conditions in rice cultivars, for example, have been recorded, ranging from a 31% yield reduction to a 41% yield gain [24,43]. Similarly, significant variation in yield response under eCO_2 conditions, ranging from 20 to 80%, was observed in soybean cultivars [44]. Further variations in yield response were observed in other studies, with yield gains of between 31 and 41% being found [24,43]. As has been seen in previous studies, in the present study, negative growth effects of eCO_2 were observed regarding vegetative biomass and grain yield. The negative yield responses may partly be associated with alterations in the shoot: root carbon allocation between the cultivars examined. Previous studies reported positive root growth effects in barley via eCO_2 conditions [45,46]. Cultivars with negative vegetative biomass accumulation under eCO_2 were allotted newly assimilated carbon, but this would preferentially take place below the ground level for the enhanced development of their root systems at the expense of the vegetative biomass [21]. A review of different experiments conducted under eCO_2 conditions listed 13 C3-plant species that exhibited reductions in vegetative biomass by up to 42% [47]. A set of more than 100 spring barley cultivars grown under eCO_2 conditions yielded negative responses comparable to the current findings [48]. In general, studies on C3 crops indicate that intraspecific yield variations under eCO_2 conditions are primarily related to changes in carbon allocation within cultivars, rather than physiological traits related to carbon assimilation [45,46]. The current study, as well as other similar studies, have found a wide range of eCO_2 responsiveness in some of the world's most important food crops, implying that selecting for eCO_2 responsiveness may ensure long-term productivity under eCO_2 conditions [18,26,49,50]. The *Lan_15, Lan_8, Lan_1, Lan_9*, and *Lan_6* variants among the landraces and the *Rel_4, Rel_5, Rel_6, Rel_7*, and *Rel_10* variants among the released cultivars are the top five highest-yielding variants due to improved grain number values under the eCO_2 condition. They represent important genetic resources for use in future barley breeding programs. Despite the overall positive correlation of genotypes/cultivars, the best genotypes under aCO_2 might not always be the best genotypes under eCO_2; thus, direct selection under eCO_2 is needed to identify the best varieties for future climates.

4. Materials and Methods

4.1. Genetic Material and CO_2 Enrichment

Thirty Ethiopian barley cultivars consisting of 15 landraces and 15 released cultivars were obtained from Holetta Agricultural Research Centre (HARC) in Ethiopia. The landraces represent dominant barley landraces that are cultivated in different parts of

Ethiopia. The released cultivars were chosen based on their diversity regarding adaptation and genetic background. They were released from 1975, are grown in different parts of the country, and differ in their traits such as grain yield (Figure A1, Tables 2 and A1). The cultivars were cultivated in six identical climate chambers (Vötsch BioLine, Balingen, Germany) in which the climatic variables could be controlled. To mimic a realistic seasonal climate within the climate chambers, the daily temperature and relative humidity mean of Holeta from the period 2008–2018, and which are registered at World Weather Online (https://www.worldweatheronline.com, accessed on 12 January 2019), was used. In total, 27 weekly climate profiles were derived from these 10-year time series, representing the main growing season in Ethiopia. The day length (12 h) and the daily temperatures (daily mean of the coldest week: 8 °C; daily mean of the warmest week: 25 °C) were adapted. The CO_2 concentration within the chambers did not follow any time course but was set to constant values of 400 ppm in three chambers (ambient concentration, aCO_2) and 550 ppm in another three chambers (elevated concentration, eCO_2).

4.2. Plant Cultivation and Measurement of Plant-Related Parameters

The polyvinyl chloride pots used in the experiment were 40 cm in height and 10.3 cm in diameter, with a total volume of 3.33 L and a surface area of 83.33 cm^2. These pots were filled with 3.3 kg of sand and standard soil (Fruhstorfer Erde LD80, Hawita GmbH, Vechta, Germany) with a 2:1 ratio. The standard soil, LD80, comprised 50% peat, 35% volcanic clay, and 15% bark humus, and it was enriched with slow-releasing fertilizers. The pH ($CaCl_2$) of the medium was 5.9, the organic matter content was 35% (loss-on-ignition), and the salt content was 1 g L^{-1} KCl. The nutrient availability of the LD80 standard medium was (mg L^{-1}) 150 N, 150 P$_2$O$_5$, and 250 K$_2$O. Per cultivar, five seeds were grown and thinned at the seedling stage in two experimental plants per pot. Once a week, pots and CO_2 treatments were rotated between chambers to avoid any potential chamber effects. Plants were watered with 500 mL at the beginning of the experiment and were regularly watered throughout with an adequate amount to avoid drought. Pots were weighed once a week and adjusted to a weight of 5 kg to monitor differences in the water consumption of plants from different CO_2 treatments over time. The total water consumption ranged between 8.6 and 9.7 L in landraces and between 8.7 and 9.8 L in released cultivars. The values of total water use (WU_T, Equation (1)), water-use efficiency of vegetative biomass (WUE_B, Equation (2)), and water-use efficiency of grain yield (WUE_G, Equation (3)) were calculated.

When the plants reached full maturity, plant height and total pot weight were measured before harvesting. Afterward, plants were harvested and separated into the vegetative biomass fractions (leaves, stems, and reproductive organs/ears). The single plant fractions were oven-dried at 30 °C (reproductive organs/ears) and 60 °C (stems and leaves) until they reached a constant weight before their dry weight was determined. The share to which single plant fractions contributed to total plant biomass was calculated and given as leaf, stem, and ear dry matter weight per plant. Grains were removed from the ears by manual threshing to determine the total grain yield, thousand-grain weight, and grain number, as well as the harvest index per plant.

$$WU_T = \frac{Total\ water\ applied\ (L)}{Plant} \tag{1}$$

$$WUE_B = \frac{Biomass\ yield\ (g)}{Total\ water\ applied\ (L)} \tag{2}$$

$$WUE_G = \frac{Grain\ yield\ (g)}{Total\ water\ applied\ (L)} \tag{3}$$

4.3. Statistical Analyses

The experiment was conducted using a randomized split-plot design with three replicates per CO_2 treatment level; the CO_2 treatment level was used as the main plot factor. The two levels of CO_2 were randomly assigned to a climate chamber, and cultivars were randomly placed in a climate chamber. Once a week, pots and CO_2 treatments were rotated between chambers to avoid any potential chamber effects. Following the experimental design, a two-way analysis of variance (ANOVA) was applied to test the significance of the main effects of genotype/cultivar and CO_2 treatments, as well as their interactions regarding both the landrace and released cultivars. In addition, the main effects of altitude and its interaction with CO_2 levels were analyzed regarding the landrace. Means were separated using Tukey HSD post hoc tests. Pearson's correlation coefficients were calculated to compare response variables and the performance of cultivars under the aCO_2 condition versus the eCO_2 condition. All the analyses were performed using the R programming language, version 4.0.1 [51].

5. Conclusions

Elevated CO_2 is beneficial to barley growth, yield, and water-use efficiency. The present study evaluated thirty Ethiopian barley cultivars and showed that eCO_2 levels provoke a significant enhancement of vegetative biomass and grain yield values. In comparison, grain yield was much more responsive to the eCO_2 condition than vegetative biomass, mainly due to a significant enhancement of the ear biomass value, grain number, and harvest index. The water-use efficiency of vegetative biomass and the water-use efficiency of grain was enhanced in future climate condition. The grain yield gain was positively associated with the high grain number and water-use efficiency of grain per plant. On average, the released cultivars benefited more from CO_2 fertilization than the landraces. However, a wide range of intraspecific variation was observed within the responses of biomass and grain yield parameters across both the landrace and released cultivars. For instance, the cultivars *Lan_15* and *Rel_4* were the highest yielding variants among the landrace and released cultivars, respectively, under the current and future CO_2 levels and represent important genetic resources for use in the future barley breeding in Ethiopia. The investigation of the interaction between cultivar types and the environment could help to better understand the thresholds for cultivars' performance under climate change conditions. Grain yield production under future climate conditions could benefit from the identification of cultivars with higher grain numbers and more efficient water use in grain. However, food security involves more than just production. Further attention is required regarding the investigation of the nutritional quality of barley cultivars under eCO_2 conditions. Moreover, the growth and stress tolerance values of Ethiopian barley cultivars in response to the interactive effects of eCO_2 conditions, warming, and drought should be examined in order to achieve better exploitation of germplasm resources under changing climatic conditions.

Author Contributions: M.W.G. conceived and designed the study; data collection was carried out by M.W.G.; M.W.G. and W.A.M. participated in the analysis of the data; M.W.G. wrote the paper, with substantial input from B.I.G.H. and W.A.M. All authors have read and agreed to the published version of the manuscript.

Funding: This publication is an output of a Ph.D. scholarship from the University of Hohenheim in the framework of the project "German-Ethiopian SDG Graduate School: Climate Change Effects on Food Security (CLIFOOD)" between the University of Hohenheim (Germany) and the Hawassa University (Ethiopia), supported by the DAAD with funds from the Federal Ministry for Economic Cooperation and Development (BMZ). Funding grant number 57316245.

Institutional Review Board Statement: Not applicable.

Informed Consent Statement: Not applicable.

Data Availability Statement: The data presented in this study are available upon request from the corresponding author.

Acknowledgments: The authors are grateful to Petra Högy, Gina Gensheimer, Annette Falter, and Jürgen Franzaring for their help during the climate chamber experiment. Additionally, special thanks go to the team members and coordinators of the CLIFOOD project. We would also like to acknowledge the support received through the DFG grant PI 377/24-1. Furthermore, we are grateful to Berhane Lakew and Holeta Agricultural Research Center for providing us with barley cultivars.

Conflicts of Interest: The authors declare no conflict of interest.

Appendix A

Figure A1. Map of origin of Ethiopian landrace cultivar collection.

Table A1. List of Ethiopian landrace cultivars and the origin of the collection.

Code	Cultivars	Region	Zone	Woreda	Latitude	Longitude	Altitude
Lan_1	215217-A	Amara	Debub	Legambo	11-39-00-N	39-00-00-E	3570
Lan_2	18330-A	Amara	Semen	Angolela	09-18-00-N	39-32-25-E	3325
Lan_3	219578-A	Amara	Semen	Debre	09-37-00-N	39-25-00-E	2690
Lan_4	243410	Amara	Semen	Ankober	09-36-00-N	39-44-00-E	2350
Lan_5	237021	Amara	Semen	Minjarna	09-10-20-N	39-20-00-E	1750
Lan_6	208816-A	Oromiya	Bale	Adaba	07-00-20-N	39-23-30-E	3500
Lan_7	237015	Oromiya	Arssi	Digeluna	07-45-00-N	39-11-00-E	2600
Lan_8	64233-C	Oromiya	Bale	Sinana	07-04-00-N	40-14-00-E	2460
Lan_9	18327	Oromiya	Semen	Leben	08-28-00-N	38-56-59-E	1642
Lan_10	216997	SNNP	Semen	Chencha	06-17-00-N	37-35-00-E	3030
Lan_11	208845	SNNP	Semen	Chencha	06-15-00-N	37-35-00-E	2850
Lan_12	234307	Tigray	Misrak Awi	Zealmbesa	14-16-00-N	39-21-00-E	3100
Lan_13	234293	Tigray	Debub Awi	Ofla	12-48-00-N	39-35-00-E	2410
Lan_14	237339	Tigray	Debub Awi	Enderta	13-30-00-N	39-28-00-E	2240
Lan_15	221325	SNNP	Semen	Chencha	06-09-00-N	37-36-00-E	2150

Table 2. List of Ethiopian released cultivars and their desired trait.

Code	Cultivars	Genetic Background/Pedigree	Year of Released	Desirable Traits of the Cultivars Other than Yield
Rel_1	Gobe	IICARDA germplasm-CBSS96Moo487T-D-1M-1Y-2M-oY	2012	
Rel_2	Cross 41/98	(50-16/3316-03)//(HB42/Alexis)	2012	High yielding, late maturing
Rel_3	EH 1493	White Sasa/Comp29//White Sasa/EH538/F2-12B-2	2012	High yielding, late maturing
Rel_4	EH1847	EH1847/F4.2P.5.2 (Beka/IBON64/91)	2011	
Rel_5	Bekji-1	EH 1293/F2-18B-11-1-14-18	2010	
Rel_6	HB-1307	EH-1700/F7. B1.63.70	2006	High yield, lodging resistant, resistant to leaf diseases (Pyrenophora teres and Rhynchosporium secalis), good biomass yield, and white seeded
Rel_7	Misccal-21	Azafran = Shyri//Gloria/Copal/3/Shyri/Grit; CMB87.643-2A	2006	High yield with good malting quality; resistance to lodging with multiple disease resistance
Rel_8	Meserach	Pure line selection- Kulumsa1/88	1998	Early maturing and tolerant to major leaf diseases (Pyrenophora teres and Rhynchosporium secalis)
Rel_9	HB-42	EIAR cross-IAR-H-81/comp29//comp14-20/coast	1984	Resistant to scalding (Rhynchosporium secalis) and good biomass yield
Rel_10	IAR/H/485	Pure line selection from local landrace in Arsi	1975	
Rel_11	Ardu 12-60B	Pure line selection from local landrace in Arsi	1986	
Rel_12	Balemi	Dominant farmers varieties in West shoa	1970	Tolerant to low soil fertility and drought, good flour quality
Rel_13	HB-1964	RECLA78//SHYRI/GRIT/3/ATAH92/GOB	2016	
Rel_14	HB-1965	Awra gebs X IBON64/91	2017	
Rel_15	HB-1966	CARDO/CHEVRON-BAR CBSS 96 WM 00019s	2017	

References

1. Prosekov, A.Y.; Ivanova, S.A. Food security: The challenge of the present. *Geoforum* **2018**, *91*, 73–77. [CrossRef]
2. Newton, A.; Johnson, S.; Gregory, P. Implications of climate change for diseases, crop yields and food security. *Euphytica* **2011**, *179*, 3–18. [CrossRef]
3. Food and Agriculture Organization of the United Nations. FAOSTAT Database. Available online: http://www.fao.org/faostat/en/#data/QC (accessed on 19 August 2019).
4. Bekele, B.; Alemayehu, F.; Lakew, B. Food Barley in Ethiopia. In *Food Barley: Importance, Uses, and Local Knowledge*; Grando, S., Macpherson, H.G., Eds.; ICARDA: Aleppo, Syria, 2005; pp. 53–82.
5. Lakew, B.; Semeane, Y.; Alemayehu, F.; Gebre, H.; Grando, S.; Van Leur, J.A.G.; Ceccarelli, S. Exploiting the diversity of barley landraces in Ethiopia. *Genet. Resour. Crop Evol.* **1997**, *44*, 109–116. [CrossRef]
6. Asfaw, Z. The Barleys of Ethiopia. In *Genes in the Field: On-farm Conservation of Crop Diversity*; Stephen, B.B., Ed.; Lewis Publishers: Boca Raton, FL, USA, 2000; ISBN 1-55250-327-5.
7. Bonman, J.; Bockelman, H.; Jackso, L.; Steffenson, B. Disease and insect resistance in cultivated barley accessions from the USDA national small grains collection. *Crop Sci.* **2005**, *45*, 1271–1280. [CrossRef]
8. Mekonnon, B.; Lakew, B.; Dessalegn, T. Genotypic Diversity and Interrelationship of Characters in Ethiopian Food Barley (*Hordeum vulgare* L.) Landraces. *Int. J. Plant Breed. Genet.* **2014**, *8*, 164–180. [CrossRef]
9. Shewayrga, H.; Sopade, P. Ethnobotany, diverse food uses, claimed health benefits and implications on conservation of barley landraces in North-Eastern Ethiopia Highlands. *J. Ethnobiol. Ethnomedicine* **2011**, *7*, 19–34. [CrossRef] [PubMed]
10. Wosene, G.A.; Berhane, L.; Bettina, I.G.H.; Karl, J.S. Ethiopian barley landraces show higher yield stability and comparable yield to improved varieties in multi-environment field trials. *J. Plant Breed. Crop Sci.* **2015**, *7*, 275–291. [CrossRef]
11. Intergovernmental Panel on Climate Change. *Mitigation of Climate Change*; Intergovernmental Panel on Climate Change: Geneva, Switzerland, 2014.
12. Meehl, G.A.; Stocker, T.F.; Collins, W.D.; Friedlingstein, P.; Gaye, A.T.; Gregory, J.M.; Zhao, Z.C. Global Climate Projections: The Physical Science Basis. Contribution of Working Group I to the Fourth Assessment Report of the Intergovernmental Panel on Climate Change. *Clim. Chang.* **2007**, *2007*, 1–10.
13. Cure, J.D.; Acock, B.; Ford, M.A.; Thorne, G.N.; Fuel, L.; Lichtenthaler, H.K.; Seneweera, S. Effect of increased atmospheric carbon dioxide concentration on plant communities. *Ann. Bot.* **1998**, *31*, 416–489.
14. Kimball, B.A.; Kobayashi, K.; Bindi, M. Responses of agricultural crops to free-air CO_2 enrichment. *Adv. Agron.* **2014**, *77*, 293–368.

15. Long, S.P.; Ainsworth, E.A.; Leakey, A.D.B.; Nösbsrger, J.; Ort, D.R. Food for thought: Lower-than-expected crop yield stimulation with rising CO_2 concentrations. *Science* **2006**, *312*, 1918–1921. [CrossRef] [PubMed]
16. Aspinwall, M.J.; Loik, M.E.; de Dios, V.R.; Tjoelker, M.G.; Payton, P.R.; Tissue, D.T. Utilizing intraspecific variation in phenotypic plasticity to bolster agricultural and forest productivity under climate change. *Plant Cell Environ.* **2015**, *38*, 1752–1764. [CrossRef] [PubMed]
17. Schmid, I.; Franzaring, J.; Müller, M.; Brohon, N.; Calvo, O.C.; Högy, P.; Fangmeier, A. Effects of CO_2 Enrichment and Drought on Photosynthesis, Growth, and Yield of an Old and a Modern Barley Cultivar. *J. Agron. Crop Sci.* **2016**, *202*, 81–95. [CrossRef]
18. Ziska, L.H.; Bunce, J.A.; Shimono, H.; Gealy, D.R.; Baker, J.T.; Newton, P.C.D.; Reynolds, M.P.; Jagadish, K.S.V.; Zhu, C.; Howden, M.; et al. Security, and climate change: On the potential to adapt global crop production by active selection to rising atmospheric carbon dioxide. *Proc. R. Soc. B Biol. Sci.* **2012**, *279*, 4097–4105. [CrossRef] [PubMed]
19. Alemayehu, F.R.; Frenck, G.; van der Linden, L.; Mikkelsen, T.N.; Jørgensen, R.B. Can barley (*Hordeum vulgare* L. s.l.) adapt to fast climate changes? A controlled selection experiment. *Genet. Resour. Crop Evol.* **2014**, *61*, 151–161. [CrossRef]
20. Clausen, S.K.; Frenck, G.; Linden, L.G.; Mikkelsen, T.N.; Lunde, C.; Jørgensen, R.B. Effects of Single and Multifactor Treatments with Elevated Temperature. CO_2 and Ozone on Oilseed Rape and Barley. *J. Agron. Crop Sci.* **2011**, *197*, 442–453. [CrossRef]
21. Mitterbauer, E.; Enders, M.; Bender, J.; Erbs, M.; Habekuß, A.; Kilian, B.; Weigel, H.J. Growth response of 98 barley (*Hordeum vulgare* L.) cultivars to elevated CO_2 and identification of related quantitative trait loci using genome-wide association studies. *Plant Breed.* **2017**, *136*, 483–497. [CrossRef]
22. Manderscheid, R.; Weigel, H.J. Photosynthetic and growth responses of old and modern spring wheat cultivars to atmospheric CO_2 enrichment. *Agric. Ecosyst. Environ.* **1997**, *64*, 65–73. [CrossRef]
23. Ziska, L.H.; Morris, C.F.; Goins, E.W. Quantitative and qualitative evaluation of selected wheat varieties released since 1903 to increasing atmospheric carbon dioxide: Can yield sensitivity to carbon dioxide be a factor in wheat performance? *Glob. Chang. Biol.* **2004**, *10*, 1810–1819. [CrossRef]
24. Shimono, H.; Okada, M.; Yamakawa, Y.; Nakamura, H.; Kobayashi, K.; Hasegawa, T. Genotypic variation in rice yield enhancement by elevated CO_2 relates to growth before heading, and not to maturity group. *J. Exp. Bot.* **2009**, *60*, 523–532. [CrossRef] [PubMed]
25. Bourgault, M.; Dreccer, M.F.; James, A.T.; Chapman, S.C. Genotypic variability in the response to elevated CO_2 of wheat lines differing in adaptive traits. *Funct. Plant Biol.* **2013**, *40*, 172–184. [CrossRef] [PubMed]
26. Ainsworth, E.A. Rice production in a changing climate: A meta-analysis of responses to elevated carbon dioxide and elevated ozone concentration. *Glob. Chang. Biol.* **2008**, *14*, 1642–1650. [CrossRef]
27. Korres, N.E.; Norsworthy, J.; Tehranchian, P.; Gitsopoulos, T.K.; Loka, D.A.; Oosterhuis, D.M.; Gealy, D.R.; Moss, S.R.; Burgos, N.R.; Miller, M.R.; et al. Cultivars to face climate change effects on crops and weeds: A review. *Agron. Sustain. Dev.* **2016**, *36*, 12. [CrossRef]
28. Fangmeier, A.; Chrost, B.; Högy, P.; Krupinska, K. CO_2 enrichment enhances flag leaf senescence in barley due to greater grain nitrogen sink capacity. *Environ. Exp. Bot.* **2000**, *44*, 151–164. [CrossRef]
29. Manderscheid, R.; Weigel, H.J. Drought stress effects on wheat are mitigated by atmospheric CO_2 enrichment. *Agron. Sustain. Dev.* **2007**, *27*, 79–87. [CrossRef]
30. Johannessen, M.M.; Mikkelsen, T.N.; Nersting, L.G.; Gullord, M.; Von Bothmer, R.; Jørgensen, R.B. Effects of increased atmospheric CO_2 on varieties of oat. *Plant Breed.* **2005**, *124*, 253–256. [CrossRef]
31. Weigel, H.J.; Manderscheid, R. Crop growth responses to free-air CO_2 enrichment and nitrogen fertilization: Rotating barley, ryegrass, sugar beet, and wheat. *Eur. J. Agron.* **2012**, *43*, 97–107. [CrossRef]
32. Högy, P.; Franzaring, J.; Schwadorf, K.; Breuer, J.; Schütze, W.; Fangmeier, A. Effects of free-air CO_2 enrichment on energy traits and seed quality of oilseed rape. *Agric. Ecosyst. Environ.* **2010**, *139*, 239–244. [CrossRef]
33. Wu, D.; Wang, X.G.X.; Bai, Y.F.; Liao, J.X. Effects of elevated CO_2 concentration on growth, water use, yield and grain quality of wheat under two soil water levels. *Agric. Ecosyst. Environ.* **2004**, *104*, 493–507. [CrossRef]
34. Högy, P.; Keck, M.; Niehaus, K.; Franzaring, J.; Fangmeier, A. Effects of atmospheric CO_2 enrichment on biomass, yield, and low molecular weight metabolites in wheat grain. *J. Cereal Sci.* **2010**, *52*, 215–220. [CrossRef]
35. Högy, P.; Fangmeier, A. Effects of elevated atmospheric CO_2 on grain quality of wheat. *J. Cereal Sci.* **2008**, *48*, 580–591. [CrossRef]
36. Ainsworth, E.A.; Davey, P.A.; Bernacchi, C.J.; Dermody, O.C.; Heaton, E.A.; Moore, D.J.; Long, S.P. A meta-analysis of elevated CO_2 effects on soybean (*Glycine max*) physiology, growth, and yield. *Glob. Chang. Biol.* **2002**, *8*, 695–709. [CrossRef]
37. Brouder, S.M.; Volenec, J.J. Impact of climate change on crop nutrient and water use efficiencies. *Physiol. Plant.* **2008**, *133*, 705–724. [CrossRef] [PubMed]
38. Robredo, A.; Pérez-López, U.; Miranda-Apodaca, J.M.; Lacuesta, A.; Mena-Petite, A.; Muñoz-Rueda, A. Elevated CO_2 reduces the drought effect on nitrogen metabolism in barley plants during drought and subsequent recovery. *Environ. Exp. Bot.* **2011**, *71*, 399–408. [CrossRef]
39. Wu, D.X.; Wang, G.X.; Bai, Y.F.J.; Liao, X.; Ren, H.X. Response of growth and water use efficiency of spring wheat to whole season CO_2 enrichment and drought. *Acta Bot. Sin.* **2002**, *44*, 1477–1483.
40. Wallace, J.S. Increasing agricultural water use efficiency to meet future food production. *Agric. Ecosyst. Environ.* **2000**, *82*, 105–119. [CrossRef]

41. Franzaring, J.; Holz, I.; Fangmeier, A. Responses of old and modern cereals to CO_2-fertilisation. *Crop Pasture Sci.* **2013**, *64*, 943–956. [CrossRef]

42. Plessl, M.; Heller, W.; Payer, H.D.; Elstner, E.F.; Habermeyer, J.; Heiser, I. Growth parameters and resistance against Drechslera teres of spring barley (*Hordeum vulgare* L. cv. Scarlett) grown at elevated ozone and carbon dioxide concentrations. *Plant Biol.* **2005**, *7*, 694–705. [CrossRef] [PubMed]

43. Ziska, L.H.; Namuco, O.; Moya, T.; Quilang, J. Growth and yield response of field-grown tropical rice to increasing carbon dioxide and air temperature. *Agron. J.* **1997**, *89*, 45–53. [CrossRef]

44. Ziska, L.H.; Bunce, J.A. Sensitivity of field-grown soybean to future atmospheric CO_2: Selection for improved productivity in the 21st century. *Funct. Plant Biol.* **2000**, *27*, 979–984. [CrossRef]

45. Manderscheid, R.; Erbs, M.; Weigel, H.J. Interactive effects of free-air CO_2 enrichment and drought stress on maize growth. *Eur. J. Agron.* **2014**, *52*, 11–21. [CrossRef]

46. Pérez-López, U.; Miranda-Apodaca, J.; Mena-Petite, A.; Muñoz-Rueda, A. Barley Growth and Its Underlying Components are Affected by Elevated CO_2 and Salt Concentration. *J. Plant Growth Regul.* **2013**, *32*, 732–744. [CrossRef]

47. Poorter, H.; Navas, M.L. Plant growth and competition at elevated CO_2: On winners, losers and functional groups. *New Phytologist.* **2003**, *157*, 175–198. [CrossRef] [PubMed]

48. Ingvordsen, C.H.; Backes, G.; Lyngkjær, M.F.; Peltonen-Sainio, P.; Jensen, J.D.; Jalli, M.; Jørgensen, R.B. Significant decrease in yield under future climate conditions: Stability and production of 138 spring barley accessions. *Eur. J. Agron.* **2015**, *63*, 105–113. [CrossRef]

49. Leakey, A.D.B.; Bishop, K.A.; Ainsworth, E.A. A multi-biome gap in understanding of crop and ecosystem responses to elevated CO_2. *Curr. Opin. Plant Biol.* **2012**, *15*, 228–236. [CrossRef]

50. Tausz, M.; Tausz-Posch, S.; Norton, R.M.; Fitzgerald, G.J.; Nicolas, M.E.; Seneweera, S. Understanding crop physiology to select breeding targets and improve crop management under increasing atmospheric CO_2 concentrations. *Environ. Exp. Bot.* **2013**, *88*, 71–80. [CrossRef]

51. R Core Team. *R: A Language and Environment for Statistical Computing*; R Foundation for Statistical Computing: Vienna, Austria, 2019.

Article

Screening for Higher Grain Yield and Biomass among Sixty Bread Wheat Genotypes Grown under Elevated CO_2 and High-Temperature Conditions

Emilio L. Marcos-Barbero, Pilar Pérez, Rafael Martínez-Carrasco, Juan B. Arellano and Rosa Morcuende *

Institute of Natural Resources and Agrobiology of Salamanca (IRNASA), Consejo Superior de Investigaciones Científicas (CSIC), 37008 Salamanca, Spain; emiliol.marcos@irnasa.csic.es (E.L.M.-B.); pili2013@gmail.com (P.P.); rafael.mcarrasco@gmail.com (R.M.-C.); juan.arellano@irnasa.csic.es (J.B.A.)
* Correspondence: rosa.morcuende@irnasa.csic.es; Tel.: +34-923-219-606

Abstract: Global warming will inevitably affect crop development and productivity, increasing uncertainty regarding food production. The exploitation of genotypic variability can be a promising approach for selecting improved crop varieties that can counteract the adverse effects of future climate change. We investigated the natural variation in yield performance under combined elevated CO_2 and high-temperature conditions in a set of 60 bread wheat genotypes (59 of the 8TH HTWSN CIMMYT collection and Gazul). Plant height, biomass production, yield components and phenological traits were assessed. Large variations in the selected traits were observed across genotypes. The CIMMYT genotypes showed higher biomass and grain yield when compared to Gazul, indicating that the former performed better than the latter under the studied environmental conditions. Principal component and hierarchical clustering analyses revealed that the 60 wheat genotypes employed different strategies to achieve final grain yield, highlighting that the genotypes that can preferentially increase grain and ear numbers per plant will display better yield responses under combined elevated levels of CO_2 and temperature. This study demonstrates the success of the breeding programs under warmer temperatures and the plants' capacity to respond to the concurrence of certain environmental factors, opening new opportunities for the selection of widely adapted climate-resilient wheat genotypes.

Keywords: climate change; elevated CO_2; high temperature; grain yield; biomass; bread wheat; genotypes

Citation: Marcos-Barbero, E.L.; Pérez, P.; Martínez-Carrasco, R.; Arellano, J.B.; Morcuende, R. Screening for Higher Grain Yield and Biomass among Sixty Bread Wheat Genotypes Grown under Elevated CO_2 and High-Temperature Conditions. *Plants* **2021**, *10*, 1596. https://doi.org/10.3390/plants10081596

Academic Editor: James Bunce

Received: 16 July 2021
Accepted: 30 July 2021
Published: 3 August 2021

1. Introduction

Climate change is considered to have detrimental effects on global food security by modifying the environmental conditions for agricultural production. The progressive increase in atmospheric CO_2 concentration and other greenhouse gases due to anthropogenic activities, such as continuous deforestation and the excessive use of fossil fuels, is the main driving force for global warming [1]. The atmospheric concentrations of CO_2 have risen since preindustrial times, and currently exceed 410 ppm [2]; this is expected to continue rising over this century, to reach levels close to 1000 ppm by 2100. Simultaneously, the global mean surface air temperature is predicted to rise by an average of 2.6–4.8 °C throughout this century, which may result in an increase in the incidence of other weather events with negative consequences for crop productivity [3]. Therefore, climate variability is threatening the current food system because crop growth and production are markedly affected by both the atmospheric concentration of CO_2 and air temperature [4].

Wheat is among the three main staple crops worldwide, covering more than 20% of the total calorie intake by the worldwide human population, and an equivalent percentage of proteins for approximately 2.5 billion people [5]. Bread wheat (*Triticum aestivum* L.) is extensively cultivated in Mediterranean regions, where the elevated temperature is likely the major yield-limiting factor [6].

One of the main effects of elevated atmospheric CO_2 on C3 crops such as wheat is an increase in the photosynthesis rate, which may thereby enhance plant growth and consequently increase grain yield. However, the initial stimulation of photosynthesis, as induced by elevated CO_2, cannot be sustained over prolonged exposure to high CO_2 concentrations, and it also results in a down-regulation of the photosynthetic capacity [7,8]. This acclimation of photosynthesis to elevated CO_2 is often found in wheat [9–14] and is associated with enhanced leaf carbohydrate content and decreased N concentration [14–17], yielding grains with lower N and protein concentration [18,19]. The acclimation of photosynthesis is particularly common in plants grown in pots with a restricted root system [20] or with low nutrient availability [12,21,22], and often reflects the plant's sink strength [23]. Environmental or genetic factors limiting the development of sink strength make plants susceptible to a greater photosynthetic acclimation, and diminish the stimulation of photosynthesis by CO_2 enrichment [7,8]. Thus, source–sink relationships have been proposed as playing a key role in the regulation of plant growth and yield response to elevated atmospheric CO_2 [23,24]. In line with this suggestion, greater growth and yield response to elevated CO_2 has been found in some older wheat cultivars when compared to modern ones, a finding which seems to be associated with the capacity of the older cultivars to produce further tillers functioning as vegetative sinks for the allocation of excess carbohydrates [25,26]. Increased tillering is not the only sink of relevance for wheat crops to deal with excess carbohydrates. The translocation of carbohydrates into grains during the reproductive growth stages is entailed in determining grain yield, and the sink strength of the maturing ears and grains becomes crucial [14,27,28].

The world is also facing a gradual rise in temperature, and most of the worldwide wheat-growing regions are undergoing temperature increases greater than are optimal, leading to alterations in plant growth, development, physiological processes, and yield [29,30]. The evaluation of the impact of global warming on six major food crops, performed by Lobell and Field (2007) [31], showed yield losses of about 40 million tonnes per year as a whole for wheat, maize and barley from 1981 to 2002, where wheat accounts for nearly half of these shortfalls (19 mt/year). This finding highlights that wheat is highly sensitive to high temperatures; the occurrence of elevated temperatures at any developmental stage can affect grain yield, although they are particularly critical at the reproductive stage and cause considerable yield losses [32,33]. Bergkamp et al. (2018) [34] reported that heat stress (35 °C) after anthesis decreased the grain-filling period and restricted the deposition of resources to grains, which led to a substantial decrease in wheat productivity by 6–51% when plants were grown in controlled conditions and by 2–27% under field conditions. Such yield losses have been linked to the direct impact of temperature on grain number and grain weight [35,36]. Further works using global- and/or country-based models showed that an increase in 1 °C above the average global surface temperatures could result in a worldwide decline in wheat yields from 4.1% to 6.4% [37,38]. Thus, the prediction of the potential impact of global warming on crop productivity is complicated because CO_2 enrichment could result in yield stimulation; and, for several crops, including wheat, this could ameliorate some of the adverse effects of higher temperatures on yields [39]. In line with this, Fitzgerald et al. (2016) [40] observed that growth under elevated atmospheric CO_2 enhanced biomass and yield in bread wheat grown in FACE facilities and, when combined with higher temperatures, it counteracted the negative impacts of temperature on grain yield.

Worldwide demand for wheat is projected to continue growing, whereas the current increase in wheat productivity is only 1.1% per year, while there is simply no increase in some regions [6]. Consequently, the actual wheat yield gain per year is insufficient to meet the demand for food from the increasing world population. Improving yield and its sustainability at a time of unprecedented climatic variability is becoming more and more urgent as the global population increases. To tackle these limitations, improved crop varieties will be required to guarantee food security. The exploitation of genetic variations in wheat responses to future climatic conditions might be a useful approach. Nevertheless, as mentioned above, there is a wealth of studies dealing with the present and predicted impact of elevated atmospheric CO_2 concentrations and temperature on the productivity of wheat under controlled and field conditions, but most of them have been carried out without considering simultaneous increases in atmospheric CO_2 and temperature, and are conducted with a single genotype or a limited range of crop germplasm [41]. The aim of this work was to assess the natural existing variation in wheat yield performance in response to combined elevated CO_2 and high temperature. We used a wide set of 60 bread wheat genotypes, including 59 genotypes of the CIMMYT heat-tolerant wheat screening nursery (8TH HTWSN) previously selected for high performance under warmer temperatures, together with the Gazul genotype, with high adaptability to the Mediterranean climate of the Salamanca region (Spain). All genotypes were grown until physiological maturity in controlled climate chambers that simulate the predicted global warming by the end of this century, in the region of Salamanca; the final biomass and grain yield components were determined. The results of our study will enable better planning for the selection of cultivars with improved adaptation to future climate change.

2. Results

2.1. Wheat Production and Grain Yield Components

Table 1 summarizes a list of descriptive statistics for wheat production and grain yield components, in response to the combined elevated atmospheric CO_2 and high temperatures across a population of 60 wheat genotypes, cultivated in climate chambers. Among genotypes, grain yield ranged from 4.63 to 10.70 g per plant, although 50% of the genotypes produced between 6.96 and 8.43 g per plant. The mean grain yield was 7.66 ± 1.12 g per plant, with genotypes 150, 74, 23, 8, and 76 as the lowest yielding genotypes, and 43, 94, 95, 61 and 41 as the highest yielding genotypes. The aboveground, stalk and chaff biomasses also ranged between 13.13−25.50, 5.99−12.21 and 1.54−4.42 g per plant, respectively, with 19.16 ± 2.51, 8.78 ± 1.20 and 2.72 ± 0.55 g per plant on average for the whole population. The mean plant height was 86.82 ± 4.88 cm per plant, ranging from 76.35 to 105.30 cm per plant. Grain number per ear (GNE) and per plant varied from 19.4 and 116.39 grains to 54.02 and 315.33 grains, respectively, with an average of 34.99 ± 5.63 grains per ear and 197.01 ± 36.05 grains per plant. The number of ears per plant ranged between 3.20 and 8.20, with a mean value of 5.67 ± 0.84 ears per plant. Grain yield per ear (GYE) varied from 0.89 to 1.96, with 1.36 ± 0.19 g per ear on average. The mean harvest index (HI) was 0.40 ± 0.03, ranging from 0.26 to 0.46. Skewness and kurtosis of variables in the population were generally located around zero, suggesting symmetry and tail weight similar to that of normal distribution. However, the skewness and kurtosis of HI accounted for −0.77 and 2.68, indicating a non-normal distribution with moderate left-skewed asymmetry. Statistically significant differences among genotypes were found for the studied traits.

Table 1. Wheat production (aboveground, stalk and chaff biomasses), grain yield components (grain yield, grain number, ear number, grain weight, grain yield per ear, grain number per ear and harvest index), and plant height of sixty wheat genotypes grown under elevated CO_2 and high-temperature conditions.

Genotype	Grain Yield (g plant^{-1})	Above-Ground (g plant^{-1})	Stalk (g plant^{-1})	Chaff (g plant^{-1})	Height (cm plant^{-1})	Grain Number (No. plant^{-1})	Ear Number (No. plant^{-1})	Grain Weight (mg plant^{-1})	GYE (g ear^{-1})	GNE (No. ear^{-1})	HI
150	5.45 ± 1.05	14.38 ± 1.64	6.82 ± 0.44	2.12 ± 0.35	80.36 ± 0.77	156.10 ± 32.80	4.10 ± 0.48	34.97 ± 0.81	1.32 ± 0.10	37.79 ± 3.34	0.38 ± 0.03
74	6.18 ± 0.68	16.29 ± 1.82	8.23 ± 1.26	1.88 ± 0.30	89.71 ± 1.21	151.79 ± 19.87	5.65 ± 0.30	40.95 ± 3.61	1.10 ± 0.12	26.98 ± 4.23	0.38 ± 0.04
23	6.42 ± 0.53	15.17 ± 1.31	6.83 ± 0.48	1.91 ± 0.35	79.86 ± 0.19	159.37 ± 12.57	4.45 ± 0.57	40.27 ± 0.42	1.45 ± 0.08	36.02 ± 2.24	0.42 ± 0.01
8	6.46 ± 1.22	17.14 ± 3.79	7.88 ± 1.72	2.80 ± 0.86	81.49 ± 3.04	161.55 ± 32.03	5.30 ± 1.16	40.04 ± 1.60	1.22 ± 0.06	30.58 ± 1.55	0.38 ± 0.01
76	6.50 ± 0.77	17.12 ± 1.72	8.46 ± 0.86	2.16 ± 0.26	87.15 ± 2.16	160.50 ± 24.84	5.40 ± 0.63	40.63 ± 1.60	1.20 ± 0.07	29.70 ± 2.47	0.38 ± 0.02
15	6.53 ± 0.85	16.11 ± 2.41	7.18 ± 1.27	2.39 ± 0.39	77.14 ± 1.38	158.58 ± 22.88	5.55 ± 0.53	41.28 ± 1.08	1.18 ± 0.09	28.51 ± 2.19	0.41 ± 0.02
73	6.59 ± 0.94	18.81 ± 1.49	9.59 ± 0.63	2.63 ± 0.23	91.38 ± 0.59	158.99 ± 15.40	4.95 ± 0.10	41.34 ± 2.39	1.33 ± 0.18	32.14 ± 3.25	0.35 ± 0.03
21	6.62 ± 0.59	16.35 ± 1.07	7.78 ± 0.28	1.94 ± 0.27	88.73 ± 1.92	163.79 ± 8.89	5.00 ± 0.52	40.44 ± 2.71	1.34 ± 0.19	32.94 ± 2.81	0.40 ± 0.01
9	6.64 ± 0.73	17.19 ± 1.80	8.18 ± 0.66	2.38 ± 0.44	88.13 ± 3.34	148.69 ± 18.14	4.95 ± 0.55	44.71 ± 1.92	1.34 ± 0.05	30.04 ± 1.74	0.39 ± 0.01
118	6.67 ± 0.23	17.19 ± 1.40	8.00 ± 0.95	2.53 ± 0.43	84.70 ± 4.05	176.82 ± 4.29	6.80 ± 0.49	37.75 ± 1.48	0.99 ± 0.08	26.07 ± 1.32	0.39 ± 0.03
81	6.75 ± 1.03	17.15 ± 2.14	8.01 ± 1.27	2.39 ± 0.07	94.69 ± 2.02	164.49 ± 28.73	5.00 ± 1.40	41.20 ± 2.23	1.40 ± 0.26	33.88 ± 5.17	0.39 ± 0.02
77	6.86 ± 0.96	17.42 ± 1.88	8.49 ± 0.63	2.07 ± 0.34	91.03 ± 4.86	168.32 ± 19.71	5.25 ± 0.64	40.66 ± 1.95	1.31 ± 0.19	32.32 ± 4.74	0.39 ± 0.02
115	6.92 ± 0.94	18.23 ± 1.64	8.64 ± 0.84	2.67 ± 0.25	87.08 ± 2.59	160.01 ± 25.82	5.55 ± 0.44	43.49 ± 2.81	1.25 ± 0.10	28.80 ± 3.72	0.38 ± 0.03
87	6.96 ± 1.43	18.14 ± 3.64	8.62 ± 1.82	2.55 ± 0.48	87.36 ± 1.30	184.34 ± 43.52	4.90 ± 0.48	38.03 ± 2.24	1.44 ± 0.35	38.03 ± 9.69	0.38 ± 0.01
29	7.04 ± 0.77	17.26 ± 1.00	7.89 ± 0.21	2.33 ± 0.07	82.15 ± 2.20	184.91 ± 16.88	4.75 ± 0.34	38.03 ± 1.02	1.48 ± 0.08	38.90 ± 1.70	0.41 ± 0.02
46	7.06 ± 0.32	18.39 ± 0.84	8.23 ± 0.57	3.11 ± 0.27	88.99 ± 3.74	147.54 ± 7.81	5.45 ± 0.10	47.84 ± 1.22	1.30 ± 0.07	27.09 ± 1.81	0.38 ± 0.02
119	7.09 ± 0.90	18.33 ± 1.63	8.84 ± 0.86	2.40 ± 0.23	92.60 ± 1.93	170.30 ± 5.50	5.10 ± 0.35	41.63 ± 5.21	1.39 ± 0.15	33.50 ± 2.44	0.39 ± 0.02
6	7.20 ± 0.75	16.94 ± 1.73	7.51 ± 0.75	2.23 ± 0.33	79.49 ± 1.19	194.79 ± 17.84	4.35 ± 0.34	36.91 ± 0.71	1.65 ± 0.08	44.78 ± 2.35	0.42 ± 0.01
110	7.25 ± 0.81	18.25 ± 1.07	8.77 ± 0.43	2.24 ± 0.06	85.64 ± 3.54	202.72 ± 19.43	5.75 ± 0.38	35.73 ± 1.16	1.26 ± 0.07	35.21 ± 1.56	0.40 ± 0.02
71	7.29 ± 0.65	18.70 ± 0.87	9.02 ± 0.57	2.39 ± 0.34	93.69 ± 1.18	169.53 ± 12.76	4.85 ± 0.34	42.97 ± 1.26	1.50 ± 0.10	35.03 ± 2.76	0.39 ± 0.03
16	7.34 ± 0.32	17.69 ± 0.36	7.94 ± 0.12	2.40 ± 0.20	83.21 ± 3.08	181.28 ± 9.93	6.35 ± 0.38	40.54 ± 1.26	1.16 ± 0.08	28.58 ± 1.32	0.42 ± 0.01
22	7.44 ± 0.79	18.10 ± 1.96	8.10 ± 1.11	2.56 ± 0.26	85.10 ± 0.27	199.14 ± 21.75	6.15 ± 0.60	37.39 ± 0.74	1.21 ± 0.07	32.38 ± 1.56	0.41 ± 0.02
13	7.49 ± 1.01	18.87 ± 2.29	8.43 ± 1.13	2.95 ± 0.32	89.96 ± 0.86	166.74 ± 23.43	5.45 ± 0.68	45.03 ± 3.37	1.38 ± 0.15	30.67 ± 3.44	0.40 ± 0.02
92	7.50 ± 0.68	18.40 ± 1.24	8.47 ± 0.43	2.43 ± 0.27	83.39 ± 1.90	199.36 ± 17.06	5.60 ± 0.37	37.61 ± 0.98	1.34 ± 0.11	35.69 ± 3.41	0.41 ± 0.01
117	7.53 ± 0.35	18.56 ± 1.17	8.42 ± 0.58	2.60 ± 0.34	89.48 ± 0.78	173.98 ± 5.18	5.65 ± 0.19	43.31 ± 1.77	1.33 ± 0.07	30.80 ± 0.52	0.41 ± 0.01
79	7.56 ± 0.91	19.75 ± 2.48	9.12 ± 1.34	3.06 ± 0.31	84.79 ± 2.56	181.24 ± 19.11	5.95 ± 0.41	41.69 ± 1.43	1.27 ± 0.14	30.47 ± 2.52	0.38 ± 0.02
33	7.60 ± 1.35	21.98 ± 0.67	10.76 ± 0.66	3.61 ± 0.47	85.28 ± 4.74	168.38 ± 37.34	5.65 ± 0.41	45.58 ± 3.37	1.36 ± 0.30	30.02 ± 7.23	0.35 ± 0.06
69	7.66 ± 1.50	17.81 ± 2.54	7.63 ± 0.89	2.51 ± 0.45	79.86 ± 2.44	193.14 ± 28.90	6.35 ± 0.85	39.48 ± 1.79	1.21 ± 0.20	30.62 ± 4.64	0.43 ± 0.03
31	7.71 ± 1.26	18.71 ± 2.69	8.27 ± 1.09	2.73 ± 0.36	87.35 ± 3.87	177.91 ± 29.82	6.50 ± 1.16	43.38 ± 2.12	1.20 ± 0.17	27.52 ± 3.13	0.41 ± 0.01
19	7.73 ± 0.66	19.04 ± 1.21	8.74 ± 0.58	2.57 ± 0.10	87.91 ± 3.11	193.25 ± 3.79	5.10 ± 0.38	40.03 ± 3.77	1.52 ± 0.11	38.04 ± 2.65	0.41 ± 0.01
114	7.75 ± 0.32	20.64 ± 1.03	10.42 ± 0.53	2.46 ± 0.37	100.68 ± 3.35	214.96 ± 15.27	5.65 ± 0.34	36.11 ± 1.47	1.37 ± 0.04	38.03 ± 0.75	0.38 ± 0.01
97	7.79 ± 0.35	19.89 ± 1.06	9.42 ± 0.50	2.68 ± 0.35	88.10 ± 3.37	213.71 ± 8.05	6.55 ± 0.19	36.49 ± 2.51	1.19 ± 0.06	32.67 ± 2.01	0.39 ± 0.01
113	7.79 ± 0.87	20.01 ± 2.47	9.91 ± 1.27	2.31 ± 0.39	90.13 ± 4.00	237.19 ± 44.64	6.40 ± 1.02	33.16 ± 2.30	1.22 ± 0.06	36.96 ± 1.32	0.39 ± 0.01
25	7.81 ± 0.47	18.72 ± 1.04	8.22 ± 0.64	2.69 ± 0.16	83.50 ± 4.10	197.07 ± 12.53	5.60 ± 0.28	39.63 ± 0.68	1.40 ± 0.10	35.23 ± 2.36	0.42 ± 0.02
18	7.83 ± 1.28	19.61 ± 2.49	8.82 ± 1.01	2.97 ± 0.61	87.00 ± 2.85	202.67 ± 31.10	5.75 ± 0.98	38.62 ± 2.16	1.37 ± 0.13	35.37 ± 2.38	0.40 ± 0.04
70	7.84 ± 1.17	19.75 ± 3.23	8.98 ± 1.72	2.93 ± 0.73	93.44 ± 3.93	220.48 ± 43.49	5.85 ± 1.23	35.89 ± 3.65	1.37 ± 0.24	37.98 ± 4.14	0.40 ± 0.04
63	7.89 ± 0.82	20.21 ± 1.54	9.23 ± 0.77	3.09 ± 0.31	85.29 ± 2.36	215.34 ± 14.60	6.00 ± 0.43	36.59 ± 1.61	1.32 ± 0.10	35.96 ± 2.46	0.39 ± 0.03
105	7.91 ± 0.89	18.50 ± 1.33	8.21 ± 0.60	2.38 ± 0.12	82.40 ± 2.10	207.79 ± 28.23	4.40 ± 0.00	38.17 ± 1.69	1.80 ± 0.20	47.22 ± 6.42	0.43 ± 0.02
88	8.00 ± 0.78	20.27 ± 1.31	9.21 ± 0.79	3.05 ± 0.23	87.35 ± 2.97	200.75 ± 19.37	5.25 ± 0.34	39.88 ± 1.76	1.53 ± 0.22	38.35 ± 4.41	0.39 ± 0.02
93	8.01 ± 1.51	20.01 ± 3.21	9.01 ± 1.47	2.99 ± 0.53	88.54 ± 1.84	192.90 ± 23.42	5.70 ± 0.60	41.28 ± 3.34	1.40 ± 0.17	33.84 ± 1.97	0.40 ± 0.03
35	8.02 ± 1.11	21.93 ± 2.59	10.30 ± 1.39	3.61 ± 0.59	87.33 ± 1.71	195.70 ± 8.82	4.90 ± 0.62	40.91 ± 4.55	1.63 ± 0.02	40.28 ± 3.92	0.37 ± 0.03
4	8.06 ± 0.63	18.18 ± 0.90	7.45 ± 0.38	2.67 ± 0.05	85.81 ± 1.84	219.29 ± 9.45	6.25 ± 0.19	36.71 ± 1.34	1.29 ± 0.08	35.09 ± 1.15	0.44 ± 0.02
26	8.12 ± 0.44	20.35 ± 1.49	9.58 ± 0.91	2.65 ± 0.24	87.34 ± 1.58	204.17 ± 6.84	6.05 ± 0.41	39.77 ± 0.95	1.34 ± 0.03	33.81 ± 1.15	0.40 ± 0.01
48	8.15 ± 0.43	20.28 ± 1.18	9.34 ± 0.76	2.79 ± 0.28	90.09 ± 4.08	185.85 ± 20.06	6.35 ± 0.47	44.08 ± 2.62	1.29 ± 0.10	29.31 ± 2.95	0.40 ± 0.01
99	8.19 ± 0.63	20.80 ± 1.14	9.62 ± 0.23	3.00 ± 0.52	88.19 ± 1.78	238.56 ± 11.01	5.50 ± 0.26	34.34 ± 2.81	1.49 ± 0.10	43.39 ± 1.40	0.39 ± 0.02
53	8.19 ± 0.72	20.25 ± 1.45	8.78 ± 0.76	3.27 ± 0.33	82.55 ± 1.55	221.46 ± 14.92	6.40 ± 0.33	36.96 ± 0.85	1.28 ± 0.12	34.65 ± 2.60	0.40 ± 0.02
59	8.24 ± 1.32	20.47 ± 2.57	9.20 ± 1.02	3.03 ± 0.33	86.99 ± 0.61	192.70 ± 29.26	5.95 ± 0.70	42.73 ± 1.07	1.38 ± 0.10	32.33 ± 1.78	0.40 ± 0.01
83	8.28 ± 0.56	21.10 ± 2.28	9.70 ± 1.19	3.12 ± 0.58	85.04 ± 7.27	216.33 ± 16.38	5.70 ± 0.84	38.31 ± 0.37	1.47 ± 0.23	38.52 ± 6.17	0.39 ± 0.02

Table 1. *Cont.*

Genotype	Grain Yield (g plant^{-1})	Above-Ground (g plant^{-1})	Stalk (g plant^{-1})	Chaff (g plant^{-1})	Height (cm plant^{-1})	Grain Number (No. plant^{-1})	Ear Number (No. plant^{-1})	Grain Weight (mg plant^{-1})	GYE (g ear^{-1})	GNE (No. ear^{-1})	HI
10	8.31 ± 0.98	19.66 ± 2.19	8.43 ± 0.97	2.92 ± 0.33	82.31 ± 2.65	226.90 ± 26.78	5.40 ± 0.78	36.71 ± 2.78	1.55 ± 0.15	42.14 ± 1.38	0.42 ± 0.02
65	8.35 ± 0.89	20.65 ± 2.87	9.13 ± 1.67	3.17 ± 0.54	91.61 ± 5.93	217.74 ± 8.02	6.30 ± 0.62	38.29 ± 2.87	1.33 ± 0.16	34.81 ± 3.61	0.41 ± 0.03
42	8.51 ± 0.65	21.34 ± 1.32	9.34 ± 0.87	3.49 ± 0.22	88.88 ± 0.71	250.79 ± 12.25	6.80 ± 0.52	33.91 ± 1.12	1.25 ± 0.09	36.96 ± 1.86	0.40 ± 0.02
107	8.56 ± 1.37	20.33 ± 3.43	9.10 ± 1.42	2.68 ± 0.74	84.05 ± 2.79	249.29 ± 47.63	5.75 ± 0.50	34.48 ± 1.63	1.49 ± 0.16	43.14 ± 5.03	0.42 ± 0.01
11	8.65 ± 0.21	20.15 ± 0.90	8.78 ± 0.72	2.72 ± 0.21	86.24 ± 2.95	232.13 ± 13.49	6.30 ± 0.62	37.33 ± 1.50	1.38 ± 0.11	36.96 ± 1.64	0.43 ± 0.02
58	8.70 ± 0.81	21.47 ± 2.01	9.85 ± 0.82	2.92 ± 0.56	86.29 ± 1.27	247.08 ± 13.25	5.90 ± 0.50	35.16 ± 1.55	1.48 ± 0.08	41.99 ± 2.19	0.41 ± 0.02
5	8.76 ± 1.17	22.09 ± 2.40	9.88 ± 0.86	3.46 ± 0.40	84.75 ± 1.54	238.93 ± 21.61	6.80 ± 0.59	36.60 ± 3.13	1.28 ± 0.09	35.13 ± 0.78	0.40 ± 0.01
43	8.97 ± 0.99	20.86 ± 1.66	8.64 ± 0.77	3.26 ± 0.13	89.44 ± 1.97	232.37 ± 31.09	5.65 ± 0.25	38.77 ± 3.25	1.58 ± 0.12	41.12 ± 5.22	0.43 ± 0.02
94	8.99 ± 0.68	21.36 ± 1.95	9.59 ± 1.18	2.79 ± 0.28	94.06 ± 3.87	226.96 ± 24.88	6.00 ± 0.59	39.71 ± 1.44	1.50 ± 0.12	37.88 ± 3.00	0.42 ± 0.02
95	9.00 ± 0.66	21.73 ± 2.01	9.59 ± 1.04	3.14 ± 0.44	85.65 ± 3.43	231.89 ± 18.72	6.15 ± 1.00	38.84 ± 1.48	1.48 ± 0.15	38.15 ± 3.88	0.41 ± 0.01
61	9.03 ± 0.71	22.21 ± 1.52	10.27 ± 1.23	2.92 ± 0.72	85.68 ± 1.24	259.49 ± 13.43	5.90 ± 0.62	34.75 ± 1.25	1.53 ± 0.09	44.23 ± 3.49	0.41 ± 0.01
41	9.68 ± 0.85	23.25 ± 1.59	9.80 ± 0.73	3.78 ± 0.22	82.73 ± 3.39	252.73 ± 27.81	7.25 ± 0.44	38.38 ± 1.53	1.33 ± 0.05	34.82 ± 2.47	0.42 ± 0.01
Mean	7.66 ± 1.12	19.16 ± 2.51	8.78 ± 1.20	2.72 ± 0.55	86.82 ± 4.88	197.01 ± 36.05	5.67 ± 0.84	39.26 ± 3.68	1.36 ± 0.19	34.99 ± 5.63	0.40 ± 0.03
Min	4.63	13.13	5.99	1.54	76.35	116.39	3.20	29.73	0.89	19.40	0.26
Max	10.70	25.50	12.21	4.42	105.30	315.33	8.20	48.93	1.96	54.02	0.46
Q1	6.96	17.58	7.95	2.36	83.80	171.31	5.00	36.87	1.24	31.23	0.39
Q3	8.43	21.03	9.65	3.04	89.75	221.81	6.20	41.23	1.48	38.22	0.42
Skewness	−0.02	0.02	0.14	0.37	0.41	0.27	0.09	0.36	0.29	0.32	−0.7
Kurtosis	−0.25	−0.23	−0.28	−0.08	0.51	−0.10	−0.06	−0.10	0.27	0.27	2.68
p-value	4.08×10^{-11}	8.82×10^{-12}	6.67×10^{-10}	8.18×10^{-17}	6.38×10^{-30}	2.85×10^{-25}	1.37×10^{-16}	2.85×10^{-28}	4.72×10^{-14}	2.32×10^{-28}	5.22×10^{-6}

GNE: grain number ear-1; GYE: grain yield ear-1; HI: harvest index; Max: maximum; Min: minimum; Q1 and Q3: first and third quartile. For each genotype, values represent the mean ± standard deviation (SD) of four replicates (n = 4). Genotypes are ranked according to the increasing mean of grain yield. Within columns, all genotypes and replicates (N = 240) were employed for the calculation of the total mean ± SD (Mean), Skewness and Kurtosis. Statistical significance (*p*-value) was calculated by multivariate analysis of variance (MANOVA).

Traits

Minimum　　　Maximum

2.2. Comparison between the CIMMYT Population and Gazul

The mean responses of the genotype 150 (Gazul) for wheat production and grain yield traits were compared with a population of 59 selected wheat genotypes belonging to the CIMMYT 8TH HTWSN collection (hereafter called the CIMMYT population (Figure 1)). The one-sample test for the comparison of the mean responses showed significant differences between Gazul and the CIMMYT population for all the traits under study. Gazul exhibited lower aboveground, chaff and stalk biomasses, grain yield, grain and ear numbers per plant, and lower GYE and HI, but a greater GNE than the CIMMYT population. The effect size was considered as large for almost all the variables, with the exception of the effect size for GYE and GNE, which were considered as small and medium-size, respectively.

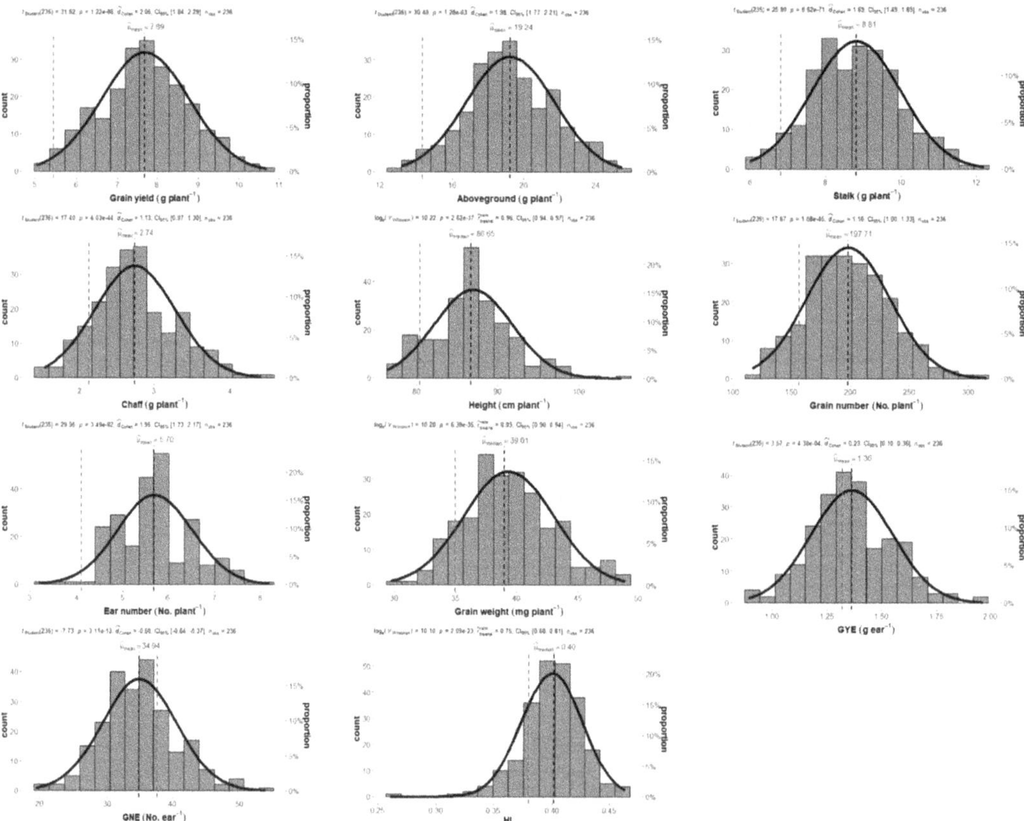

Figure 1. Histograms with a one-sample test of wheat production and grain yield components for 59 wheat genotypes, from CIMMYT grown under elevated CO_2 and high-temperature conditions. *GNE*: grain number ear^{-1}; *GYE*: grain yield ear^{-1}; *HI*: harvest index. Each plot shows the distribution of frequencies with a superimposed normal curve for the selected trait among a population of 59 wheat genotypes belonging to CIMMYT and their four replicates (n_{obs} = 236). Statistics ($t_{student}$ and $V_{Wilcoxon}$), p-value (p) and the size of the effect (d_{Cohen} and $r_{Wilcoxon}$) with confidence intervals (CI) for a one-sample test are added. The black dotted line represents the mean or median of population, according to the test employed (Student's t-test or Wilcoxon test). The blue dotted line represents the mean value for the four replicates (n = 4) of genotype 150 (Gazul). Differences were considered statistically significant at $p < 0.05$.

2.3. Correlation Network and Coefficient Matrix

To study the relationship among the traits under study in the population of the 60 wheat genotypes, both a correlation matrix and a correlation network were performed (Figure 2; Table S1). Positive correlations were found among the aboveground, stalk, chaff, grain yield, grain number and ear number traits, with very high correlations of aboveground biomass to stalk biomass (0.90), grain yield (0.88), chaff biomass (0.77) and grain number (0.76), as well as between grain yield and grain number (0.85). GYE and GNE were also positively correlated with each other (0.78), and both were correlated, to a lesser extent (0.11−0.61), with aboveground, stalk and chaff biomasses, grain yield and grain number traits. Furthermore, they were also negatively correlated with ear numbers (−0.44 and −0.29, respectively). Grain weight showed no correlation with chaff biomass and very low negative correlation with HI, stalk and aboveground biomasses, grain yield and ear number (ranged from −0.02 to −0.16), but it was negatively correlated with GNE (−0.54) and grain number per plant (−0.58). HI showed positive correlations with grain yield, GYE, grain number and GNE, but negative correlations with plant height and stalk and chaff biomasses, all of them lower than |0.45|. Plant height was in turn positively correlated with grain weight and with stalk and aboveground biomasses (0.28−0.15).

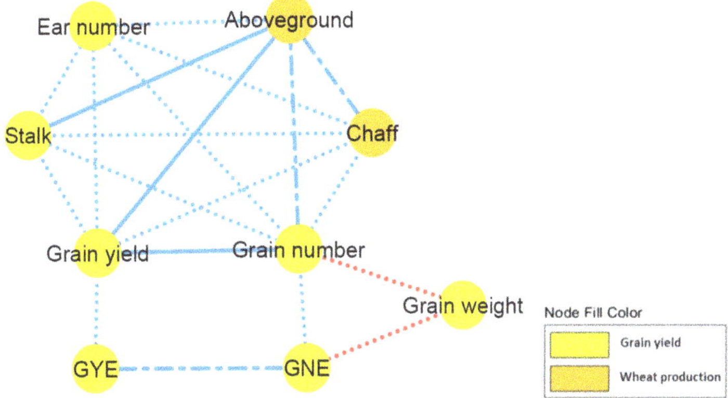

Figure 2. Correlation network for the wheat production and grain yield components of sixty wheat genotypes grown under elevated CO_2 and high-temperature conditions. *GNE*: grain number ear^{-1}; *GYE*: grain yield ear^{-1}. The different traits (nodes) were classified by colours according to their grain yield or wheat production nature (see legend). Edges stand for a Spearman's correlation $r \geq$ |0.45|, split up as dot (|0.65| $> r \geq$ |0.45|), dash and dot (|0.85| $> r \geq$ |0.65|) or solid (|1| $> r \geq$ |0.85|) line types. Blue edges indicate a positive correlation, whereas red lines implicate negative correlation.

2.4. Principal Component and Hierarchical Clustering Analyses

The natural variation in plant biomass and grain yield within the 60 wheat genotypes was further investigated using principal component analysis (PCA). Graphs of variables and individuals from PCA are shown in Figure 3a,b. Almost all the variables under study were positively correlated with the first dimension of the PCA (Table 2). Grain yield, grain number and aboveground biomass per plant were the most commonly contributing eigenvectors (18.85, 18.07 and 18.01, respectively) and correlated variables (0.95, 0.93 and 0.93) with the first dimension of the PCA. The stalk and chaff biomasses, GNE and ear number were also highly correlated variables, with correlation values ranging between 0.74 and 0.52, followed by GYE, HI and plant height. The grain weight was the only variable that was negatively correlated with the first dimension of the PCA (−0.49). Weaker positive correlations were found for the second dimension of the PCA, while the number of negatively correlated variables increased, together with their intensity. GNE, GYE and HI were the most contributing variables (21.58, 14.96 and 14.24) and the highest negatively correlated ones (-0.73, -0.61 and -0.59) with this second dimension. Grain number per plant was also negatively correlated with the second dimension (-0.21), although it barely contributed to this dimension (1.81). Ear number, grain weight and stalk biomass were the most positively correlated variables ($0.53-0.46$), followed by the height of the plant and the chaff and aboveground biomasses. The grain yield barely correlated with this second dimension (0.02).

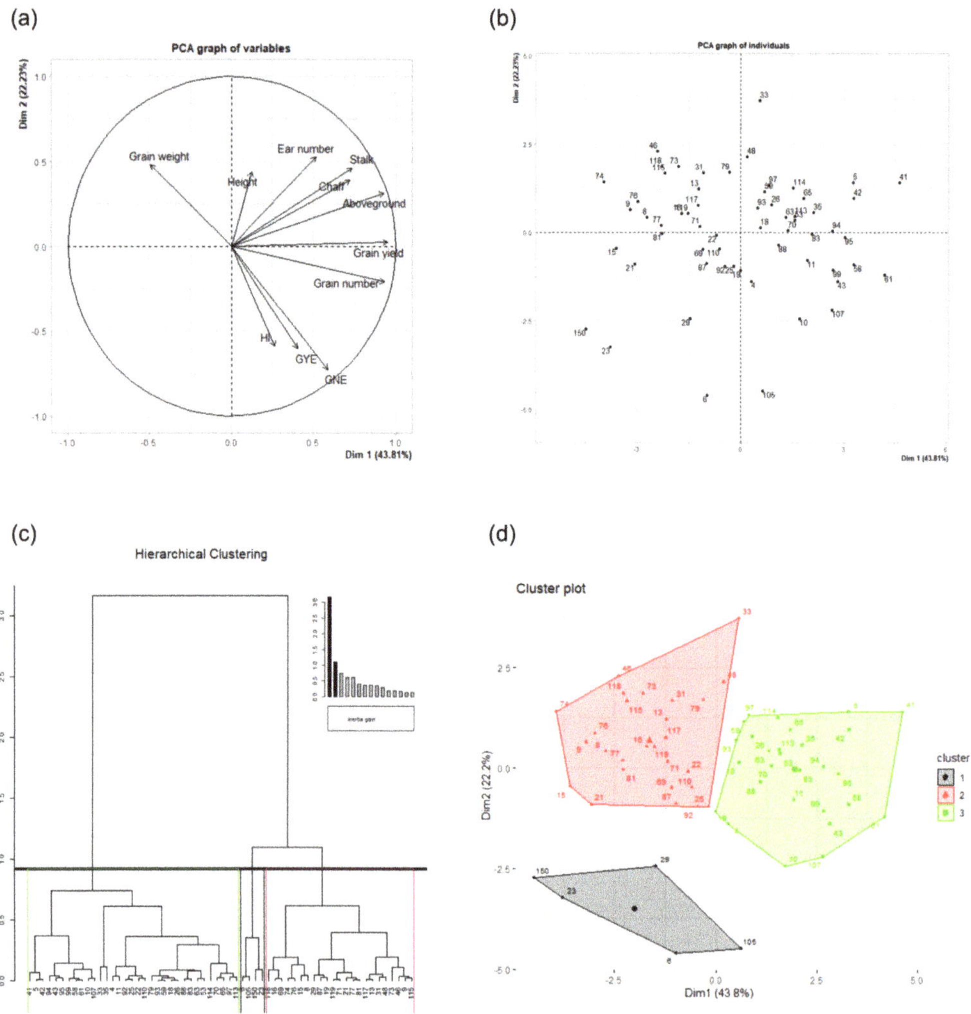

Figure 3. Hierarchical clustering based on principal component analysis (PCA) for the 60 wheat genotypes grown under elevated CO_2 and high-temperature conditions. *GNE*: grain number ear^{-1}; *GYE*: grain yield ear^{-1}; *HI*: harvest index. (**a**) The *variables plot* represents the correlation of traits with the PCA axes. (**b**) The *individuals plot* exhibits the position of genotypes in the PCA. (**c**) *Hierarchical clustering* of genotypes based on their distribution on the PCA. (**d**) *Cluster plot* of genotypes distributed in the PCA. PCA and hierarchical clustering were performed using individuals as the mean of the four replicates ($n = 4$) for each genotype.

Table 2. Correlations and contributions between the original variables with the first two dimensions of the principal component analysis.

	Dim. 1				Dim. 2		
Continuous Variables	Corr.	Cos^2	Contr.	Continuous Variables	Corr.	Cos^2	Contr.
Grain yield	0.95	0.91	18.85	Ear number	0.53	0.28	11.39
Grain number	0.93	0.87	18.07	Grain weight	0.48	0.23	9.44
Aboveground	0.93	0.87	18.01	Stalk	0.46	0.21	8.69
Stalk	0.74	0.55	11.33	Height	0.44	0.19	7.80
Chaff	0.72	0.52	10.88	Chaff	0.39	0.15	6.09
GNE	0.59	0.35	7.18	Aboveground	0.31	0.10	4.00
Ear number	0.52	0.27	5.52	Grain yield	0.02	0.00	0.02
GYE	0.40	0.16	3.39	Grain number	−0.21	0.04	1.81
HI	0.27	0.07	1.48	HI	−0.59	0.35	14.24
Height	0.13	0.02	0.33	GYE	−0.61	0.37	14.96
Grain weight	−0.49	0.24	4.96	GNE	−0.73	0.53	21.58

Dim.: dimension; *Contr.*: contribution; *Corr.*: correlation; *GNE*: grain number ear^{-1}; *GYE*: grain yield ear^{-1}; *HI*: harvest index; *Corr.* indicates the correlation between each variable and the dimension. The squared correlation (Cos^2) values between the variables and the dimensions are used to estimate the quality of the representation. *Cont.* expresses the contributions, in percentages, of each variable when accounting for the variability in the dimension.

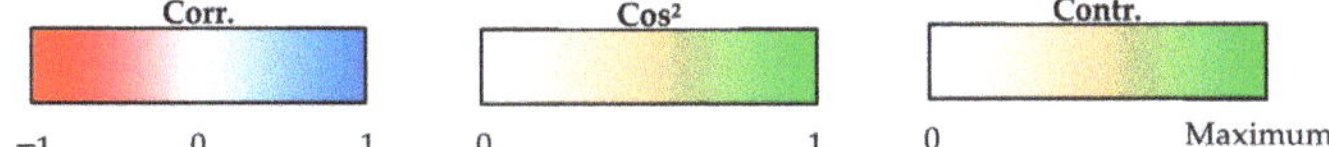

Hierarchical clustering on principal components (HCPC) for the 60 wheat genotypes was performed, based on the PCA results. Cutting of the tree defined three major clusters (clusters 1, 2 and 3; Figure 3c). Grouped by clusters, the distribution of the genotypes in the PCA is shown in Figure 3d, whereas Figure 4 shows boxplots of wheat production and grain yield components with a one-way analysis of variance (ANOVA) for clusters. Cluster 3 included the 28 genotypes most related to grain yield, aboveground biomass and grain number, whereas the 27 genotypes from cluster 2 were mainly associated with grain weight. Cluster 1 was composed of 5 genotypes, including genotype 150 (Gazul).

The one-way ANOVA found statistically significant differences for wheat production and grain yield components when the three clusters were compared. Cluster 3 showed significantly higher grain yield and grain number per plant than clusters 2 and 1, whereas no differences between the latter two were found. Aboveground, stalk and chaff biomasses were also significantly greater for cluster 3 than for clusters 2 and 1, as well as for cluster 2 than for cluster 1. Cluster 2 showed higher grain weight than clusters 1 or 3 but lower HI, with non-significant differences between clusters 1 and 3 for any variable. Furthermore, cluster 2 also had lower GYE and GNE than clusters 1 and 3, whereas cluster 1 had the highest. Finally, cluster 1 showed the lowest ear number and plant height, with no differences between clusters 2 and 3. All the traits under study exhibited a large effect ranging between 0.18−0.57, except for HI, which showed a middle-sized effect (0.06).

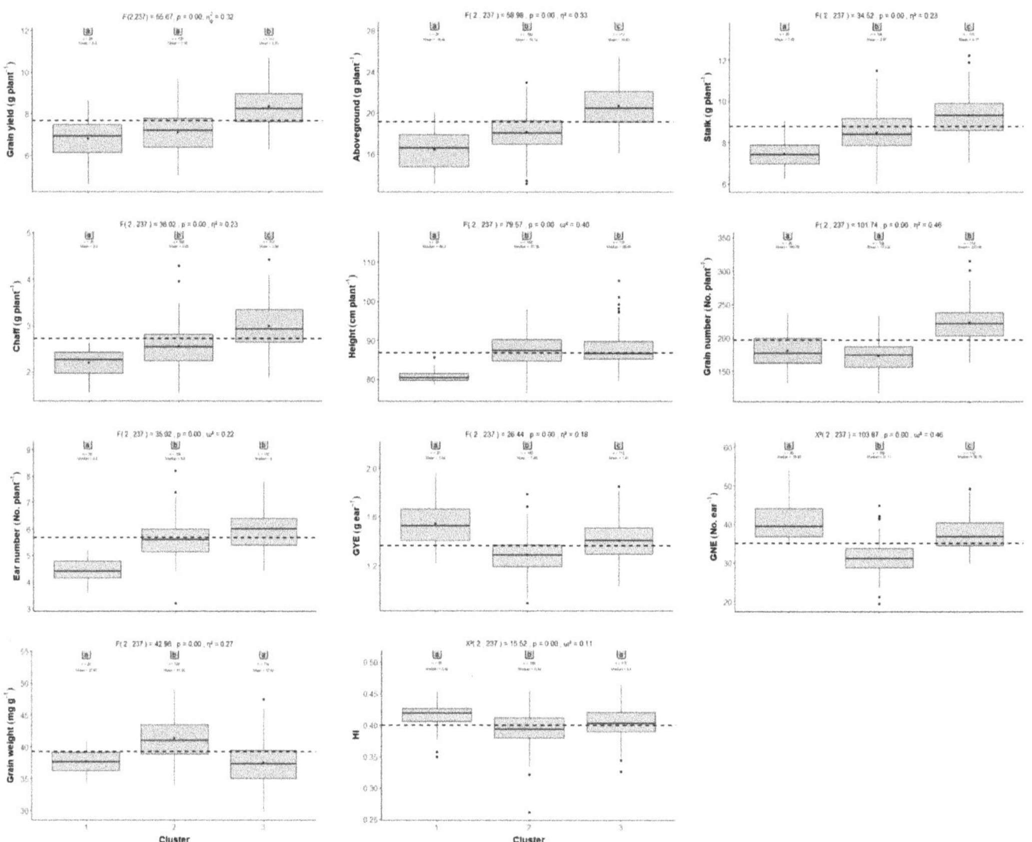

Figure 4. Boxplot with ANOVA of wheat production and grain yield components for the three defined clusters of the 60 wheat genotypes grown under elevated CO_2 and high-temperature conditions. *GNE*: grain number ear^{-1}; *GYE*: grain yield ear^{-1}; *HI*: harvest index. Black dots represent the mean of the replicates per cluster (Cluster 1, $n = 20$; Cluster 2, $n = 108$; Cluster 3, $n = 112$). The black dotted line represents the mean for all the genotypes and replicates ($N = 240$). Statistics (*F* and X^2), degrees of freedom, *p*-value (*p*) and size of the effect (η^2 and ω^2) for one-way ANOVA are added. Among columns, numbers followed by the same letter are not significantly different at $p < 0.05$ for post-hoc tests.

2.5. Days from Sowing to Ear Emergence, Anthesis and Maturity

Growth duration varied among genotypes. Days from sowing until plants reached ear emergence, anthesis and maturity growth stages ranged between 57−68, 60−69 and 92−102 days after sowing, respectively (Table S2). Negative correlations were found of the average number of days until plants reached these growth stages to wheat production and grain yield components, although all of them were under | 0.45 | (Table S3). Weak positive correlations were found among growth durations and GYE, GNE, HI and plant height.

3. Discussion

Climate change will inevitably affect crop development and productivity, increasing uncertainty regarding food production. In recent years, more attention has been paid to the risks of concomitant increases in atmospheric CO_2 concentrations and temperature for plant performance, and both factors should be assessed together to build up a realistic picture of how global warming will impact the productivity of food crops. In this context, the potential to exploit genetic variation becomes an essential strategy for the selection of crops with better adaptation to future climate conditions. So far, the effects of elevated CO_2 when combined with increasing temperatures on wheat yield have rarely been investigated under field conditions or enclosure facilities, and most of those studies have used either a single genotype or a very limited number of cultivars [4,40–45]. Here, the genotypic variation in grain yield is investigated across a set of 60 bread wheat genotypes, using controlled environmental chambers as an approach to perform a more accurate simulation of predicted daily and seasonal variations of temperature during the wheat-growing cycle in the region of Salamanca (Spain) by the end of the century. Among the 60 wheat genotypes under study (Table 1), there were several genotypes for which the range of values in biomass production and yield-related traits were more than double the average values achieved by the lowest-ranking genotypes, showing evidence for variability in the performance of the 60 genotypes under combined elevated CO_2 and high-temperature conditions. This implies that the selection of genotypes achieving greater productivity in a future climate scenario is feasible, as is in accordance with previous works by the authors regarding bread wheat [45], and of others regarding durum wheat [44]. It is worth noting that the CIMMYT genotypes exhibited higher biomass and grain yield when compared to Gazul (Figure 1), a result that can be interpreted as the former having better adaptability to the environment. In spite of the higher grain number per ear observed in Gazul, this was insufficient to compensate for the reduction in ear number per plant and grain weight, thus resulting in lower grain yield and biomass. Taken together, these findings suggest that the CIMMYT genotypes performed better than the Gazul genotype under combined elevated CO_2 and high-temperature conditions, and also provide evidence for the success of breeding programs conducted under warmer temperature environments.

Likewise, we observed a strong positive correlation of grain yield with aboveground biomass and grain number (Figure 2; Table S1), as well as a highly positive correlation of grain yield with ear number and stalk and chaff biomasses; all of them, in turn, correlated with each other. These results suggest that increased grain yield was attained by a higher grain number due to a larger number of ears, but not from an increase in grain weight, as is in agreement with previous studies in wheat grown under elevated CO_2 [18,46–48]. Data from the chamber experiments [46] also reported increased grain numbers per ear, but not the FACE experiments [49]. As is consistent with our finding, the grain number per square meter was the most important yield component, accounting for the effects of elevated CO_2 and temperature in wheat and rice [4]. Altogether, these findings suggest that tillering capacity, and hence the increased number of fertile tillers, is an important factor in yield response to elevated CO_2 [26,50], especially under higher temperatures and dryland conditions [40,44]. In a study comprising 20 wheat genotypes grown in glasshouses and controlled environment chambers, it was also found that CO_2 enrichment stimulated tillering, regardless of the cultivars [51]. For nine durum wheat varieties grown hydroponically, Sabella et al. (2020) [44] observed a higher yield for the old cultivar, Cappelli, when compared to modern cultivars under elevated CO_2 and high temperatures, and this was associated with a larger number of ears per plant rather than to changes within the ears. These data resemble previous studies showing a greater response to elevated CO_2 by older wheat varieties than the more recently released ones, because of the higher plasticity of the former in tiller production [26,52]. Evaluating plant developmental responses to CO_2 enrichment has underlined the importance of sink strength at the whole plant level, both as a mechanism to increase plant growth and to avoid photosynthetic acclimation [23]. Enhanced leaf carbohydrate content caused by higher photosynthesis rates under CO_2 enrichment might be efficiently used for the development of further sinks, such as new tillers, to distribute the photoassimilates [53]. Consequently, this ability to develop new tillers when an extra carbon source is made available is in accordance with the previous suggestion that ensuring adequate sink strength is essential for maximizing the response to elevated CO_2.

Furthermore, we observed that grain weight was negatively correlated with grain number per plant and per ear (Figure 2; Table S1), indicative of the change in size distribution toward smaller grains, as previously shown in wheat plants grown under elevated CO_2 [18]. This finding is in contrast to the slight shifts toward a larger grain size, as reported by Högy et al. (2013) [54], and the increase in grain size found in three wheat cultivars by Panozzo et al. (2014) [55], although most previous results from FACE experiments showed slight changes in grain weight or that it simply remained unchanged [48,49]. Such contradictory results in wheat grain weight could be associated with the co-influence of other factors related to differences in growing conditions [18,56], such as water availability [47,57]. Increased temperature adversely impacts grain development, owing to several limitations in assimilate supply, the length of the grain-filling period, and the starch biosynthesis and deposition rates, together leading to smaller grains [32,58]. Wheat is most sensitive to abrupt heat stress around flowering, and becomes more tolerant to higher temperatures during the grain-filling stage [59]. In line with this observation, Wollenweber et al. (2003) [60] found distinct effects on grain weight according to the timing of heat application. Weichert et al. (2017) [61] reported the adverse impact of a post-anthesis heatwave on wheat grain yield, being responsible for reduced grain size in the last term. In addition, it is worthy of highlighting that central spikelets and proximal florets flower earlier and receive priority in the deposition of assimilates, and thus tend to have a larger grain size when compared to distal spikelets and florets [58,62]. Higher temperatures may induce upper and lower spikelets and distal florets to abort or develop constitutively smaller grains; this may, therefore, bring together a greater heterogeneity in the crop grain size and a shift in the distribution toward smaller grain size [58]. Even genetic differences among cultivars may contribute to the inconsistent results on wheat grain under elevated CO_2 [50]. Among wheat genotypes, a negative relationship between grains per square meter and grain weight has frequently been observed [63]. In fact, with the advent of the semidwarf cultivars, there was a trend toward a decrease in grain weight, accompanied by more grains in the distal positions in spikelets having a lower potential grain weight [64]. These observations may explain the negative relationship between grain weight and yield components, including the grain number per plant and per ear, as observed in our study, which is not related to competition among grains for assimilates but to the increasing contribution of grains of low potential weight that are, thereby, smaller [65,66].

From the hierarchical clustering analysis (Figure 3c), we inferred that the wheat genotypes were grouped into three clusters with similar variation sources in each of them (Figure 3d), although differences in biomass and yield-related traits among clusters were identified in line with the PCA analysis (Figure 3a,b). These findings suggest that there were contrasting yield strategies among the wheat genotypes under the studied environmental conditions. Despite the larger number of grains per ear exhibited by genotypes of cluster 1, and the improved grain weight of those of cluster 2, the highest yielding genotypes were found within cluster 3, in association with the highest aboveground, stalk and chaff biomasses, grain number, and ears per plant (Figure 4). This suggests that genotypes that can preferentially increase these yield-related traits will lead to better yield response under combined elevated CO_2 and high temperature. In line with this observation, most of the previous studies worldwide have shown that grain yield progress in wheat was mostly correlated with grain number per square meter rather than grain weight [67–69]. Similarly, in Spain, the bread wheat genetic improvement on yield from 1930 to 2000 was also achieved by an increase in grain number, while grain weight was unchanged [70]. In fact, the introduction of semidwarf cultivars throughout the 1960s increased grains per square meter by increasing assimilate allocation to the ear during the pre-flowering period, allowing more floret primordia to become fertile florets [71]. Therefore, past genetic gains in bread wheat yield have been widely linked to an increase in harvest index and a decline in plant height [72]. Overall, improved semidwarf high-yielding bread wheat varieties from Mexico and European countries, mostly France and Italy, have played a major role in yield improvements in Spain [73], where the scarcity of rain is frequent in spring when temperatures rise rapidly. Thus, the incorporation of drought tolerance has always been one of the main goals of Spanish wheat-breeding programs, because drought and high temperature are the prevalent stresses during the reproductive stages constraining wheat production in Spain [74]. During the 1990s, the private seed sector successfully introduced and marketed 74 bread wheat varieties, including Gazul, which is currently cultivated in Spain [73]. Tillering is regulated genotypically, but it is also influenced by the environment [75]. In this context, tillering reduction has been proposed as a suitable feature under terminal drought stress because it decreases soil water use before anthesis [76]. Improved tiller economy could enhance the partitioning of assimilates to the ear, leading to increased grain yield per ear, mainly resulting from higher grain number per ear, but decreased grain per square meter [76,77]. In accordance with these findings, cluster 1 grouped the lowest-yielding genotypes, driven by a decreased number of productive tillers with improved ear fertility but decreased grain yield. The fact that genotype 150 (Gazul) was closely related to four CIMMYT genotypes highlights the important contribution of CIMMYT germplasm in the release of Spanish varieties during the last century [70,73], as reported by the seed company (Limagrain Iberica SA) involved in the marketing of improved bread varieties. It is interesting, however, that the genotypes grouped into cluster 2 seem to sustain grain yield production by higher grain weight, to compensate for reduced grain numbers per ear. Therefore, these intermediate yielding genotypes, when compared with the lowest and highest yielding genotypes, had different attributes with heavier grains rather than higher grain numbers per square meter. In agreement with these results, it has been found that the grain yield progress of CIMMYT advanced lines from 1977 to 2008 was driven by higher grain weight, rather than more grains per square meter [78]. Other earlier studies have also reported that grain weight has contributed to yield progress [79,80].

Breeding programs have focused on developing germplasm adapted to different crop-producing geographic regions and, in Mediterranean areas, the earliness of heading has been recognized as an adaptive trait of modern cultivars aiming to escape drought/heat terminal stress [77,81]. Indeed, Lopes et al. (2012) [78] reported that, in warmer environments, the reduction in days until heading displayed in high-yielding spring wheat advanced lines allowed an escape from drought/heat episodes at the most sensitive stage of grain setting. In our study, there were differences among genotypes in the number of days from sowing to ear emergence, which could partly explain the better performance under the studied conditions of the early-heading genotypes when compared to the late-heading ones (Table S2). Nevertheless, genotype 150 (Gazul) performed poorly under combined elevated CO_2 and high temperature, although it was one of the earliest-heading genotypes. Hence, other features should be considered, because phenology does not appear to be the only cause of variability in grain yield (Table S3). It must be highlighted that one of the highest yielding genotypes in our study (genotype 5) has also exhibited good and stable performance under warm temperatures in previous studies [82].

4. Materials and Methods

4.1. Plant Material and Experimental Conditions

The experiment was carried out with 60 bread wheat genotypes (*Triticum aestivum* L.). Fifty-nine genotypes were selected to represent the germplasm of the CIMMYT heat-tolerance wheat screening nursery (8TH HTWSN), together with the Gazul genotype (referred to as genotype 150), with a high yield and adaptability to the Mediterranean climate of the Salamanca region [10,83,84], as well as being identified as a genetic source of excellent bread-making quality owing to the high protein content and specific glutenin and gliadin quality of their grain [73] (Table S4). The genotypes of the 8TH HTWSN were previously selected for high performance in warmer temperatures [85]. Seeds were sown in 5-L pots, with a density of five plants per pot after emergence. Pots were filled with 1.2 kg of a mixture of peat:perlite (4:1), and 4 g of both KNO_3 and KH_2PO_4 were added to each pot, with the peat providing enough provision of other nutrients [86]. Pots were placed in two, 3.6-m length × 4.8-m width × 2.4-m height, controlled environment chambers maintained on a 16-/8-hour light/dark regime with an irradiance of 400 μmol m^{-2} s^{-1}

at the top of the canopy, provided by a combination of blue- plus red-peak fluorescent lamps, and relative humidity of 40%/60% day/night. The atmospheric CO_2 concentration was set at 700 μmol mol^{-1} by injecting pure CO_2 [87,88]. The temperature was 4 °C above current temperatures, simulating the daily and seasonal oscillations of typical temperatures in the natural environments of the Salamanca region [45]. Four different sections were established to reproduce the daily temperature oscillations: night, and the initial, central and final parts of the photoperiod. These temperatures were increased by three levels reproducing the natural seasonal oscillations throughout wheat development, when at least 50% of the plants reached ear emergence and anthesis, respectively. The experiment was conducted in a completely randomized design, with four replicates/pot per each of the studied genotypes. Once germinated, 1200 plants were grown in 240 pots. Water was supplied during crop development three times per week to maintain pot field capacity, and the pots were rotated twice a week in order to avoid edge effects.

4.2. Evaluation of Phenological Traits

In the 60 bread wheat genotypes, through periodic observation (three times per week) the following traits were evaluated: the number of days from sowing to ear emergence (days until heading) was recorded when 50% of the plants per each of the studied genotypes had fully emerged ears; days from sowing to anthesis were determined when 50% of ears showed extruded anthers along the head; the days from anthesis to maturity were also evaluated. After anthesis, plant height was determined as the distance from the ground to the ear tip, excluding awns, of the main stem of two plants per pot per each of the studied genotypes.

4.3. Harvesting and Yield Component Determination

At physiological maturity, the aboveground plant parts were harvested from each pot and divided into stalks (stems and leaves) and ears. Grains and chaff components were separated from the ears by manual threshing. The number of ears and grains per plant and per ear were determined, and the dry weights for the stalk, chaff, and the grain yield per plant and per ear were recorded after drying in an oven at 60 °C over 48 h. The grain weight was estimated as the quotient between the grain yield and the grain number per plant. The harvest index (HI) was also determined as the ratio of grain yield to total aboveground biomass.

4.4. Statistical Analysis

The normal distribution of all parameters under study was tested using the Shapiro–Wilk test from the function *tapply* of the *base* package in the statistical software *R* [89]. Descriptive statistics of variables were measured using the function *describBy* from the *psych* package [90]. A multiple analysis of variance (MANOVA) was performed to determine differences for each studied trait among the 60 genotypes employed, and their four replicates per genotype ($N = 240$), using the function *manova* of the package *stats* [89]. A one-sample test was carried out for comparison of the wheat production and grain yield components between the average values of the Gazul genotype ($n = 4$) and the 59 wheat genotype population belonging to CIMMYT ($n = 236$). A PCA was applied to the genotype by a trait (GxT) matrix of means, as described in Driever et al. (2014) [5]. The principal components performed were employed as a pre-process for genotype clustering (hierarchical clustering on principal components (HCPC)) as suggested by Husson et al. (2010) [91]. Clusters grouped a different number of genotypes, with four replicates per genotype (Cluster 1: 5 genotypes, $n = 20$; Cluster 2: 27 genotypes, $n = 108$; Cluster 3: 28 genotypes, $n = 112$). The test of Levene for homogeneity of variance was conducted to determine homoscedasticity among clusters in the traits under study, using the package *DescTools* [92]. For comparisons of the traits among clusters, a one-way ANOVA for independent measures was performed. Table S5 schematizes decisions taken to the choice of the best one-sample or one-way ANOVA test, together with the interpretation for the values of the effect size. Correlation networks were performed with the packages *psych* and *reshape2* [91,93], along with the software "Cytoscape" [94], using a threshold for the Spearman's correlation values of r $\geq$ |0.45|. Correlation coefficients for wheat production and grain yield components were also calculated, using the 60 genotypes employed and their replicates ($N = 240$), whereas, for correlations of these traits with the days to heading, only the mean values for each genotype were employed. For the whole study, differences were considered significant at $p < 0.05$.

5. Conclusions

Crops are predicted to be exposed more frequently to high temperatures in the future as a consequence of climate change. Therefore, the exploration of variability in wheat productivity under combined elevated CO_2 and high temperatures is essential for the selection of genotypes that are better adapted to future global warming. The dissection of yield-related traits in our study discloses that the bread wheat genotypes employed diverse strategies to achieve final grain yield under elevated CO_2 levels combined with a high temperature, most probably as a consequence of the contrasting breeding strategies used to improve grain yield throughout the 20th century. Increased grain yield was driven by grain numbers and ears per plant, rather than per grain weight. The CIMMYT genotypes performed better than the Gazul genotype under combined elevated CO_2 and high temperature, providing evidence for the success of the breeding programs under warmer temperature environments. Further research is needed to investigate whether the highest yielding genotypes identified in the present study will be able to maintain their yield advantage, regardless of nutrient availability or drought stress, combined with high temperature, for the selection of widely adapted climate-resilient wheat genotypes.

Supplementary Materials: The following are available online at https://www.mdpi.com/article/10.3390/plants10081596/s1, Table S1: Correlation coefficient matrix for wheat production and grain yield components. Table S2: Average days from sowing to ear emergence, anthesis and maturity developmental stages. Table S3: Correlation coefficient matrix for average wheat production and grain yield components, with days from sowing to ear emergence, anthesis and maturity developmental stages. Table S4: Catalogue of the 60 wheat genotypes employed in the present study. Table S5: Statistical design with tests employed.

Author Contributions: R.M.-C., P.P. and R.M. conceived and designed the experiment. R.M.-C., P.P., J.B.A., E.L.M.-B. and R.M. contributed to the experimental work and conducted plant growth analyses. P.P. and E.L.M.-B. compiled the data, while E.L.M.-B. carried out the statistical analysis and description of results. R.M. wrote the paper. All authors have read and agreed to the published version of the manuscript.

Funding: This research was supported by the Spanish National R&D&I Plan of the Ministry of Economy and Competitiveness (Projects AGL2013-41363-R and AGL2016-79589-R) and Ministry of Science and Innovation (Project PID2019-107154RB-100), as well as the regional government, the Junta de Castilla y León (Projects CSI083U16 and CSI260P20), with European Regional Development Fund, ERDF. Project "CLU-2019-05—IRNASA/CSIC Unit of Excellence", funded by the Junta de Castilla y León and co-financed by the European Union (ERDF "Europe drives our growth"). E.L. Marcos-Barbero was the recipient of a pre-doctoral contract from the Junta de Castilla y León (E-37-2017-0066125).

Institutional Review Board Statement: Not applicable.

Informed Consent Statement: Not applicable.

Data Availability Statement: Data are available upon request from the corresponding authors.

Acknowledgments: The technical cooperation of A. Verdejo and M.A. Boyero in control of plant growth and harvesting of plant material is acknowledged. We also acknowledge CIMMYT for providing the seeds of the nursery 8TH HTWSN. The support of the publication fee by the CSIC Open Access Publication Support Initiative, through its Unit of Information Resources for Research (URICI), is acknowledged as well.

Conflicts of Interest: The authors declare no conflict of interest.

References

1. *IPCC Climate Change 2014: Synthesis Report. Contribution of Working Groups I, II and III to the Fifth Assessment Report of the Intergovernmental Panel on Climate Change*; IPCC: Geneva, Switzerland, 2014.
2. NOAA-ESRL Trends in atmospheric carbon dioxide. Available online: https://www.esrl.noaa.gov/gmd/ccgg/trends/index.html (accessed on 1 July 2021).
3. Lobell, D.B.; Schlenker, W.; Costa-Roberts, J. Climate trends and global crop production since 1980. *Science* **2011**, *333*, 616–620. [CrossRef]
4. Cai, C.; Yin, X.; He, S.; Jiang, W.; Si, C.; Struik, P.C.; Luo, W.; Li, G.; Xie, Y.; Xiong, Y.; et al. Responses of wheat and rice to factorial combinations of ambient and elevated CO_2 and temperature in FACE experiments. *Glob. Chang. Biol.* **2016**, *22*, 856–874. [CrossRef]
5. Driever, S.M.; Lawson, T.; Andralojc, P.J.; Raines, C.A.; Parry, M.A.J. Natural variation in photosynthetic capacity, growth, and yield in 64 field-grown wheat genotypes. *J. Exp. Bot.* **2014**, *65*, 4959–4973. [CrossRef]
6. Abdelrahman, M.; Burritt, D.J.; Gupta, A.; Tsujimoto, H.; Tran, L.-S.P. Heat stress effects on source–sink relationships and metabolome dynamics in wheat. *J. Exp. Bot.* **2020**, *71*, 543–554. [CrossRef]
7. Long, S.P.; Ainsworth, E.A.; Rogers, A.; Ort, D.R. Rising atmospheric carbon dioxide: Plants FACE the future. *Annu. Rev. Plant Biol.* **2004**, *55*, 591–628. [CrossRef]
8. Ainsworth, E.A.; Rogers, A. The response of photosynthesis and stomatal conductance to rising [CO_2]: Mechanisms and environmental interactions. *Plant Cell Environ.* **2007**, *30*, 258–270. [CrossRef]
9. Pérez, P.; Morcuende, R.; Martín del Molino, I.; Martínez-Carrasco, R. Diurnal changes of Rubisco in response to elevated CO_2, temperature and nitrogen in wheat grown under temperature gradient tunnels. *Environ. Exp. Bot.* **2005**, *53*, 13–27. [CrossRef]
10. Gutiérrez, D.; Gutiérrez, E.; Pérez, P.; Morcuende, R.; Verdejo, A.L.; Martinez-Carrasco, R. Acclimation to future atmospheric CO_2 levels increases photochemical efficiency and mitigates photochemistry inhibition by warm temperatures in wheat under field chambers. *Physiol. Plant.* **2009**, *137*, 86–100. [CrossRef] [PubMed]
11. Pérez, P.; Alonso, A.; Zita, G.; Morcuende, R.; Martínez-Carrasco, R. Down-regulation of Rubisco activity under combined increases of CO_2 and temperature minimized by changes in Rubisco kcat in wheat. *Plant Growth Regul.* **2011**, *65*, 439–447. [CrossRef]
12. Vicente, R.; Pérez, P.; Martínez-Carrasco, R.; Usadel, B.; Kostadinova, S.; Morcuende, R. Quantitative RT-PCR platform to measure transcript levels of C and N metabolism-related genes in durum wheat: Transcript profiles in elevated [CO_2] and high temperature at different levels of n supply. *Plant Cell Physiol.* **2015**, *56*, 1556–1573. [CrossRef] [PubMed]
13. Martínez-Carrasco, R.; Pérez, P.; Morcuende, R. Interactive effects of elevated CO_2, temperature and nitrogen on photosynthesis of wheat grown under temperature gradient tunnels. *Environ. Exp. Bot.* **2005**, *54*, 49–59. [CrossRef]
14. Aranjuelo, I.; Cabrera-Bosquet, L.; Morcuende, R.; Avice, J.C.; Nogués, S.; Araus, J.L.; Martínez-Carrasco, R.; Pérez, P. Does ear C sink strength contribute to overcoming photosynthetic acclimation of wheat plants exposed to elevated CO_2? *J. Exp. Bot.* **2011**, *62*, 3957–3969. [CrossRef]
15. Taub, D.R.; Wang, X. Why are nitrogen concentrations in plant tissues lower under elevated CO_2? *A critical examination of the hypotheses. J. Integr. Plant Biol.* **2008**, *50*, 1365–1374. [CrossRef]
16. Carlisle, E.; Myers, S.; Raboy, V.; Bloom, A. The effects of inorganic nitrogen form and CO_2 concentration on wheat yield and nutrient accumulation and distribution. *Front. Plant Sci.* **2012**, *3*, 1–13. [CrossRef] [PubMed]
17. Leakey, A.D.B.; Ainsworth, E.A.; Bernacchi, C.J.; Rogers, A.; Long, S.P.; Ort, D.R. Elevated CO_2 effects on plant carbon, nitrogen, and water relations: Six important lessons from FACE. *J. Exp. Bot.* **2009**, *60*, 2859–2876. [CrossRef]
18. Högy, P.; Wieser, H.; Köhler, P.; Schwadorf, K.; Breuer, J.; Franzaring, J.; Muntifering, R.; Fangmeier, A. Effects of elevated CO_2 on grain yield and quality of wheat: Results from a 3-year free-air CO_2 enrichment experiment. *Plant Biol.* **2009**, *11*, 60–69. [CrossRef] [PubMed]
19. Erbs, M.; Manderscheid, R.; Jansen, G.; Seddig, S.; Pacholski, A.; Weigel, H.J. Effects of free-air CO_2 enrichment and nitrogen supply on grain quality parameters and elemental composition of wheat and barley grown in a crop rotation. *Agric. Ecosyst. Environ.* **2010**, *136*, 59–68. [CrossRef]
20. ARP, W.J. Effects of source-sink relations on photosynthetic acclimation to elevated CO_2. *Plant. Cell Environ.* **1991**, *14*, 869–875. [CrossRef]
21. Del Pozo, A.; Pérez, P.; Morcuende, R.; Alonso, A.; Martínez-Carrasco, R. Acclimatory responses of stomatal conductance and photosynthesis to elevated CO_2 and temperature in wheat crops grown at varying levels of N supply in a Mediterranean environment. *Plant Sci.* **2005**, *169*, 908–916. [CrossRef]
22. Del Pozo, A.; Pérez, P.; Gutiérrez, D.; Alonso, A.; Morcuende, R.; Martínez-Carrasco, R. Gas exchange acclimation to elevated CO_2 in upper-sunlit and lower-shaded canopy leaves in relation to nitrogen acquisition and partitioning in wheat grown in field chambers. *Environ. Exp. Bot.* **2007**, *59*, 371–380. [CrossRef]
23. Ainsworth, E.A.; Rogers, A.; Nelson, R.; Long, S.P. Testing the "source-sink" hypothesis of down-regulation of photosynthesis in elevated [CO_2] in the field with single gene substitutions in Glycine max. *Agric. For. Meteorol.* **2004**, *122*, 85–94. [CrossRef]

24. Tausz-Posch, S.; Norton, R.M.; Seneweera, S.; Fitzgerald, G.J.; Tausz, M. Will intra-specific differences in transpiration efficiency in wheat be maintained in a high CO_2 world? *A FACE study. Physiol. Plant.* **2013**, *148*, 232–245. [CrossRef]
25. Ziska, L.H.; Morris, C.F.; Goins, E.W. Quantitative and qualitative evaluation of selected wheat varieties released since 1903 to increasing atmospheric carbon dioxide: Can yield sensitivity to carbon dioxide be a factor in wheat performance? *Glob. Chang. Biol.* **2004**, *10*, 1810–1819. [CrossRef]
26. Ziska, L.H. Three-year field evaluation of early and late 20th century spring wheat cultivars to projected increases in atmospheric carbon dioxide. *F. Crop. Res.* **2008**, *108*, 54–59. [CrossRef]
27. Schynder, H. The role of carbohydrate storage and redistribution in the source-sink relations of wheat and barley during grain filling—A review. *New Phytol.* **1993**, *123*, 233–245. [CrossRef]
28. Yang, J.; Zhang, J. Grain filling of cereals under soil drying. *New Phytol.* **2006**, *169*, 223–236. [CrossRef]
29. Cossani, C.M.; Reynolds, M.P. Physiological traits for improving heat tolerance in wheat. *Plant Physiol.* **2012**, *160*, 1710–1718. [CrossRef]
30. Djanaguiraman, M.; Narayanan, S.; Erdayani, E.; Prasad, P.V.V. Effects of high temperature stress during anthesis and grain filling periods on photosynthesis, lipids and grain yield in wheat. *BMC Plant Biol.* **2020**, *20*, 1–12. [CrossRef]
31. Lobell, D.B.; Field, C.B. Global scale climate–crop yield relationships and the impacts of recent warming. *Environ. Res. Lett.* **2007**, *2*, 014002. [CrossRef]
32. Farooq, M.; Bramley, H.; Palta, J.A.; Siddique, K.H.M. Heat stress in wheat during reproductive and grain-filling phases. *CRC. Crit. Rev. Plant Sci.* **2011**, *30*, 491–507. [CrossRef]
33. Prasad, P.V.V.; Djanaguiraman, M. Response of floret fertility and individual grain weight of wheat to high temperature stress: Sensitive stages and thresholds for temperature and duration. *Funct. Plant Biol.* **2014**, *41*, 1261–1269. [CrossRef]
34. Bergkamp, B.; Impa, S.M.; Asebedo, A.R.; Fritz, A.K.; Jagadish, S.V.K. Prominent winter wheat varieties response to post-flowering heat stress under controlled chambers and field based heat tents. *F. Crop. Res.* **2018**, *222*, 143–152. [CrossRef]
35. Barnabás, B.; Jäger, K.; Fehér, A. The effect of drought and heat stress on reproductive processes in cereals. *Plant Cell Environ.* **2008**, *31*, 11–38. [CrossRef] [PubMed]
36. Dupont, F.M.; Hurkman, W.J.; Vensel, W.H.; Tanaka, C.; Kothari, K.M.; Chung, O.K.; Altenbach, S.B. Protein accumulation and composition in wheat grains: Effects of mineral nutrients and high temperature. *Eur. J. Agron.* **2006**, *25*, 96–107. [CrossRef]
37. Asseng, S.; Ewert, F.; Martre, P.; Rötter, R.P.; Lobell, D.B.; Cammarano, D.; Kimball, B.A.; Ottman, M.J.; Wall, G.W.; White, J.W.; et al. Rising temperatures reduce global wheat production. *Nat. Clim. Chang.* **2015**, *5*, 143–147. [CrossRef]
38. Liu, B.; Asseng, S.; Müller, C.; Ewert, F.; Elliott, J.; Lobell, D.B.; Martre, P.; Ruane, A.C.; Wallach, D.; Jones, J.W.; et al. Similar estimates of temperature impacts on global wheat yield by three independent methods. *Nat. Clim. Chang.* **2016**, *6*, 1130–1136. [CrossRef]
39. Hatfield, J.L.; Dold, C. Agroclimatology and wheat production: Coping with climate change. *Front. Plant Sci.* **2018**, *9*, 1–5. [CrossRef]
40. Fitzgerald, G.J.; Tausz, M.; O'Leary, G.; Mollah, M.R.; Tausz-Posch, S.; Seneweera, S.; Mock, I.; Löw, M.; Partington, D.L.; Mcneil, D.; et al. Elevated atmospheric [CO_2] can dramatically increase wheat yields in semi-arid environments and buffer against heat waves. *Glob. Chang. Biol.* **2016**, *22*, 2269–2284. [CrossRef]
41. Ainsworth, E.A.; Long, S.P. 30 years of free-air carbon dioxide enrichment (FACE): What have we learned about future crop productivity and its potential for adaptation? *Glob. Chang. Biol.* **2021**, *27*, 27–49. [CrossRef] [PubMed]
42. Sánchez De La Puente, L.; Pérez Pérez, P.; Martínez-Carrasco, R.; Morcuende Morcuende, R.; Martín Del Molino, I.M. Action of elevated CO_2 and high temperatures on the mineral chemical composition of two varieties of wheat. *Agrochimica* **2000**, *44*, 221–230.
43. Högy, P.; Kottmann, L.; Schmid, I.; Fangmeier, A. Heat, wheat and CO_2: The relevance of timing and the mode of temperature stress on biomass and yield. *J. Agron. Crop Sci.* **2019**, *205*, 608–615. [CrossRef]
44. Sabella, E.; Aprile, A.; Negro, C.; Nicolì, F.; Nutricati, E.; Vergine, M.; Luvisi, A.; De Bellis, L. Impact of climate change on durum wheat yield. *Agronomy* **2020**, *10*, 793. [CrossRef]
45. Marcos-Barbero, E.L.; Pérez, P.; Martínez-Carrasco, R.; Arellano, J.B.; Morcuende, R. Genotypic variability on grain yield and grain nutritional quality characteristics of wheat grown under elevated CO_2 and high temperature. *Plants* **2021**, *10*, 1043. [CrossRef]
46. Wang, L.; Feng, Z.; Schjoerring, J.K. Effects of elevated atmospheric CO2 on physiology and yield of wheat (Triticum aestivum L.): A meta-analytic test of current hypotheses. *Agric. Ecosyst. Environ* **2013**, *178*, 57–63. [CrossRef]
47. Erice, G.; Sanz-Sáez, Á.; González-Torralba, J.; Méndez-Espinoza, A.M.; Urretavizcaya, I.; Nieto, M.T.; Serret, M.D.; Araus, J.L.; Irigoyen, J.J.; Aranjuelo, I. Impact of elevated CO_2 and drought on yield and quality traits of a historical (Blanqueta) and a modern (Sula) durum wheat. *J. Cereal Sci.* **2019**, *87*, 194–201. [CrossRef]
48. Tausz-Posch, S.; Seneweera, S.; Norton, R.M.; Fitzgerald, G.J.; Tausz, M. Can a wheat cultivar with high transpiration efficiency maintain its yield advantage over a near-isogenic cultivar under elevated CO_2? *F. Crop. Res.* **2012**, *133*, 160–166. [CrossRef]
49. Högy, P.; Keck, M.; Niehaus, K.; Franzaring, J.; Fangmeier, A. Effects of atmospheric CO_2 enrichment on biomass, yield and low molecular weight metabolites in wheat grain. *J. Cereal Sci.* **2010**, *52*, 215–220. [CrossRef]
50. Tausz-Posch, S.; Dempsey, R.W.; Seneweera, S.; Norton, R.M.; Fitzgerald, G.; Tausz, M. Does a freely tillering wheat cultivar benefit more from elevated CO_2 than a restricted tillering cultivar in a water-limited environment? *Eur. J. Agron.* **2015**, *64*, 21–28. [CrossRef]

51. Bourgault, M.; Dreccer, M.F.; James, A.T.; Chapman, S.C. Genotypic variability in the response to elevated CO_2 of wheat lines differing in adaptive traits. *Funct. Plant Biol.* **2013**, *40*, 172–184. [CrossRef]
52. Tausz, M.; Tausz-Posch, S.; Norton, R.M.; Fitzgerald, G.J.; Nicolas, M.E.; Seneweera, S. Understanding crop physiology to select breeding targets and improve crop management under increasing atmospheric CO_2 concentrations. *Environ. Exp. Bot.* **2013**, *88*, 71–80. [CrossRef]
53. Makino, A.; Tadahiko, M. Photosynthesis and plant growth at elevated levels of CO_2. *Plant Cell Physiol.* **1999**, *40*, 999–1006. [CrossRef]
54. Högy, P.; Brunnbauer, M.; Koehler, P.; Schwadorf, K.; Breuer, J.; Franzaring, J.; Zhunusbayeva, D.; Fangmeier, A. Grain quality characteristics of spring wheat (Triticum aestivum) as affected by free-air CO_2 enrichment. *Environ. Exp. Bot.* **2013**, *88*, 11–18. [CrossRef]
55. Panozzo, J.F.; Walker, C.K.; Partington, D.L.; Neumann, N.C.; Tausz, M.; Seneweera, S.; Fitzgerald, G.J. Elevated carbon dioxide changes grain protein concentration and composition and compromises baking quality. *A FACE study. J. Cereal Sci.* **2014**, *60*, 461–470. [CrossRef]
56. Högy, P.; Fangmeier, A. Effects of elevated atmospheric CO_2 on grain quality of wheat. *J. Cereal Sci.* **2008**, *48*, 580–591. [CrossRef]
57. Fernando, N.; Panozzo, J.; Tausz, M.; Norton, R.M.; Neumann, N.; Fitzgerald, G.J.; Seneweera, S. Elevated CO_2 alters grain quality of two bread wheat cultivars grown under different environmental conditions. *Agric. Ecosyst. Environ.* **2014**, *185*, 24–33. [CrossRef]
58. Nuttall, J.G.; O'Leary, G.J.; Panozzo, J.F.; Walker, C.K.; Barlow, K.M.; Fitzgerald, G.J. Models of grain quality in wheat—A review. *F. Crop. Res.* **2017**, *202*, 136–145. [CrossRef]
59. Tashiro, T.; Wardlaw, I. The response to high temperature shock and humidity changes prior to and during the early stages of grain development in wheat. *Funct. Plant Biol.* **1990**, *17*, 551. [CrossRef]
60. Wollenweber, B.; Porter, J.R.; Schellberg, J. Lack of interaction between extreme high-temperature events at vegetative and reproductive growth stages in wheat. *J. Agron. Crop Sci.* **2003**, *189*, 142–150. [CrossRef]
61. Weichert, H.; Högy, P.; Mora-Ramirez, I.; Fuchs, J.; Eggert, K.; Koehler, P.; Weschke, W.; Fangmeier, A.; Weber, H. Grain yield and quality responses of wheat expressing a barley sucrose transporter to combined climate change factors. *J. Exp. Bot.* **2017**, *68*, 5511–5525. [CrossRef]
62. González, F.G.; Slafer, G.A.; Miralles, D.J. Floret development and survival in wheat plants exposed to contrasting photoperiod and radiation environments during stem elongation. *Funct. Plant Biol.* **2005**, *32*, 189–197. [CrossRef]
63. Foulkes, M.J.; Slafer, G.A.; Davies, W.J.; Berry, P.M.; Sylvester-Bradley, R.; Martre, P.; Calderini, D.F.; Griffiths, S.; Reynolds, M.P. Raising yield potential of wheat. III. Optimizing partitioning to grain while maintaining lodging resistance. *J. Exp. Bot.* **2011**, *62*, 469–486. [CrossRef]
64. Waddington, S.R.; Ransom, J.K.; Osmanzai, M.; Saunders, D.A. Improvement in the yield potential of bread wheat adapted to northwest mexico 1. *Crop Sci.* **1986**, *26*, 698–703. [CrossRef]
65. Acreche, M.M.; Slafer, G.A. Grain weight response to increases in number of grains in wheat in a Mediterranean area. *F. Crop. Res.* **2006**, *98*, 52–59. [CrossRef]
66. Miralles, D.J.; Slafer, G.A. Individual grain weight responses to genetic reduction in culm length in wheat as affected by source-sink manipulations. *F. Crop. Res.* **1995**, *43*, 55–66. [CrossRef]
67. Sayre, K.D.; Rajaram, S.; Fischer, R.A. Yield potential progress in short bread wheats in northwest Mexico. *Crop Sci.* **1997**, *37*, 36–42. [CrossRef]
68. Shearman, V.J.; Sylvester-Bradley, R.; Scott, R.K.; Foulkes, M.J. Physiological processes associated with wheat yield progress in the UK. *Crop Sci.* **2005**, *45*, 175–185. [CrossRef]
69. Xiao, Y.G.; Qian, Z.G.; Wu, K.; Liu, J.J.; Xia, X.C.; Ji, W.Q.; He, Z.H. Genetic gains in grain yield and physiological traits of winter wheat in shandong province, china, from 1969 to 2006. *Crop Sci.* **2012**, *52*, 44–56. [CrossRef]
70. Sanchez-Garcia, M.; Royo, C.; Aparicio, N.; Martín-Sánchez, J.A.; Álvaro, F. Genetic improvement of bread wheat yield and associated traits in Spain during the 20th century. *J. Agric. Sci.* **2013**, *151*, 105–118. [CrossRef]
71. Miralles, D.J.; Katz, S.D.; Colloca, A.; Slafer, G.A. Floret development in near isogenic wheat lines differing in plant height. *F. Crop. Res.* **1998**, *59*, 21–30. [CrossRef]
72. Canevara, M.G.; Romani, M.; Corbellini, M.; Perenzin, M.; Borghi, B. Evolutionary trends in morphological, physiological, agronomical and qualitative traits of Triticum aestivum L. cultivars bred in Italy since 1900. *Eur. J. Agron.* **1994**, *3*, 175–185. [CrossRef]
73. Royo, C.; Briceño-Felix, G.A. Spanish wheat pool. In *The World Wheat Book: A History of Wheat Breeding*; Bojean, A.P., Angus, W.J., van Ginkel, M., Eds.; Lavoisier Publishing Inc.: Paris, France, 2011; pp. 121–154.
74. Araus, J.L.; Slafer, G.A.; Royo, C.; Serret, M.D. Breeding for yield potential and stress adaptation in cereals. *CRC. Crit. Rev. Plant Sci.* **2008**, *27*, 377–412. [CrossRef]
75. Dreccer, M.F.; Chapman, S.C.; Rattey, A.R.; Neal, J.; Song, Y.; Christopher, J.T.; Reynolds, M. Developmental and growth controls of tillering and water-soluble carbohydrate accumulation in contrasting wheat (*Triticum aestivum* L.) genotypes: Can we dissect them? *J. Exp. Bot.* **2013**, *64*, 143–160. [CrossRef] [PubMed]
76. Duggan, B.L.; Richards, R.A.; van Herwaarden, A.F.; Fettell, N.A. Agronomic evaluation of a tiller inhibition gene (tin) in wheat. I. Effect on yield, yield components, and grain protein. *Aust. J. Agric. Res.* **2005**, *56*, 169. [CrossRef]

77. Reynolds, M.; Foulkes, M.J.; Slafer, G.A.; Berry, P.; Parry, M.A.J.; Snape, J.W.; Angus, W.J. Raising yield potential in wheat. *J. Exp. Bot.* **2009**, *60*, 1899–1918. [CrossRef]
78. Lopes, M.S.; Reynolds, M.P.; Manes, Y.; Singh, R.P.; Crossa, J.; Braun, H.J. Genetic yield gains and changes in associated traits of CIMMYT spring bread wheat in a "Historic" set representing 30 years of breeding. *Crop Sci.* **2012**, *52*, 1123–1131. [CrossRef]
79. Morgounov, A.; Zykin, V.; Belan, I.; Roseeva, L.; Zelenskiy, Y.; Gomez-Becerra, H.F.; Budak, H.; Bekes, F. Genetic gains for grain yield in high latitude spring wheat grown in Western Siberia in 1900–2008. *F. Crop. Res.* **2010**, *117*, 101–112. [CrossRef]
80. Zheng, T.C.; Zhang, X.K.; Yin, G.H.; Wang, L.N.; Han, Y.L.; Chen, L.; Huang, F.; Tang, J.W.; Xia, X.C.; He, Z.H. Genetic gains in grain yield, net photosynthesis and stomatal conductance achieved in Henan Province of China between 1981 and 2008. *F. Crop. Res.* **2011**, *122*, 225–233. [CrossRef]
81. Del Pozo, A.; Matus, I.; Serret, M.D.; Araus, J.L. Agronomic and physiological traits associated with breeding advances of wheat under high-productive Mediterranean conditions. *The case of Chile. Environ. Exp. Bot.* **2014**, *103*, 180–189. [CrossRef]
82. Mondal, S.; Dutta, S.; Crespo-Herrera, L.; Huerta-Espino, J.; Braun, H.J.; Singh, R.P. Fifty years of semi-dwarf spring wheat breeding at CIMMYT: Grain yield progress in optimum, drought and heat stress environments. *F. Crop. Res.* **2020**, *250*, 107757. [CrossRef]
83. Gutiérrez, E.; Gutiérrez, D.; Morcuende, R.; Verdejo, A.L.; Kostadinova, S.; Martinez-Carrasco, R.; Pérez, P. Changes in leaf morphology and composition with future increases in CO_2 and temperature revisited: Wheat in field chambers. *J. Plant Growth Regul.* **2009**, *28*, 349–357. [CrossRef]
84. Gutiérrez, D.; Morcuende, R.; Del Pozo, A.; Martínez-Carrasco, R.; Pérez, P. Involvement of nitrogen and cytokinins in photosynthetic acclimation to elevated CO_2 of spring wheat. *J. Plant Physiol.* **2013**, *170*, 1337–1343. [CrossRef] [PubMed]
85. Gourdji, S.M.; Mathews, K.L.; Reynolds, M.; Crossa, J.; Lobell, D.B. An assessment of wheat yield sensitivity and breeding gains in hot environments. *Proc. R. Soc. B Biol. Sci.* **2013**, *280*. [CrossRef] [PubMed]
86. Córdoba, J.; Molina-Cano, J.L.; Pérez, P.; Morcuende, R.; Moralejo, M.; Savé, R.; Martínez-Carrasco, R. Photosynthesis-dependent/independent control of stomatal responses to CO_2 in mutant barley with surplus electron transport capacity and reduced SLAH3 anion channel transcript. *Plant Sci.* **2015**, *239*, 15–25. [CrossRef] [PubMed]
87. Vicente, R.; Pérez, P.; Martínez-Carrasco, R.; Gutiérrez, E.; Morcuende, R. Nitrate supply and plant development influence nitrogen uptake and allocation under elevated CO_2 in durum wheat grown hydroponically. *Acta Physiol. Plant.* **2015**, *37*, 114. [CrossRef]
88. Vicente, R.; Pérez, P.; Martínez-Carrasco, R.; Feil, R.; Lunn, J.E.; Watanabe, M.; Arrivault, S.; Stitt, M.; Hoefgen, R.; Morcuende, R. Metabolic and transcriptional analysis of durum wheat responses to elevated CO_2 at low and high nitrate supply. *Plant Cell Physiol.* **2016**, *57*, 2133–2146. [CrossRef] [PubMed]
89. *R Core Team R: A Language and Environment for Statistical Computing*; R Foundation for Statistical Computing: Vienna, Austria, 2019.
90. Revelle, W. Package "psych"—Procedures for Psychological, Psychometric and Personality Research. R Package Version 2.1.3. Available online: https://cran.r-project.org/package=psych (accessed on 1 April 2021).
91. Husson, F.; Josse, J.; Pages, J. Principal component methods-hierarchical clustering-partitional clustering: Why would we need to choose for visualizing data. *Appl. Math. Dep.* **2010**, 1–17.
92. Signorell, A. DescTools: Tools for Descriptive Statistics. R package version 0.99.42. Available online: https://cran.r-project.org/package=DescTools (accessed on 1 August 2021).
93. Wickham, H. Reshaping Data with the reshape Package. *J. Stat. Softw.* **2007**, *21*, 1–20. [CrossRef]
94. Shannon, P. Cytoscape: A software environment for integrated models of biomolecular interaction networks. *Genome Res.* **2003**, *13*, 2498–2504. [CrossRef] [PubMed]

Article

Elevated Temperature Induced Adaptive Responses of Two Lupine Species at Early Seedling Phase

Sigita Jurkonienė [1,*] , Jurga Jankauskienė [1] , Rima Mockevičiūtė [1] , Virgilija Gavelienė [1] ,
Elžbieta Jankovska-Bortkevič [1] , Iskren Sergiev [2] , Dessislava Todorova [2] and Nijolė Anisimovienė [1]

[1] Nature Research Centre, Institute of Botany, Akademijos Str. 2, 08412 Vilnius, Lithuania;
jurga.jankauskiene@gamtc.lt (J.J.); rima.mockeviciute@gamtc.lt (R.M.); virgilija.gaveliene@gamtc.lt (V.G.);
elzbieta.jankovska@gamtc.lt (E.J.-B.); n.anisimoviene@gmail.com (N.A.)

[2] Bulgarian Academy of Sciences, Institute of Plant Physiology and Genetics, Acad. G. Bonchev Str. 21,
1113 Sofia, Bulgaria; iskren@bio21.bas.bg (I.S.); dessita@bio21.bas.bg (D.T.)

* Correspondence: sigita.jurkoniene@gamtc.lt

Abstract: This study aimed to investigate the impact of climate warming on hormonal traits of invasive and non-invasive plants at the early developmental stage. Two different lupine species—invasive *Lupinus polyphyllus* Lindl. and non-invasive *Lupinus luteus* L.—were used in this study. Plants were grown in climate chambers under optimal (25 °C) and simulated climate warming conditions (30 °C). The content of phytohormone indole-3-acetic acid (IAA), ethylene production and the adaptive growth of both species were studied in four-day-old seedlings. A higher content of total IAA, especially of IAA-amides and transportable IAA, as well as higher ethylene emission, was determined to be characteristic for invasive lupine both under optimal and simulated warming conditions. It should be noted that IAA-L-alanine was detected entirely in the invasive plants under both growth temperatures. Further, the ethylene emission values increased significantly in invasive lupine hypocotyls under 30 °C. Invasive plants showed plasticity in their response by reducing growth in a timely manner and adapting to the rise in temperature. Based on the data of the current study, it can be suggested that the invasiveness of both species may be altered under climate warming conditions.

Keywords: early growth stage; ethylene; IAA conjugates; indole-3-acetic acid; invasiveness; lupine seedlings; simulated conditions; warming simulation

Citation: Jurkonienė, S.;
Jankauskienė, J.; Mockevičiūtė, R.;
Gavelienė, V.; Jankovska-Bortkevič,
E.; Sergiev, I.; Todorova, D.;
Anisimovienė, N. Elevated
Temperature Induced Adaptive
Responses of Two Lupine Species at
Early Seedling Phase. *Plants* **2021**, *10*,
1091. https://doi.org/10.3390/
plants10061091

Academic Editor: James Bunce

Received: 27 April 2021
Accepted: 27 May 2021
Published: 29 May 2021

Publisher's Note: MDPI stays neutral
with regard to jurisdictional claims in
published maps and institutional affiliations.

1. Introduction

One of the most cited indicators of global climate change is the increase in global temperature. The impact of this factor on various plant growth and development processes has already been comprehended [1–4]. It is suggested that different non-native—alien—plant species may respond differently to the same climate changes and become invasive species—a threat to native species and biodiversity [2,3]. Model studies to date indicate that elevated temperature may increase invasion risk by accelerating physiological processes of alien species [5–7]. Therefore, the expected global warming with a predicted 5 °C rise during the 21st century may lead alien plant species to become invaders [1,2,5,8]. However, there is also evidence that the growth of plants may be adversely affected by temperature stress caused by warming [9]. These contradictory data on the global warming issue encourage us to study the adaptive response of alien plants to the rise in temperature by 5 °C.

Research on the molecular mechanisms determining the invasiveness of plants has only just begun [10,11]. The knowledge of plant physiological responses to climate warming will help mitigate the impact of future environmental conditions on plants. Currently, there is no consensus among researchers regarding physiological-biochemical traits of plants that could determine invasiveness [10–13]. A special role of phytohormones in regulating invasiveness should be considered due to their involvement in plant growth, development and processes of response to environmental factors [14–18].

The role of indole-3-acetic acid (IAA) as a principal phytohormone coordinating developmental processes in response to environmental signals has been recognized [17,19–22]. Several studies have shown plant phenotypic adaptations of elongated hypocotyls and activation of IAA biosynthesis in certain plant tissues under elevated temperature [23,24]. It is suggested that growth responses to such conditions are partially regulated through the phytohormone IAA signaling pathway [25–27]. It is thought that the distribution of IAA and regulation of its content in cells and tissues could be important for the modulation of plant high-temperature response [25,28,29]. A direct connection between the IAA pathway and high temperature-induced adaptive growth has been shown in *Arabidopsis* hypocotyls [26,27]. However, in contrast, high temperature reduced IAA levels in the reproductive tissues of *Arabidopsis*, barley [25,28] and pepper [30]. Elkinawy [31] employed the method for cotyledon excision from four-day-old *Lupinus albus* intact seedlings and experimentally demonstrated the impact of IAA from cotyledons of lupines on the content of IAA in axial organs (hypocotyls). Additionally, it was stated that IAA synthesis de novo occurs in storage tissues (cotyledons) of the seeds during germination, post-germination and early developmental stages, mostly. A small amount of IAA can be synthesized in axial organs (hypocotyls) four days after seed swelling [32,33]. Thus, IAA level and turnover should be analyzed both in cotyledons and hypocotyls. The plant hormone ethylene has also been defined as a modulator of plant growth and development [34–36], though this is not its only role. Studies have shown that it enables plants to adapt to elevated temperature [18,37,38]. Increased ethylene emission reduced the growth of plants until the high temperature was removed [16,39,40].

Meanwhile, climate warming-induced adaptive responses of invasive plants are still not well studied. It is unclear how they depend on the IAA state and content in tissues and ethylene emission. The discovery of these characteristics may be carried out using a comparative study of genetically related plants [41–44]. Young intact lupine seedlings can be suitable objects to test the impact of increased temperature on IAA content and ethylene emission [44,45]. Such a model can provide a new insight into the hormonal traits of plant invasiveness onset detection and analysis under future climate warming scenarios. Thus, two different lupine species were tested in the current study.

We hypothesized that climate warming may affect the plant hormone regulatory system—the factor determining plant growth and development. Thus, the goal of the current study was to evaluate the adaptive responses of two lupine species under simulated climate warming conditions at the early phase of growth.

2. Results

2.1. Germination and Growth under 25 °C and 30 °C

The simulated climate warming conditions were found to differently affect seed germination and resulted in diverse growth responses of invasive and non-invasive lupines (Figure 1, Table 1). The data of the current study show that simulated 5 °C warming conditions had no significant effect on the seed germination percentage of non-invasive lupine. However, the seed germination of invasive lupine was 5% lower under 30 °C than under optimal conditions. The growth of non-invasive lupine was more intensive under 30 °C than under 25 °C. The weight of hypocotyls increased by 30%. The weight of roots and cotyledons was higher as well. On the other hand, elevated temperature (30 °C) resulted in slower growth of invasive lupine. The weight of hypocotyls and roots was about 40% and of cotyledons about 9% lower than that of plants grown under 25 °C (Figure 1, Table 1).

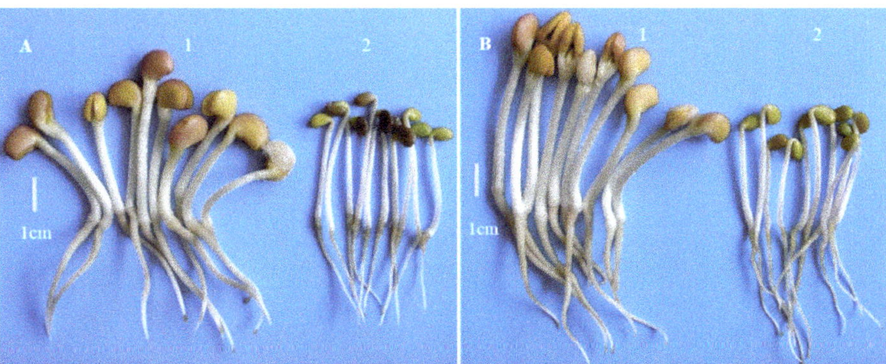

Figure 1. Four-day-old seedlings of non-invasive (*L. luteus*) (1) and invasive (*L. polyphyllus*) lupines (2) grown at 25 °C (**A**) and 30 °C (**B**). Scale bar, 1 cm.

Table 1. Effect of temperatures of 25 °C and 30 °C on germination and growth parameters of four-day-old seedlings of two lupine species.

Species	Temperature, °C	Germination, %	Cotyledons	Fresh Mass, g Hypocotyls	Roots
L. polyphyllus	25	36.65 ± 1.03 a	0.906 ± 0.15 n.s.	1.04 ± 0.16 a	0.30 ± 0.01 a
	30	31.28 ± 1.18 b	0.825 ± 0.19 n.s.	0.61 ± 0.11 b	0.15 ± 0.05 b
L. luteus	25	39.82 ± 1.36 n.s.	2.229 ± 0.30 n.s.	1.12 ± 0.21 a	0.21 ± 0.06 n.s.
	30	39.24 ± 1.54 n.s.	2.271 ± 0.29 n.s.	1.46 ± 0.22 b	0.27 ± 0.06 n.s.

Values are mean ± SD of three experiments with five replicates in each. Different lowercase letters indicate significant difference ($p < 0.05$) between mean values at 25 °C and 30 °C for each lupine species; n.s.—non-significant difference.

2.2. IAA Content under 25 °C and 30 °C

The analysis of IAA status in cotyledons of plants grown under 25 °C revealed that the content of free IAA reached about 21% of the total IAA in invasive lupine and about 14% in non-invasive lupine. The content of this transportable IAA was about twice higher in cotyledons of invasive plants than in non-invasive plants (Figure 2).

The major part of total IAA in cotyledons of both tested lupine species was in a bound form. The content of these reversible low-molecular mass complexes (IAA-esters and amides) reached at least 70% of the total IAA content (Figure 2). The content of IAA-amides and IAA-esters was higher in cotyledons of invasive lupine; however, the proportions of these IAA conjugates were equivalent in both lupine species. The amount of IAA conjugates was 30% higher in cotyledons of invasive lupine. This led us to predict that seedlings of invasive lupine would be provided with a higher amount of free IAA. A negligible and almost similar part of total IAA (9–13%) was identified as IAA catabolites (irreversibly degraded) in cotyledons of both tested lupine species (Figure 2).

Meanwhile, the content of free IAA was 26% higher in hypocotyls of invasive lupine than in non-invasive lupine (Figure 3). The amount of IAA conjugates, especially of IAA-amides, was also greater in hypocotyls of *L. polyphyllus* (by about 23%). The content of IAA catabolites was 31% higher in hypocotyls of invasive lupine (Figure 3).

Two IAA-amides—indole-3-acetyl-L-aspartic acid (IAA-Asp) and indole-3-acetyl-L-glutamic acid (IAA-Glu)—were identified in cotyledons and hypocotyls of non-invasive lupine, and three—IAA-Asp, IAA-Glu and indole-3-acetyl-L-alanine (IAA-Ala)—were identified in cotyledons and hypocotyls of invasive lupine (Tables 2 and 3). Both lupine species contained one IAA-ester-type conjugate—IAA complex with glucose (IAA-Glc)—and one IAA catabolite—2-oxindole-3-acetic acid (Ox-IAA) (Tables 2 and 3). The results on the 30 °C temperature effect on the total IAA content in tested plants reveal that the total IAA content was 20% higher in invasive than non-invasive lupine (Figures 2 and 3).

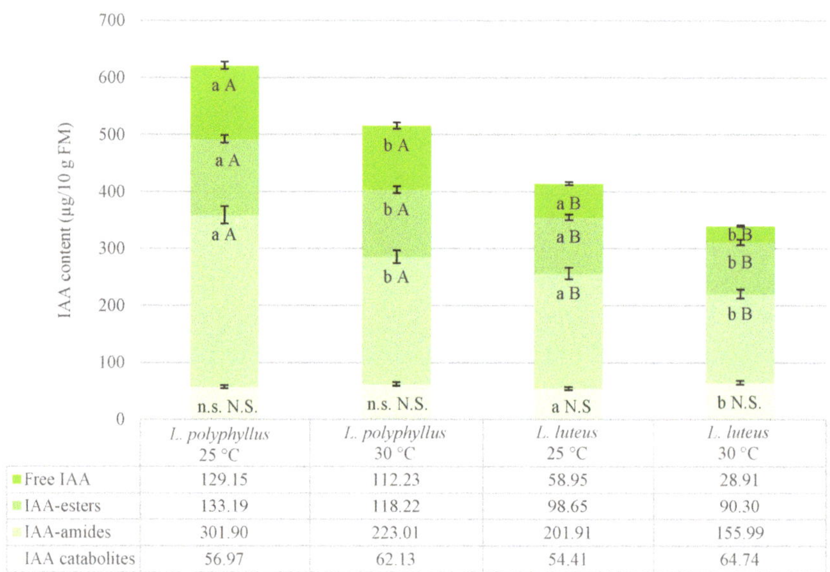

	L. polyphyllus 25 °C	*L. polyphyllus* 30 °C	*L. luteus* 25 °C	*L. luteus* 30 °C
Free IAA	129.15	112.23	58.95	28.91
IAA-esters	133.19	118.22	98.65	90.30
IAA-amides	301.90	223.01	201.91	155.99
IAA catabolites	56.97	62.13	54.41	64.74

Figure 2. IAA content in cotyledons of seedlings grown at 25 °C and 30 °C. Vertical bars represent the total IAA content. Sub-bars depict the content of IAA compounds. Values are mean ± SD of three experiments with five replicates in each. Different uppercase letters indicate significant difference ($p < 0.05$) between mean values of two lupine species grown under the same temperature; N.S.—non-significant difference. Significant difference between mean values at 25 °C and 30 °C for each lupine species is marked with different lowercase letters; n.s.—non-significant difference.

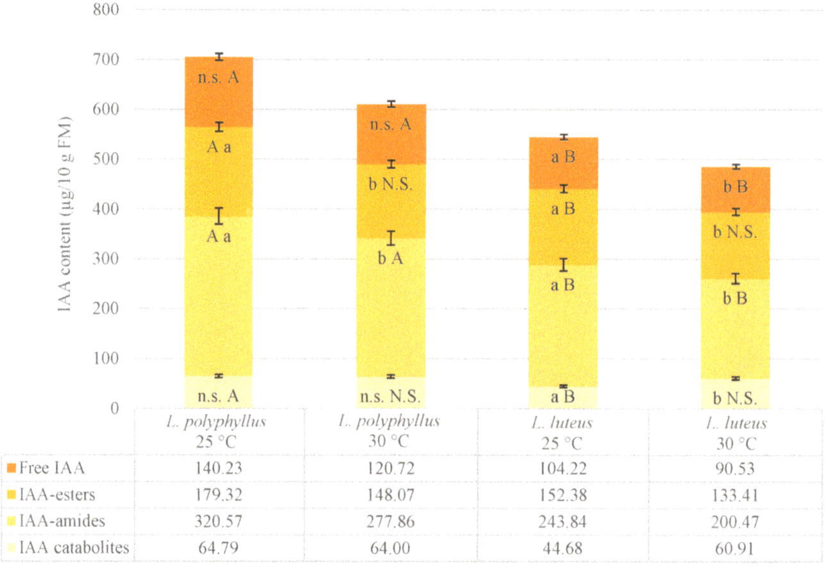

	L. polyphyllus 25 °C	*L. polyphyllus* 30 °C	*L. luteus* 25 °C	*L. luteus* 30 °C
Free IAA	140.23	120.72	104.22	90.53
IAA-esters	179.32	148.07	152.38	133.41
IAA-amides	320.57	277.86	243.84	200.47
IAA catabolites	64.79	64.00	44.68	60.91

Figure 3. IAA content in hypocotyls of seedlings grown at 25 °C and 30 °C. Vertical bars represent the total IAA content. Sub-bars depict the content of IAA compounds. Values are mean ± SD of three experiments with five replicates in each. Different uppercase letters indicate significant difference ($p < 0.05$) between mean values of two lupine species grown under the same temperature; N.S.—non-significant difference. Significant difference between mean values of each lupine species at 25 °C and 30 °C is marked with different lowercase letters; n.s.—non-significant difference.

Table 2. The comparison of IAA content in cotyledons of four-day-old seedlings of two lupine species grown at 25 °C versus 30 °C.

IAA Form		Changes in Amount (%)	
		L. polyphyllus	*L. luteus*
Free IAA	IAA	↓ 13.10 ± 0.85 *	↓ 50.96 ± 3.46 *
IAA-esters	IAA-Glc	↓ 12.70 ± 1.23 *	↓ 8.40 ± 2.03 *
	IAA-Glu	↓ 18.20 ± 1.16 *	↓ 18.33 ± 1.23 *
IAA-amides	IAA-Asp	↓ 38.57 ± 2.65 *	↓ 22.98 ± 2.60 *
	IAA-Ala	↓ 6.88 ± 0.29 *	Non-detected
IAA catabolites	Ox-IAA	↑ 9.03 ± 0.28	↑ 18.17 ± 1.88 *

Values are mean ± SD of three experiments with five replicates in each. ↑ and ↓—increase and decrease in content, respectively. *—significant difference ($p < 0.05$) between mean values at 25 °C and 30 °C for each lupine species.

Table 3. The comparison of IAA content in hypocotyls of four-day-old seedlings of two lupine species grown at 25 °C versus 30 °C.

IAA Form		Changes in Amount (%)	
		L. polyphyllus	*L. luteus*
Free IAA	IAA	↓ 13.90 ± 0.90	↓ 13.13 ± 0.46 *
IAA-esters	IAA-Glc	↓ 17.29 ± 1.15 *	↓ 12.42 ± 0.33 *
	IAA-Glu	↓ 12.55 ± 0.23 *	↓ 16.46 ± 0.57 *
IAA-amides	IAA-Asp	↓ 27.23 ± 1.85 *	↓ 21.17 ± 1.46 *
	IAA-Ala	↑ 11.10 ± 0.44 *	Non-detected
IAA catabolites	Ox-IAA	↓ 2.77 ± 0.30	↑ 36.32 ± 2.98 *

Values are mean ± SD of three experiments with five replicates in each. ↑ and ↓—increase and decrease in content, respectively. *—significant difference ($p < 0.05$) between mean values at 25 °C and 30 °C for each lupine species.

The content of free IAA decreased by 13% in cotyledons and hypocotyls of *L. polyphyllus* under elevated temperature (Figures 2 and 3, Tables 2 and 3). The same temperature regime had a similar effect on the free IAA level in hypocotyls and a more substantial effect (reduced twice) in cotyledons of *L. luteus*. The content of IAA-esters (IAA-Glc and IAA-Glu) under 30 °C decreased in both species. The decrease in IAA-Glc content was greater in the invasive plants (Figure 2, Table 2). The major decrease in IAA-Glu content (up to 18%) was in cotyledons of both species. The IAA-amide composition changed unevenly in both lupines under 30 °C (Tables 2 and 3). A major, more than 20% decrease was detected in IAA-Asp content in tissues of both tested lupine organs, especially in cotyledons of *L. polyphyllus* (by 38%). It must be noted that the IAA-Ala complex was detected in seedlings of invasive lupine only (Tables 2 and 3). Its content in cotyledons was lower than in hypocotyls, and the 30 °C temperature affected it differently. IAA-Ala content increased by 11% (70.84 ± 5.54 µg/10 g of fresh mass at 25 °C and 78.68 ± 1.81 µg/10 g of fresh mass at 30 °C) in hypocotyls and decreased by 7% in cotyledons under simulated warming conditions.

No significant changes under simulated 5 °C warming were detected in the content of IAA catabolite Ox-IAA in invasive lupine. However, it increased by 18% in cotyledons and by 36% in hypocotyls of non-invasive lupine (Tables 2 and 3).

2.3. Ethylene Production under 25 °C and 30 °C

Comparative analysis of ethylene production in hypocotyls of both lupine species under 25 °C showed significant differences. Ethylene emission in hypocotyls of invasive lupine was 29% higher than that in non-invasive lupine (Figure 4). However, no significant differences were found in ethylene emission from cotyledons of both investigated lupine species.

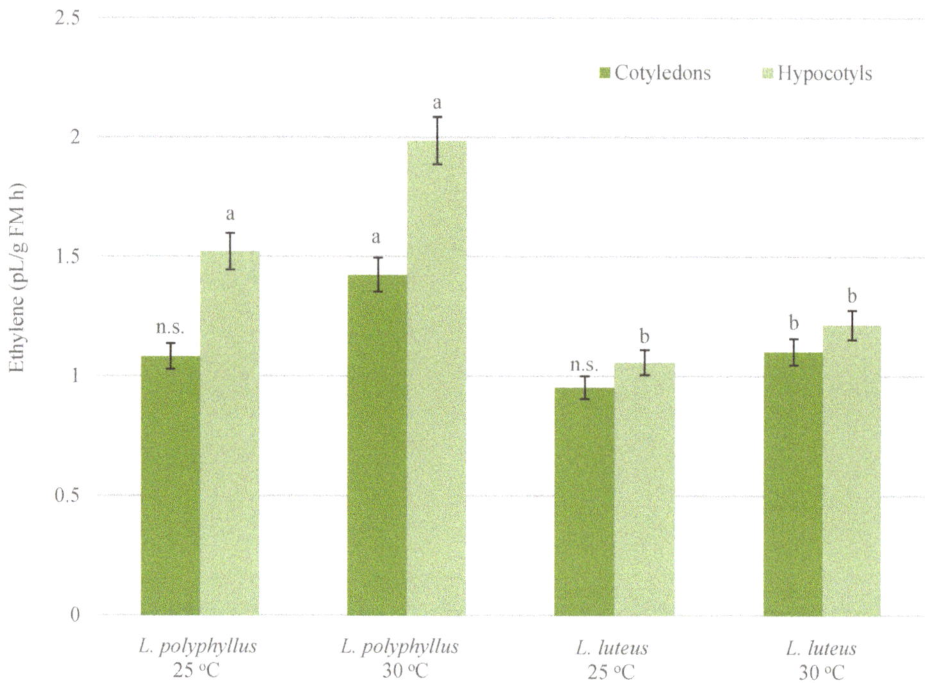

Figure 4. Ethylene production in seedlings grown at 25 °C and 30 °C. Values are mean ± SD of three experiments with five replicates in each. Different letters indicate significant difference ($p < 0.05$) between two lupine species grown under the same temperature; n.s.—non-significant difference.

The elevated temperature significantly (up to 30%) increased ethylene production in both tested organs of invasive lupine. The production of ethylene in hypocotyls and cotyledons of non-invasive lupine was less intensive. This research reveals statistically significant differences in ethylene emission in tissues of invasive and non-invasive lupines under simulated warming conditions. Emission values in hypocotyls and cotyledons of invasive lupine were found to be higher by 38% and 21%, respectively, as compared to non-invasive lupine tissues (Figure 4).

3. Discussion

L. polyphyllus originates from North America [38,41]. It is one of the most common alien plants in Europe. It has been planted as a fodder crop and as an ornamental plant and is now widely naturalized. It is one of the seven most aggressive invasive plant species in Lithuania [46]. *L. polyphyllus* changes meadow and sand communities and eliminates uncompetitive native plants [47]. *L. luteus* originates from the Mediterranean region of Southern Europe. It is cultivated as a fodder and cover crop and is non-invasive in temperate regions.

Invasive and non-invasive species differ in many traits [48,49]. In general, invasive alien plants have broad environmental tolerance and are usually characterized by fast growth [7,8,48]. It is thought that elevated temperature may increase invasion risk by accelerating physiological processes and growth by increasing the competitive ability of invasive species [5–7,49,50]. On the other hand, there is evidence that climate warming may cause declines in populations of invasive plants [2,5,51]. Our data show that the temperature of 30 °C is not optimal for the growth of invasive lupine in the early stages of development as well as for non-invasive lupine as compared to the growth at 25 °C (Table 1, Figure 1). Furthermore, the data of the current study show that the seed germination of invasive lupine was 5% lower under 30 °C (Table 1). On the other hand, the seed germination of non-invasive lupine remained the same at 30 °C. This agrees with the data on *L. polyphyllus* germination at high summer temperatures in the studies of other authors [52]. Therefore, the tendency to climate warming seems to be not beneficial for the germination and growth of invasive *L. polyphyllus*.

Numerous studies have shown the role of phytohormones in plant reactions to environmental changes. The important regulators of adaptive plant growth responses to environmental stresses are IAA and ethylene [38,53]. Their content in invasive and non-invasive lupines was analyzed in the current study. Analysis of the IAA state in cotyledons and hypocotyls of both lupine species showed that IAA homeostasis was maintained through both key IAA metabolic pathways—reversible conjugation and catabolism through non-decarboxylative oxidation (Figures 2 and 3). This has also been shown in early phases of growth in other plants [33,54–57] Previous studies have shown that ox-IAA is a major primary IAA catabolite in higher plants, which has low biological activity and is important in the regulation of IAA homeostasis and stress response mechanisms [32,43,58]. Data of the current study show that 8–19% of the IAA content in hypocotyls and cotyledons of both tested lupine species can be catabolized through the oxidative catabolic pathway.

The data of this experiment show a decrease in the total amount of IAA in cotyledons and hypocotyls of both tested lupine species grown at 30 °C, compared to plants grown at 25 °C (Figures 2 and 3). The studies of IAA turnover in cotyledons at the early phases of growth of invasive and non-invasive lupines showed that the 5 °C temperature increase (from 25 up to 30 °C) affected the hydrolysis of IAA reversible complexes (IAA-amides and IAA-esters) in both lupine species (Figures 2 and 3). Earlier, it was shown that IAA-Asp, IAA-Glu and IAA-Glc are common in higher plants, and various physiological roles in growing parts are attributed to them [17,54,59]. However, the contribution of these conjugates in particular developmental pathways as well as in specific invasive and non-invasive plant responses to temperature changes (e.g., simulated climate warming) is unknown. Some studies have presented evidence that IAA conjugates may be involved in abiotic stress tolerance [17,60]. The mutant cell line of henban with impaired IAA-Asp biosynthesis dies at 33 °C to which the wild type is resistant [61]. Our results show a significant decrease in the IAA-Asp amount at 30 °C in both lupine species, especially in cotyledons of invasive lupine (by 38.57%). This could indicate that the 5 °C warming effect induced temperature stress in *L. polyphyllus*. Nevertheless, the IAA-Ala complex has been detected in several plants, though its role has been poorly investigated [54,58]. There are few data suggesting that IAA-Ala could be related to plant growth inhibition [17,62,63]. Similar results were obtained in the current study with *L. polyphyllus* seedlings at 30 °C. The IAA-Ala complex was found only in invasive lupine. Moreover, the proportion in the content of IAA-amides was greater at 30 °C compared to 25 °C (Tables 2 and 3).

Our results show that the content of free IAA and IAA conjugates (except IAA catabolites) was higher in *L. polyphyllus* than that of *L. luteus* at both temperatures (Figure 2). The content of free IAA under 30 °C was about three-fold higher in cotyledons and 33% higher in hypocotyls of invasive lupine than in non-invasive lupine. The higher level of free IAA in hypocotyls of invasive lupine seedlings may have resulted from the maintenance of the level of IAA in cotyledons (Tables 2 and 3). The higher amount of IAA can be transported from cell to cell, interact with specific receptors, moderate IAA inducible gene expression and participate in growth and development processes [27,33,64]. The role of IAA in the post-germination growth period was obvious in both lupine species under simulated warming conditions (Tables 1–3). This is in agreement with the data on the role of IAA under changing environmental conditions, including global warming, obtained by other authors [14,15,17,27,53].

Studies have shown that the plant hormone ethylene participates in numerous aspects of plant development. The content of ethylene can be modified by biotic and abiotic factors [18,37,38]. The production of ethylene increases in response to temperature stress and enables plants to reach a high level of plasticity and to adapt to environmental changes [16,38,40]. These studies have shown that invasive plants are more adaptable to changing environmental conditions (e.g., increased temperature). They are more sensitive to environmental stress and reduce growth processes in a timely manner. In our study, the growth reduction of hypocotyls of invasive lupine at 30 °C could be triggered by elevated ethylene production. These growth adaptations enable plants to minimize the risk of heat damage and enhance evaporative leaf cooling for optimal plant growth [65]. Results of the current study show that elevated temperature has an impact on ethylene emission in cotyledons and hypocotyls at early phases of development. It was increased in both lupine species under warming conditions. However, the rise in ethylene production in invasive lupine was higher (Figure 4).

The diversity of the ethylene function is thought to be achieved in combination with other phytohormones, e.g., IAA [16,35,66,67]. A significant alteration of ethylene emission in cotyledons of invasive lupine may be related to significant changes in the utilization of IAA resources and a higher level of free IAA. An increase in ethylene emission and a decrease in IAA content in seedlings of invasive lupine at 30 °C were observed in this study. These results show the possible link between phytohormones IAA and ethylene and the manifestation of invasiveness under simulated climate warming.

The obtained data show the possible role of phytohormones IAA and ethylene in adaptive responses of invasive plants to simulated warming at the early developmental phase. It was determined that the level of IAA conjugates (especially of IAA-amides) and transportable IAA, as well as ethylene emission, was higher in invasive than in non-invasive lupine. The decreased level of transportable IAA and IAA conjugates and increased ethylene emission were detected under elevated temperature in seedlings of both species. These changes coincided with the slower growth of *L. polyphyllus*. However, the growth of *L. luteus* was stimulated under elevated temperature conditions. Additionally, the IAA-amide IAA-Ala was found in invasive lupine only. The data of the current study show that the hormonal traits of invasive and non-invasive lupines were altered by elevated temperature at the early seedling phase. It can be suggested that the invasive properties of these plants may be changed by global warming scenarios.

4. Materials and Methods

4.1. Plant Material and Treatments

Seeds of two lupine species (*L. polyphyllus* Lindl. and *L. luteus* L.) were used in the study. One hundred seeds of each lupine species (with five replications) were soaked in distilled water and grown in climate chambers (Climacell, Czech Republic) at 90% relative humidity in the dark at two different temperatures: at 25 °C (optimal temperature for lupine) and 30 °C (simulated climate warming temperature) [49]. The germination of soaked seeds was monitored (Figure A1), and germinated seeds were counted. The germination percentage was calculated.

Following germination, seedlings were grown for four days under previous conditions at 25 °C and 30 °C.

Cotyledons and hypocotyls of four-day-old invasive and non-invasive lupines (Figure 1) were separated and weighed after being washed with sterile distilled water for determination of fresh mass and assays.

4.2. Indole-3-acetic Acid Assay

4.2.1. Extraction and Hydrolysis of IAA

IAA compounds were extracted from the samples of plant material by grinding with 80% methanol, containing 1 mg/L antioxidant butylated hydroxytoluene at a ratio of 1:10 (*w/v*), using a porcelain mortar with a pestle and incubated for 16 h at 4 °C in the dark using a Multi-Pulse Vortexer device (Glass-Col, Terre Haute, IN, USA). The extracts were separated from the residue by filtration through 0.2 μm pore size membrane filters (Whatman, Maidstone, UK), and after removing phenolic compounds by poly(vinylpolypyrrolidone) (PVP40) (Sigma, Neustadt, Germany) at concentration of 0.5%, they were concentrated using a vacuum evaporator (IKA RV-10 Basic, Germany) until dry. Following ether and ethylacetate extractions of the IAA compounds, extracts were purified from peptides and small biomolecules (molecular weight > 700–1500) on a Sephadex G-10 or G-15 column. Indole compounds were detected using specific coloring Salkowski and Ehrlich reagents. The alkaline hydrolysis method was applied for IAA-ester and IAA-amide quantification according to free IAA release following IAA-ester complex hydrolysis at 1 N NaOH 30 °C for 30 min and IAA-amide complex hydrolysis at 7 N NaOH 100 °C for 1 h [32,44,54,57,68].

4.2.2. Chromatographic Isolation and Quantitative Estimation of IAA

Thin-layer chromatography (TLC) and high-performance liquid chromatography (HPLC) were used to separate the individual indole compounds [48,54,68]. The IAA compounds were separated on Alugram SIL G/UV 24 TLC plates (Macherey-Nagel, Düren, Germany), detected under UV light λ − 254 nm and identified by comparing their RF values with synthetic standards. The absorption spectra (λ range: from 220 to 320 nm) of IAA and other indoles were generally identified using a UV–Vis spectrophotometer, Specord 210 PLUS (Analytik Jena GmbH, Jena, Germany). Amounts of indole compounds were calculated (µg/10 g of fresh mass) using the calibration curve.

Final identification of IAA and its metabolites was performed by HPLC analysis according to the procedures described by Kowalczyk and Sandberg [69], using a Shimadzu PROMINENCE LC-20 series system (Shimadzu Technologies, Kyoto, Japan). The samples were separated in a reversed phase column, YMC-PackPro18 (YMC CO, Japan), with a particle diameter of 3 µm. The linear gradient of eluent A (methanol) was from 1 to 95% (*v/v*) in eluent B (water acidified by 1% acetic acid (*v/v*)), at a flow rate of 0.6 mL/min, a time span of over 45 min and an oven temperature of 30 °C. IAA and its metabolites were identified by co-elution with authentic standards. IAA and IAA metabolites (five IAA-amides, one IAA-ester and three IAA catabolites) were used as standards to evaluate the composition and quantity of IAA compounds. (Figure 5). The standard compounds were purchased from OlChemIm Ltd. (Olomouc, Czech Republic) and Sigma-Aldrich (Germany).

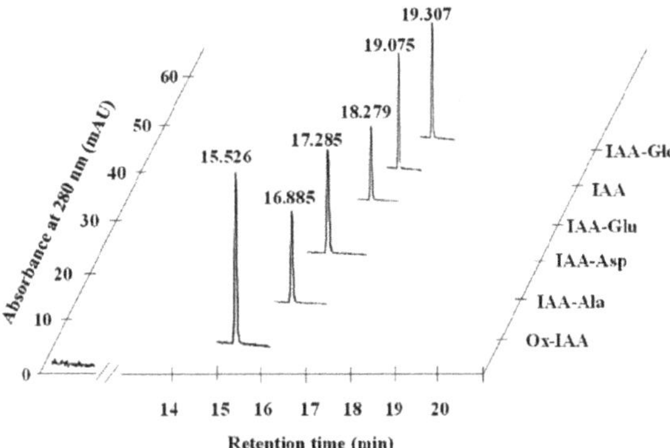

Figure 5. Separation of standards (1 pmol) of IAA and IAA conjugates under HPLC analysis. Ox-IAA—2-oxindole-3-acetic acid; IAA-Ala—indole-3-acetyl-L-alanine; IAA-Asp—indole-3-acetyl-L-aspartic acid; IAA-Glu—indole-3-acetyl-L-glutamic acid; IAA—indole-3-acetic acid; IAA-Glc—IAA complex with glucose.

4.3. Ethylene Assay

Ethylene production was determined in freshly harvested samples of hypocotyls and cotyledons. Samples of a known mass were placed in 30 mL glass vials sealed with a rubber stopper [70]. After 24 h of incubation in the dark at the same temperatures in which plants were grown (at 25 °C and 30 °C), 1 mL of head gas was sampled from each vial, and the ethylene content was measured using a FOCUS GC (Thermo Fischer Scientific, Italy) gas chromatographer, equipped with a flame ionization detector and a stainless-steel matrix 80/100 column, PROPAC R (Sigma-Aldrich, USA). The carrier gas was helium. The temperatures of the column, injector and detector were 90 °C, 110 °C and 150 °C, respectively. Ethylene contents were expressed in picolitres evolved per gram of tissue per hour (pL/g h). Ethylene standard (Alltech, Germany) was used to quantify the content of ethylene in the samples.

4.4. Statistical Analysis

The data presented are mean values $\pm$ standard deviation (SD) of three experiments with five replicates in each. Germination test was performed in three independent experiments with five replications. Each replicate included one hundred seeds of each lupine species. All analytical data are expressed on a fresh mass basis. The data were statistically analyzed using analysis of variance (ANOVA) and tested for significant mean differences ($p < 0.05$) using Tukey's test. Statistical analyses were performed with SPSS Statistics v. 17.0 (SPSS Inc., Chicago, IL, USA) software.

5. Conclusions

The obtained data show the role of phytohormones IAA and ethylene in the adaptive response of invasive L. *polyphyllus* and non-invasive L. *luteus* to simulated warming at the early developmental phase.

The germination rate of L. *polyphyllus* decreased under elevated 30 °C temperature; nevertheless, the rate of germination of L. *luteus* remained the same comparing to the optimal 25 °C temperature.

The simulated warming resulted in decreased growth of L. *polyphyllus* at the early seedling phase as compared to L. *luteus*.

A higher amount of total IAA, especially of IAA-amides, and transportable IAA, as well as a higher amount of ethylene emission, was characteristic for L. *polyphyllus* under both temperatures in comparison to L. *luteus*.

Additionally, the IAA-amide IAA-Ala was found in invasive L. *polyphyllus* only.

A higher supply of IAA in cotyledons of L. *polyphyllus* seedlings at the early phases of development was observed due to the intensive hydrolysis of IAA-amides.

The decreased level of transportable free IAA and IAA conjugates and increased ethylene emission were detected under elevated temperature in seedlings of both species. These changes coincided with the slower growth of L. *polyphyllus*.

The data of the current study show that the hormonal traits of both tested lupine species were altered by the elevated temperature at the early seedling phase.

Author Contributions: Conceptualization, N.A., I.S. and S.J.; methodology, J.J., E.J.-B. and V.G.; formal analysis, S.J. and D.T.; investigation, R.M., V.G., J.J. and D.T.; data curation, S.J. and J.J.; writing—original draft preparation, S.J.; writing—review and editing, E.J.-B., R.M., I.S. and J.J.; visualization, R.M. and E.J.-B.; supervision, S.J. All authors have read and agreed to the published version of the manuscript.

Funding: This research was supported by the Nature Research Centre R&D III programme.

Data Availability Statement: The data supporting reported results can be found in scientific reports of the Laboratory of Plant Physiology of Institute of Botany of Nature Research Centre, where archived datasets generated during the study are included.

Acknowledgments: We thank R. Paškauskas, J. Labokas and the technical staff of the Nature Research Centre subdivision Open Access Centre for their assistance with GC and HPLC. We also thank the staff of the Laboratory of Plant Physiology of the Nature Research Centre for support and help provided.

Conflicts of Interest: The authors declare no conflict of interest.

Appendix A

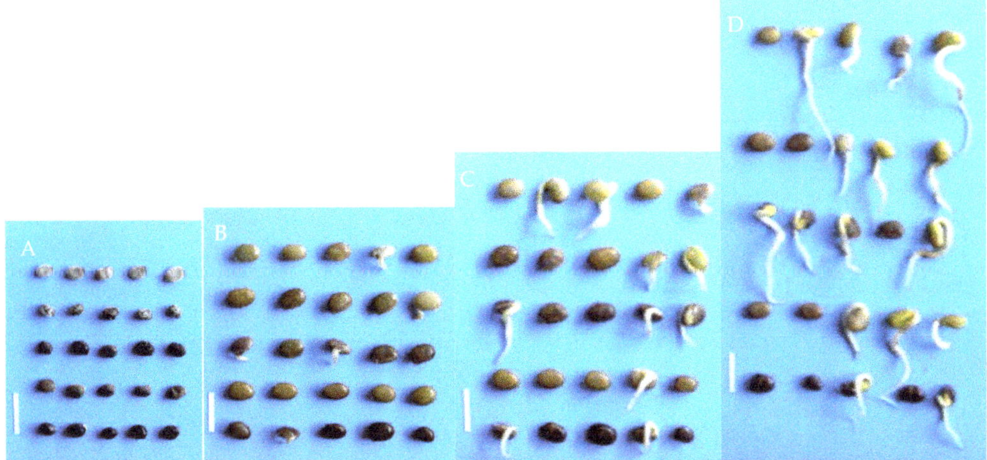

Figure A1. Development of *L. polyphyllus* seedlings under 25 °C temperature: 0 h (**A**), 24 h (**B**), 48 h (**C**) and 72 h (**D**). Scale bar, 1 cm.

References

1. Zhang, Q.; Zhang, Y.; Peng, S.; Zobel, K. Climate Warming May Facilitate Invasion of the Exotic Shrub Lantana camara. *PLoS ONE* **2014**, *9*, e105500. [CrossRef]
2. Chen, B.-M.; Gao, Y.; Liao, H.-X.; Peng, S.-L. Differential responses of invasive and native plants to warming with simulated changes in diurnal temperature ranges. *AoB PLANTS* **2017**, *9*, plx028. [CrossRef]
3. Liu, Y.; Oduor, A.M.O.; Zhang, Z.; Manea, A.; Tooth, I.M.; Leishman, M.R.; Xu, X.; van Kleunen, M. Do invasive alien plants benefit more from global environmental change than native plants? *Glob. Chang. Biol.* **2017**, *23*, 3363–3370. [CrossRef]
4. Stephens, K.L.; Dantzler-Kyer, M.E.; Patten, M.A.; Souza, L. Differential responses to global change of aquatic and terrestrial invasive species: Evidences from a meta-analysis. *Ecosphere* **2019**, *10*, e02680. [CrossRef]
5. Verlinden, M.; De Boeck, H.J.; Nijs, I. Climate warming alters competition between two highly invasive alien plant species and dominant native competitors. *Weed Res.* **2014**, *54*, 234–244. [CrossRef]
6. Kleinbauer, I.; Dullinger, S.; Peterseil, J.; Essl, F. Climate change might drive the invasive tree Robinia pseudacacia into nature reserves and endangered habitats. *Biol. Conserv.* **2010**, *143*, 382–390. [CrossRef]
7. He, W.-M.; Li, J.-J.; Peng, P.-H. A Congeneric Comparison Shows That Experimental Warming Enhances the Growth of Invasive Eupatorium adenophorum. *PLoS ONE* **2012**, *7*, e35681. [CrossRef] [PubMed]
8. Hellmann, J.J.; Byers, J.; Bierwagen, B.G.; Dukes, J.S. Five Potential Consequences of Climate Change for Invasive Species. *Conserv. Biol.* **2008**, *22*, 534–543. [CrossRef]
9. White, T.A.; Campbell, B.D.; Kemp, P.D.; Hunt, C.L. Sensitivity of three grassland communities to simulated extreme tem-perature and rainfall events. *Glob. Chang. Biol.* **2000**, *6*, 671–684. [CrossRef]
10. Wahid, A.; Gelani, S.; Ashraf, M.; Foolad, M. Heat tolerance in plants: An overview. *Environ. Exp. Bot.* **2007**, *61*, 199–223. [CrossRef]
11. Xu, C.; Ge, Y.; Wang, J. Molecular basis underlying the successful invasion of hexaploid cytotypes of Solidago canadensis L.: Insights from integrated gene and miRNA expression profiling. *Ecol. Evol.* **2019**, *9*, 4820–4852. [CrossRef]
12. Drenovsky, R.E.; Khanasova, A.; James, J.J. Trait convergence and plasticity among native and invasive species in re-source-poor environments. *Am. J. Bot.* **2012**, *99*, 629–639. [CrossRef] [PubMed]
13. Darginavičienė, J.; Jurkonienė, S. Characteristics of transmembrane proton transport in the cells of Lupinus polyphyllus. *Open Life Sci.* **2013**, *8*, 461–469. [CrossRef]
14. Teale, W.; Ditengou, F.A.; Dovzhenko, A.; Li, X.; Molendijk, A.; Ruperti, B.; Paponov, I.; Palme, K. Auxin as a Model for the Integration of Hormonal Signal Processing and Transduction. *Mol. Plant.* **2008**, *1*, 229–237. [CrossRef]
15. Vanneste, S.; Friml, J. Auxin: A Trigger for Change in Plant Development. *Cell* **2009**, *136*, 1005–1016. [CrossRef] [PubMed]
16. Djanaguiraman, M.; Prasad, P.V.V. Ethylene production under high temperature stress causes premature leaf senescence in soybean. *Funct. Plant. Biol.* **2010**, *37*, 1071–1084. [CrossRef]
17. Ludwig-Müller, J. Auxin conjugates: Their role for plant development and in the evolution of land plants. *J. Exp. Bot.* **2011**, *62*, 1757–1773. [CrossRef] [PubMed]

18. Wu, Y.-S.; Yang, C.-Y. Ethylene-mediated signaling confers thermotolerance and regulates transcript levels of heat shock factors in rice seedlings under heat stress. *Bot. Stud.* **2019**, *60*, 1–12. [CrossRef]

19. Enders, T.A.; Strader, L.C. Auxin activity: Past, present, and future. *Am. J. Bot.* **2015**, *102*, 180–196. [CrossRef]

20. Yamamuro, C.; Zhu, J.K.; Yang, Z. Epigenetic modifications and plant hormone action. *Mol. Plant* **2016**, *9*, 57–70. [CrossRef]

21. Wani, S.H.; Kumar, V.; Saroj, V.S.; Sah, K. Phytohormones and their metabolic engineering for abiotic stress tolerance in crop plants. *Crop J.* **2016**, *4*, 162–176. [CrossRef]

22. Šimura, J.; Antoniadi, I.; Široká, J.; Tarkowská, D.; Strnad, M.; Ljung, K.; Novák, O. Plant hormonomics: Multiple phytohormone profiling by targeted metabolomics. *Plant Physiol.* **2018**, *177*, 476–489. [CrossRef]

23. Gray, W.M.; Östin, A.; Sandberg, G.; Romano, C.P.; Estelle, M. High temperature promotes auxin-mediated hypocotyl elongation in Arabidopsis. *Proc. Natl. Acad. Sci. USA* **1998**, *95*, 7197–7202. [CrossRef] [PubMed]

24. Koini, M.A.; Alvey, L.; Allen, T.; Tilley, C.A.; Harberd, N.P.; Whitelam, G.C.; Franklin, K.A. High Temperature-Mediated Adaptations in Plant Architecture Require the bHLH Transcription Factor PIF4. *Curr. Biol.* **2009**, *19*, 408–413. [CrossRef]

25. Sakata, T.; Yagihashi, N.; Atsushi, H.; Higashitani, A. Tissue-specific auxin signaling in response to temperature fluctuation. *Plant. Signal. Behav.* **2010**, *5*, 1510–1512. [CrossRef] [PubMed]

26. Sun, J.; Qi, L.; Li, Y.; Chu, J.; Li, C. PIF4–Mediated Activation of YUCCA8 Expression Integrates Temperature into the Auxin Pathway in Regulating Arabidopsis Hypocotyl Growth. *PLoS Genet.* **2012**, *8*, e1002594. [CrossRef]

27. Sakata, T.; Oshino, T.; Miura, S.; Tomabechi, M.; Tsunaga, Y.; Higashitani, N.; Miyazawa, Y.; Takahashi, H.; Watanabe, M.; Higashitani, A. Auxins reverse plant male sterility caused by high temperatures. *Proc. Natl. Acad. Sci. USA* **2010**, *107*, 8569–8574. [CrossRef] [PubMed]

28. Lee, H.-J.; Jung, J.-H.; Llorca, L.C.; Kim, S.-G.; Lee, S.; Baldwin, I.T.; Park, C.-M. FCA mediates thermal adaptation of stem growth by attenuating auxin action in Arabidopsis. *Nat. Commun.* **2014**, *5*, 5473. [CrossRef]

29. Higashitani, A. High temperature injury and auxin biosynthesis in microsporogenesis. *Front. Plant. Sci.* **2013**, *4*, 47. [CrossRef] [PubMed]

30. Huberman, M.; Riov, J.; Aloni, B.; Goren, R. Role of ethylene biosynthesis and auxin content and transport in high temper-ature-induced abscission of pepper reproductive organs. *J. Plant Growth Regul.* **1997**, *16*, 129–135. [CrossRef]

31. Elkinawy, M. Physiological significance of indoleacetic acid and factors determining its level in cotyledons of Lupinus albus during germination and growth. *Physiol. Plant.* **1982**, *54*, 302–308. [CrossRef]

32. Merkys, A.; Anisimovienė, N.; Darginavičienė, J.; Maksimov, G. Principles of IAA action in plants. *Biologija* **2003**, *4*, 28–31.

33. Zazimalova, E.; Napier, R.M. Points of regulation for auxin action. *Plant. Cell Rep.* **2003**, *21*, 625–634. [CrossRef] [PubMed]

34. Wang, K.L.-C.; Li, H.; Ecker, J.R. Ethylene Biosynthesis and Signaling Networks. *Plant. Cell* **2002**, *14*, S131–S151. [CrossRef] [PubMed]

35. Iqbal, N.; Khan, N.A.; Ferrante, A.; Trivellini, A.; Francini, A.; Khan, M.I.R. Ethylene Role in Plant Growth, Development and Senescence: Interaction with Other Phytohormones. *Front. Plant. Sci.* **2017**, *8*, 475. [CrossRef] [PubMed]

36. Jankovska-Bortkevič, E.; Gavelienė, V.; Jankauskienė, J.; Mockevičiūtė, R.; Koryznienė, D.; Jurkonienė, S. Response of winter oilseed rape to imitated temperature fluctuations in autumn-winter period. *Environ. Exp. Bot.* **2019**, *166*, 103801. [CrossRef]

37. Vandenbussche, F.; Van Der Straeten, D. The role of ethylene in plant growth and development. *Annu. Plant Rev.* **2018**, *4*, 219–241.

38. Dubois, M.; Broeck, L.V.D.; Inzé, D. The Pivotal Role of Ethylene in Plant Growth. *Trends Plant. Sci.* **2018**, *23*, 311–323. [CrossRef] [PubMed]

39. Larkindale, J.; Huang, B. Thermotolerance and antioxidant systems in Agrostis stolonifera: Involvement of salicylic acid, ab-scisic acid, calcium, hydrogen peroxide, and ethylene. *J. Plant Physiol.* **2004**, *161*, 405–413. [CrossRef]

40. Bita, C.E.; Gerats, T. Plant tolerance to high temperature in a changing environment: Scientific fundamentals and production of heat stress-tolerant crops. *Front. Plant. Sci.* **2013**, *4*, 273. [CrossRef]

41. Aniszewski, T.; Kupari, M.H.; Leinonen, A.J. Seed number, seed size and seed diversity in Washington lupine (*Lupinus poly-phyllus* Lindl.). *Ann. Bot.* **2001**, *87*, 77–82. [CrossRef]

42. Murray, B.; Phillips, M. Temporal introduction patterns of invasive alien plant species to Australia. *NeoBiota* **2012**, *13*, 1–14. [CrossRef]

43. Sõber, V.; Ramula, S. Seed number and environmental conditions do not explain seed size variability for the invasive herb Lupinus polyphyllus. *Plant Ecol.* **2013**, *214*, 883–892. [CrossRef]

44. Musatenko, L.I.; Vedenicheva, N.P.; Vasyuk, V.A.; Generalova, V.N.; Martyn, G.I.; Sytnik, K.M. Phytohormones in Seedlings of Maize Hybrids Differing in Their Tolerance to High Temperatures. *Russ. J. Plant. Physiol.* **2003**, *50*, 444–448. [CrossRef]

45. Nicolás, J.I.L.; Acosta, M.; Sánchez-Bravo, J. Variation in indole-3-acetic acid transport and its relationship with growth in etiolated lupin hypocotyls. *J. Plant. Physiol.* **2007**, *164*, 851–860. [CrossRef]

46. Gudzinskas, Z. Conspectus of alien plant species of Lithuania. 10. Fabaceae. *Bot. Lith.* **1999**, *5*, 103–114.

47. Fremstad, E. NOBANIS—Invasive Alien Species Fact Sheet—Lupinus polyphyllus. Available online: https://www.nobanis.org/globalassets/speciesinfo/l/lupinus-polyphyllus/lupinus-polyphyllus.pdf (accessed on 15 January 2020).

48. Wolkovich, E.M.; Cleland, E.E. Phenological niches and the future of invaded ecosystems with climate change. *AoB PLANTS* **2014**, *6*, 013. [CrossRef]

49. Wang, R.L.; Zeng, R.S.; Peng, S.L.; Chen, B.M.; Liang, X.T.; Xin, X.W. Elevated temperature may accelerate invasive expansion of the liana plant Ipomoea cairica. *Weed Res.* **2011**, *51*, 574–580. [CrossRef]

50. Pyšek, P.; Jarošík, V.; Hulme, P.E.; Pergl, J.; Hejda, M.; Schaffner, U.; Vilà, M. A global assessment of invasive plant impacts on resident species, communities and ecosystems: The interaction of impact measures, invading species' traits and environment. *Glob. Chang. Biol.* **2012**, *18*, 1725–1737. [CrossRef]
51. Williams, A.L.; Wills, K.E.; Janes, J.K.; Schoor, J.K.V.; Newton, P.C.; Hovenden, M.J. Warming and free-air CO2 enrichment alter demographics in four cooccurring grassland species. *New Phytol.* **2007**, *176*, 365–374. [CrossRef] [PubMed]
52. Elliott, C.W.; Fischer, D.G.; Leroy, C.J. Germination of Three NativeLupinusSpecies in Response to Temperature. *Northwest Sci.* **2011**, *85*, 403–410. [CrossRef]
53. De Wit, M.; Lorrain, S.; Fankhauser, C. Auxin-mediated plant architectural changes in response to shade and high temperature. *Physiol. Plant.* **2013**, *151*, 13–24. [CrossRef] [PubMed]
54. Bajguz, A.; Piotrowska, A. Conjugates of auxin and cytokinin. *Phytochemistry* **2009**, *70*, 957–969. [CrossRef] [PubMed]
55. Staswick, P.E.; Serban, B.; Rowe, M.; Tiryaki, I.; Maldonado, M.T.; Maldonado, M.C.; Suza, W. Characterization of an Ara-bidopsis enzyme family that conjugates amino acids to indole-3-acetic acid. *Plant Cell* **2005**, *17*, 616–627. [CrossRef]
56. Korasick, D.; Enders, T.A.; Strader, L.C. Auxin biosynthesis and storage forms. *J. Exp. Bot.* **2013**, *64*, 2541–2555. [CrossRef] [PubMed]
57. Ljung, K. Auxin metabolism and homeostasis during plant development. *Development* **2013**, *140*, 943–950. [CrossRef]
58. Pěnčík, A.; Simonovik, B.; Petersson, S.V.; Henyková, E.; Simon, S.; Greenham, K.; Zhang, Y.; Kowalczyk, M.; Estelle, M.; Zazímalová, E.; et al. Regulation of auxin homeostasis and gradients in Arabidopsis roots through the formation of the indole-3-acetic acid catabolite 2-oxindole-3-acetic acid. *Plant Cell* **2013**, *25*, 3858–3870. [CrossRef]
59. Ljun, K.; Hul, A.K.; Kowalczyk, M.; Marchant, A.; Celenza, J.; Cohen, J.D.; Sandberg, G. Biosynthesis, conjugation, catabo-lism and homeostasis of indole-3-acetic acid in Arabidopsis thaliana. *Plant Mol. Biol.* **2002**, *49*, 249–272. [CrossRef]
60. Junghans, U.; Polle, A.; Düchting, P.; Weiler, E.; Kuhlmann, B.; Gruber, F.; Teichmann, T. Adaptation to high salinity in poplar involves changes in xylem anatomy and auxin physiology. *Plant Cell Environ.* **2006**, *29*, 1519–1531. [CrossRef]
61. Oetiker, J.H.; Aeschbacher, G. Temperature-Sensitive Plant Cells with Shunted Indole-3-Acetic Acid Conjugation. *Plant. Physiol.* **1997**, *114*, 1385–1395. [CrossRef]
62. LeClere, S.; Tellez, R.; Rampey, R.A.; Matsuda, S.P.T.; Bartel, B. Characterization of a Family of IAA-Amino Acid Conjugate Hydrolases from Arabidopsis. *J. Biol. Chem.* **2002**, *277*, 20446–20452. [CrossRef]
63. Rampey, R.A.; LeClere, S.; Kowalczyk, M.; Ljung, K.; Sandberg, G.; Bartel, B. A Family of Auxin-Conjugate Hydrolases That Contributes to Free Indole-3-Acetic Acid Levels during Arabidopsis Germination. *Plant. Physiol.* **2004**, *135*, 978–988. [CrossRef]
64. Teale, W.D.; Paponov, I.; Palme, K. Auxin in action: Signalling, transport and the control of plant growth and development. *Nat. Rev. Mol. Cell Biol.* **2006**, *7*, 847–859. [CrossRef] [PubMed]
65. Crawford, A.J.; McLachlan, D.H.; Hetherington, A.; Franklin, K.A. High temperature exposure increases plant cooling capacity. *Curr. Biol.* **2012**, *22*, R396–R397. [CrossRef] [PubMed]
66. Muday, G.K.; Rahman, A.; Binder, B.M. Auxin and ethylene: Collaborators or competitors? *Trends Plant. Sci.* **2012**, *17*, 181–195. [CrossRef]
67. Hao, D.; Sun, X.; Ma, B.; Zhang, J.-S.; Guo, H. Ethylene. In *Hormone Metabolism and Signaling in Plants*; Elsevier Academic Press: London, UK, 2017; pp. 203–241.
68. Porfírio, S.; da Silva, M.D.R.G.; Peixe, A.; Cabrita, M.J.; Azadi, P. Current analytical methods for plant auxin quantification—A review. *Anal. Chim. Acta* **2016**, *902*, 8–21. [CrossRef] [PubMed]
69. Kowalczyk, M.; Sandberg, G. Quantitative Analysis of Indole-3-Acetic Acid Metabolites in Arabidopsis. *Plant. Physiol.* **2001**, *127*, 1845–1853. [CrossRef] [PubMed]
70. Child, R.D.; Chauvaux, N.; John, K.; Van Onckelen, H.; Ulvskov, P. Ethylene biosynthesis in oilseed rape pods in relation to pod shatter. *J. Exp. Bot.* **1998**, *49*, 829–838. [CrossRef]

Article

Yield, Physiological Performance, and Phytochemistry of Basil (*Ocimum basilicum* L.) under Temperature Stress and Elevated CO$_2$ Concentrations

T. Casey Barickman [1,*], Omolayo J. Olorunwa [1], Akanksha Sehgal [2], C. Hunt Walne [2], K. Raja Reddy [2] and Wei Gao [3]

[1] North Mississippi Research and Extension Center, Mississippi State University, Verona, MS 38879, USA; ojo26@msstate.edu

[2] Department of Plant and Soil Sciences, Mississippi State University, Mississippi State, MS 39762, USA, as5002@msstate.edu (A.S.); chw148@msstate.edu (C.H.W.); krreddy@pss.msstate.edu (K.R.R.)

[3] USDA UVB Monitoring and Research Program, Natural Resource Ecology Laboratory, Department of Ecosystem Science and Sustainability, Colorado State University, Fort Collins, CO 80523, USA; wei.gao@colostate.edu

* Correspondence: t.c.barickman@msstate.edu; Tel.: +1-(662)-566-2201

Citation: Barickman, T.C.; Olorunwa, O.J.; Sehgal, A.; Walne, C.H.; Reddy, K.R.; Gao, W. Yield, Physiological Performance, and Phytochemistry of Basil (*Ocimum basilicum* L.) under Temperature Stress and Elevated CO$_2$ Concentrations. *Plants* **2021**, *10*, 1072. https://doi.org/10.3390/plants 10061072

Academic Editor: James Bunce

Received: 8 April 2021
Accepted: 22 May 2021
Published: 27 May 2021

Abstract: Early season sowing is one of the methods for avoiding yield loss for basil due to high temperatures. However, basil could be exposed to sub-optimal temperatures by planting it earlier in the season. Thus, an experiment was conducted that examines how temperature changes and carbon dioxide (CO$_2$) levels affect basil growth, development, and phytonutrient concentrations in a controlled environment. The experiment simulated temperature stress, low (20/12 °C), and high (38/30 °C), under ambient (420 ppm) and elevated (720 ppm) CO$_2$ concentrations. Low-temperature stress prompted the rapid closure of stomata resulting in a 21% decline in net photosynthesis. Chlorophylls and carotenoids decreased when elevated CO$_2$ interacted with low-temperature stress. Basil exhibited an increase in stomatal conductance, intercellular CO$_2$ concentration, apparent quantum yield, maximum photosystem II efficiency, and maximum net photosynthesis rate when subjected to high-temperature stress. Under elevated CO$_2$, increasing the growth temperature from 30/22 °C to 38/30 °C markedly increased the antioxidants content of basil. Taken together, the evidence from this research recommends that varying the growth temperature of basil plants can significantly affect the growth and development rates compared to increasing the CO$_2$ concentrations, which mitigates the adverse effects of temperature stress.

Keywords: Genovese cultivar; photosynthesis; stomatal conductance; chlorophyll; carotenoids; antioxidant defense metabolites

1. Introduction

Climate change remains an important challenge affecting the attainment of global food security as it negatively impacts the growth and development of crops. Several studies have demonstrated higher atmospheric carbon dioxide (CO$_2$) concentrations, extreme temperature conditions, and other extreme weather events as evidence of climate change [1,2]. Global atmospheric CO$_2$ is rising (above 415 ppm in 2020). It is projected by climate models to reach the range of 540 to 970 ppm by 2100 because of human activities, declining carbon sinks, and natural global cycles [3,4]. Recent climate models have also predicted that global air temperature may experience increments in the range of 1.5 and 4.5 °C in the next century due to the increasing levels of atmospheric CO$_2$ and other greenhouse gases at an alarming rate [1,5]. Atmospheric CO$_2$ and temperature are critical in the photosynthesis, physiological, and developmental processes that occur in many crops, especially C3 crops [6,7]. Thus, anticipated increasing atmospheric CO$_2$ and temperature will affect several crops' growth and development, including basil (*Ocimum basilicum*). Increased environmental fluctuations resulting from climate change have been forecasted in many agricultural regions [8], and this poses a bane to achieving sustainable food production. Hence, it is pertinent to understand the mechanisms associated with basil's response to elevated atmospheric CO_2 and temperature stress to managing future production.

Basil is an important herbaceous aromatic plant with a noteworthy contribution to enhancing cuisine nutrition, healthy living, and landscape aesthetics. Globally, a large proportion of high-quality basil is cultivated for its essential oil, dry leaves, and flowers [9,10]. Studies have revealed different curative properties of basil, such as lowering blood pressures and fevers, reducing glucose and cholesterol levels in the blood, suppressing muscle spasms and inflammation, and strengthening the body's natural activity [9].

In general, basil is widely adapted and grown throughout the globe. However, it is most suitable for warmer temperatures [11,12]. The optimum temperature for basil growth is in the range of 25 and 30 °C, while the minimum temperature at which basil can survive is 10.9 °C [11,13]. Recent evidence suggests that temperature significantly affects the growth and development of basil plants. For example, Chang et al. [11] indicated that increasing the growth temperature of basil to 30 °C culminated in the maximum rate of net photosynthesis (P_n), transpiration rate (E), and stomatal conductance (g_s), with resultant benefits on basil yields. Though basil is considered heat-tolerant, a temperature above 38 °C has been noted to cause detrimental effects on the yield, especially during the reproductive stages of development [11,14]. Temperature stress perturbs plant metabolism due to its damaging impact on the binding affinity and structure of proteins and enzymes, thereby resulting in the build-up of undesirable toxic intermediates, disjoining of diverse reactions, and increased levels of reactive oxygen species (ROS) [15]. Thus, many farmers usually resort to growing basil plants early in the growing season to avoid harmful environmental stress factors such as heat and drought, which can cause the plants to produce elevated ROS in tissues. Elevated levels of ROS can have a devastating effect on the growth, development, and production of phytonutrients. However, early-season planting increases the risk of exposing basil crops to low-temperature stress conditions.

Basil is sensitive to low temperatures, mostly below 10 °C, resulting in damage to growth and developmental processes [16]. Chilling causes brown discoloration of interveinal leaf areas, increased leaf blade thickening, decreased plant growth, reduced postharvest shelf life, and deterioration of quality and marketability [11,16]. Many studies show that cold temperature stress can also have harmful effects on the physiological traits of basils. Kalisz et al. [17] demonstrated that cold temperatures have negative impacts on basil P_n, E, and g_s, which impaired plant growth and development. Additionally, the negative impact was identified to decrease photosystem II (PSII) activity (F_v'/F_m') and variable-to-initial chlorophyll fluorescence (F_v/F_o) after a 16-day low-temperature treatment [17].

Moreover, basil leaves subjected to low-temperature stress were identified to experience chlorosis due to rapid degradation of carotenoid, chlorophyll, and antioxidant content [11,18]. It is important to note that the alteration of basil leaf pigments due to chilling stress could damage the photosynthetic apparatus, with detrimental impacts on plant growth and development. Previous studies have revealed that basil's greenness is an indicator of chlorophyll content, usually used by consumers to determine its productivity [19]. Therefore, further research is required to quantify basil's physiological response to low-temperature stress to promote the breeding of cold-tolerant genotypes of basil.

Several studies have indicated that increasing CO_2 concentrations positively impact plant growth and development, primarily because of the significant role CO_2 plays in respiration and photosynthesis [6,20]. Al Jaouni et al. [2] reported that biomass production increased by 40% along with the photosynthetic and respiratory rate of basil, which significantly improved by 80% when atmospheric CO_2 was increased from 360 to 620 ppm. The improved photosynthetic rate was attributed to the role of elevated atmospheric CO_2 in repressing the oxygenation reaction of Rubisco, leading to improved carbon gains [20]. Previous studies by Gillig et al. [21] also found that basil grown under 1500 ppm of CO_2

had a significant increase in their chlorophyll concentrations due to the accumulation of large grains of starch and non-structural carbohydrates. The levels of antioxidant defense metabolites, such as fumarate, glutamine, glutathione (GSH), ascorbic acid (ASC), phylloquinone (vitamin K1), anthocyanins (Anth), and most flavonoids and minerals were significantly improved by elevated CO_2 [2]. Hence, the beneficial impacts of elevated atmospheric CO_2 can counteract the adverse impacts of low- and high-temperature stress on basil physiology and phytochemistry. Even more than that, the benefits of improved growth rate in basil due to elevated CO_2 could facilitate harvesting early, which will aid more crop cycles each year and contribute positively to food security.

Studies have been conducted to find out the influence of elevated CO_2 and temperature stress on basil's growth and photosynthesis. However, there has been a lack of information on the interactive impacts of CO_2 and temperature on basil plants' physiological and phytochemical performance. Hence, this study aims to evaluate the interactive impacts of elevated CO_2 and temperature stress on photosynthesis parameters, carotenoids, and chlorophylls content of basil plants. Moreover, we evaluated important biochemical parameters acting as enzymatic and nonenzymatic antioxidant defense metabolites responsible for cellular osmotic adjustments in stressed plants.

2. Results

2.1. Gas Exchange Parameters

The results showed that temperature and its interaction with CO_2 significantly ($p < 0.001$) affected the P_n of the basil plant (Table 1). However, elevated CO_2 had no significant effect ($p > 0.05$) on P_n. Under ambient CO_2, high-temperature stress increased P_n by 72%, while low-temperature stress decreased the P_n by 38%. Correspondingly, under elevated CO_2, high-temperature stress increased the P_n of basil by 45%, while low-temperature stress decreased the P_n by 21%, compared to the control treatments.

Table 1. Interactive effects of temperature stress and CO_2 on net photosynthesis (P_n), stomatal conductance to water vapor (g_s), intercellular CO_2 concentration (C_i), electron transport rate (ETR), leaf transpiration rate (E), intercellular/ambient CO_2 ratio (C_i/C_a), and the maximal quantum yield of photosystem II photochemistry (F_v'/F_m'), of basil plants. Measurements were taken on the fourth/fifth fully expanded leaf of plants grown without temperature stress (Control), with low-temperature stress, and high-temperature stress at 420 and 720 ppm of CO_2 concentration between 37 and 38 days of treatment.

Treatment	P_n ($\mu mol \cdot m^2 \cdot s^{-1}$)	g_s ($mol \cdot m^2 \cdot s^{-1}$)	C_i ($\mu mol \cdot m^2 \cdot s^{-1}$)	ETR ($\mu mol\ m^{-2} \cdot s^{-1}$)	E ($mmol \cdot m^2 \cdot s^{-1}$)	C_i/C_a [1]	F_v'/F_m'
420 PPM							
Control	24.48 c	0.38 b	295.09 d	187.33 ab	6.79 c	0.70 b	0.47 b
High Temperature	42.22 a	0.71 a	303.47 d	205.94 a	15.66 a	0.72 b	0.52 a
Low Temperature	15.20 d	0.14 c	237.64 e	146.46 bc	1.91 d	0.54 c	0.42 c
720 PPM							
Control	31.51 b	0.31 b	530.71 b	184.97 ab	6.67 c	0.74 b	0.51 a
High Temperature	35.51 b	0.63 a	597.80 a	183.19 abc	14.39 b	0.83 a	0.49 ab
Low Temperature	19.46 d	0.13 c	444.50 c	130.16 c	1.63 d	0.62 c	0.47 b
Treatment [2,3]	***	***	***	*	***	***	**
CO_2	NS	NS	***	NS	NS	**	NS
Trt *CO_2	***	NS	*	NS	NS	NS	*

[1] The measured intercellular CO_2/ambient CO_2 of LI-6400XT leaf cuvette. [2] SE— standard error of the mean; P_n = 1.5044; g_s = 0.03683; C_i = 15.4158; ETR = 17.876; E = 0.3881; C_i/C_a = 0.02751; F_v'/F_m' = 0.01425. [3] NS represents non-significant $p > 0.05$; *, **, *** represent significance levels at $p \leq 0.05$, $p \leq 0.01$, and $p \leq 0.001$ respectively; within columns, values followed by the same letters are not significantly different.

Additionally, temperature treatments significantly impacted the g_s and E of basil plants (Table 1). Specifically, the g_s and E of the basil plant were significantly increased due to exposure to high-temperature stress. Under low-temperature stress, g_s and E

were reduced by 63% and 72%, respectively. Although elevated CO_2 had no significant effects ($p > 0.05$) on g_s and E, there was a decreasing trend of g_s and E at elevated CO_2 (Table 1). In contrast, the C_i concentrations of basil plants were significantly affected by both temperature and CO_2 ($p < 0.001$), with significantly higher C_i at elevated CO_2 as compared to the ambient CO_2.

The C_i/C_a was significantly different from temperature and CO_2 stresses, but there was no difference from the control treatment when CO_2 interacted with temperature stress. The photosynthetic ETR of basil reduced by 22% when exposed to low-temperature treatment and increased by 10% under high-temperature treatment compared to the control treatment. Similarly, the quantum efficiency (F_v'/F_m') was affected by temperature stress and its interaction with CO_2 (Table 1). The F_v'/F_m' was decreased at low-temperature stress at ambient CO_2, whereas F_v'/F_m' increased when interacted with elevated CO_2 at both low- and high-temperature stresses.

2.2. Chlorophyll Content and Total Carotenoids

Compared to the control treatments, low-temperature stress when interacted with elevated CO_2 caused a loss of pigment content in basil leaves by decreasing Chl a, b, and total chlorophyll (a + b) content by 1%, 12%, and 2%, respectively (Figure 1). While high-temperature stress at elevated CO_2 significantly increased Chl a, b, and total chlorophyll (a + b) by 35%, 18%, and 33%, respectively. However, temperature stress only had significant effects ($p > 0.001$) on the total xanthophyll of the basil plant (Table 2). At ambient CO_2, the total xanthophyll content was higher under heat stress, whereas it was lower under cold stress than in the control treatments. Compared to the control treatment, the carotenoid cycle (violaxanthin (V) + antheraxanthin (A) + zeaxanthin (Z)) decreased significantly under both low- and high -temperature stresses. Analogous results were observed for ZA/ZAV at elevated CO_2. Elevated CO_2 and its interaction with temperature stress significantly decreased the proportion of both lutein and neoxanthin when compared to the control treatments. However, temperature stress and its interaction with CO_2 had no significant ($p < 0.05$) effects on the β-carotene. β-carotene was atypically much lower (24%) under elevated CO_2 than in the ambient CO_2 in the present study. The concentration of violaxanthin revealed no effects with temperature or elevated CO_2 treatments.

Table 2. Interactive effects of temperature stress and CO_2 on carotenoids concentration of basil leaf tissue. Leaf samples were taken from basil plants grown without temperature stress (control), with low-temperature stress and high-temperature stress at 420 and 720 ppm of CO_2 concentration between 37 and 38 days of treatment.

	Concentration ($\mu g \cdot g^{-1}$ Dry Mass)							
Treatment	Neo [a]	Viol	Anth	Zea	Lut	β-car	Total Xan	ZA/ZAV [b]
				420 ppm				
Control	276.43 a	204.25 a	68.76 ab	163.94 a	793.09 a	509.66 a	436.95 a	0.53 a
High Temperature	252.16 abc	215.89 a	58.23 b	101.84 bc	710.64 ab	464.48 ab	375.97 bc	0.43 c
Low Temperature	264.59 ab	239.62 a	74.12 a	161.19 a	669.04 b	506.33 a	474.93 a	0.50 ab
				720 ppm				
Control	222.36 c	208.03 a	74.29 a	157.75 a	561.09 c	384.95 bc	440.08 a	0.53 a
High Temperature	265.46 ab	226.79 a	43.14 c	78.38 c	687.31 b	390.65 bc	348.31 c	0.35 d
Low Temperature	235.20 bc	236.26 a	62.88 ab	119.79 b	552.75 c	342.73 c	418.93 ab	0.44 bc
Treatment [c,d]	NS	NS	***	***	NS	NS	***	***
CO_2	*	NS	NS	*	***	***	NS	*
Trt *CO_2	*	NS	*	NS	*	NS	NS	NS

[a] Neo—neoxanthin; Vio—violaxanthin; Anth—antheraxanthin; Zea—zeaxanthin; Lut—lutein; β-car—beta carotene; Xan—xanthophylls. [b] Xanthophyll cycle ratio = zeaxanthin to antheraxanthin/zeaxanthin to antheraxanthin to violaxanthin. [c] The standard error of mean was: Neo—14.10; Vio—14.55; Anth—7.17; Zea—11.27; Lut—35.85; Bcar—32.55; Total Xan—19.09; ZA/ZAV—0.025. [d] NS represents non-significant $p > 0.05$; *, *** represent significance levels at $p \leq 0.05$ and $p \leq 0.001$, respectively; within columns, values followed by the same letters are not significantly different.

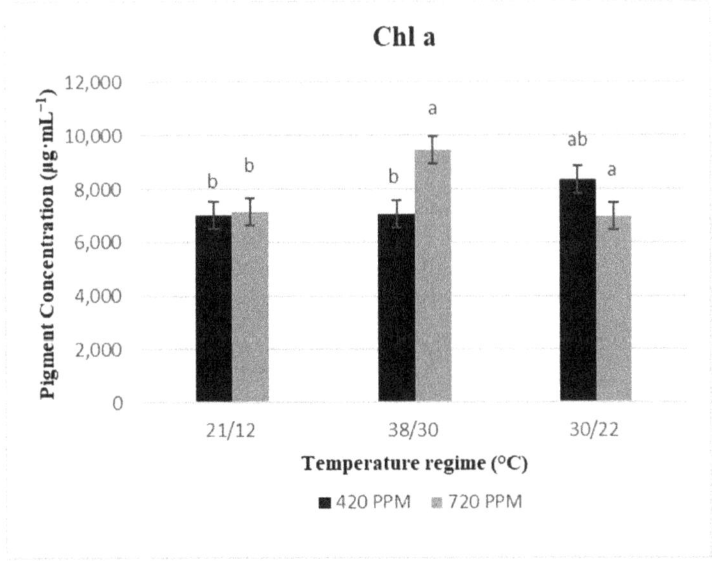

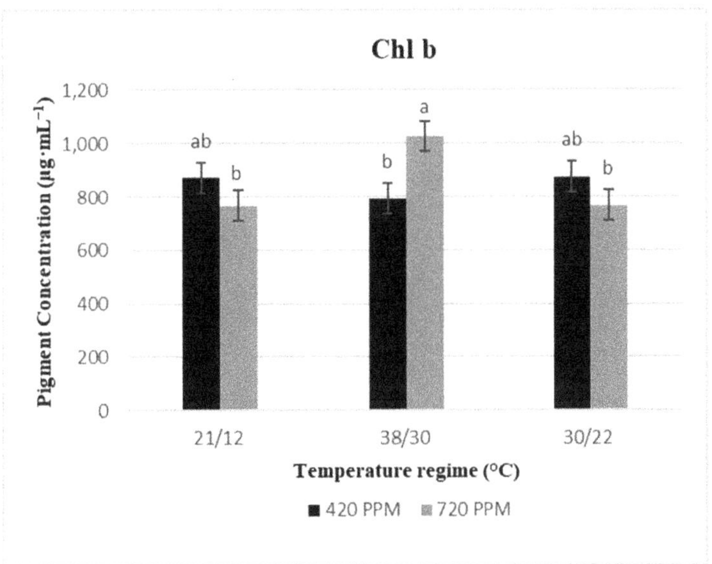

Figure 1. *Cont.*

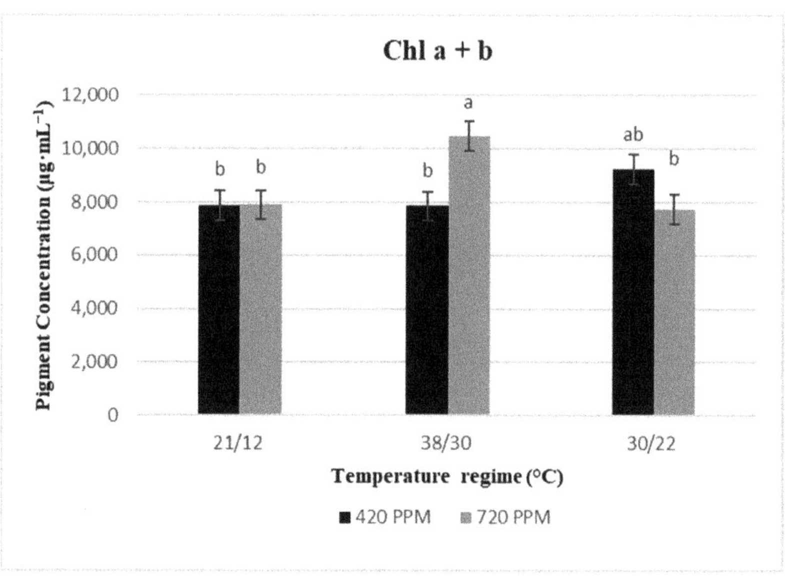

Figure 1. Chlorophyll a (Chl a), and chlorophyll b (Chl b), and total chlorophyll (Chl a + b), concentrations of basil plants under no temperature stress (Control), low-temperature stress, and high-temperature stress at 420 and 720 ppm of CO_2 concentration. The standard error mean was Chl a = 506.33, Chl b = 57.19, and Chl a + b = 547.56. Different low case letters indicate significant difference at $p = < 0.05$ by least significant difference.

2.3. Total Phenolics

Conflicting with the pattern of changes of chlorophyll contents, total phenolics reduced (7%) in basil plants subjected to the interactions of high-temperature stress and elevated CO_2 (Figure 2), whereas a significant increase of 10% was observed under low temperature at elevated CO_2.

2.4. Epicular Wax

The basil plants showed a significant reduction in leaf wax content when subjected to both low- and high-temperature stresses and elevated CO_2 (Figure 3). However, there was no interaction effect between CO_2 and temperature treatments on basil leaf wax content.

2.5. Antioxidant and Oxidative Parameters

Interactions between temperature and CO_2 significantly affected the content of MDA and GSH only (Table 3). Basil grown at low temperature under elevated CO_2 significantly increased the MDA content by 150%, whereas the MDA content decreased by 43% under elevated CO_2 at high temperature. In contrast, the total GSH levels of basil were markedly increased by 43% under high-temperature stress at elevated CO_2. It decreased by 2% when subjected to low-temperature treatment at elevated CO_2 related to the control. There was only a small difference between the GSH content under low-temperature stress at ambient CO_2 and low-temperature stress at elevated CO_2. Compared to the control treatments, the SOD content increased significantly under elevated CO_2 at low and high-temperature stresses. Additionally, elevated CO_2 alone was discovered to increase the ASC and TRE content of basil substantially by 89% and 41%, respectively, compared to the control treatments. However, temperature and elevated CO_2 treatments and their interactions had no significant effects on H_2O_2 content.

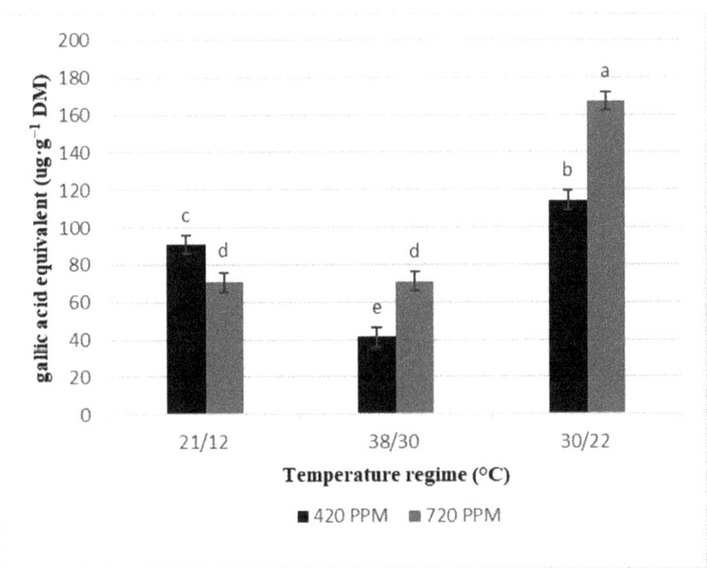

Figure 2. Total phenolic of basil leaf tissue subjected to no temperature stress (Control), low-temperature stress, and high-temperature stress at 420 and 720 ppm of CO_2 concentration. Total phenolic content is presented as gallic acid equivalent concentration ug·g^{-1} dry mass (DM). The standard error mean was total phenolic = 5.133. Different low case letters indicate significant difference at $p = < 0.05$ by least significant differences.

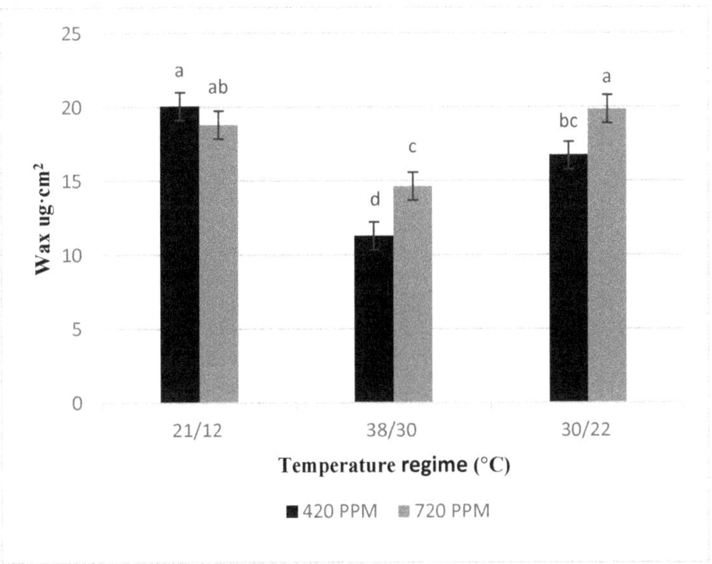

Figure 3. Average epicuticular wax content for basil plants grown without temperature stress (Control), with low-temperature stress and high-temperature stress at 420 and 720 ppm of CO_2 concentration after 34 days of treatment. The standard error mean for wax was 0.9463. Different low case letters indicate a significant difference at $p = < 0.05$ by the least significant difference.

Table 3. Interactive effects of temperature stress and CO_2 on metabolites of basil leaf tissues. Leaf samples were taken from basil plants grown without temperature stress (Control), with low-temperature stress and high-temperature stress at 420 and 720 ppm of CO_2 concentration between 37 and 38 days of treatment.

	Concentration					
	$nmol\cdot g^{-1}$ DM	$\mu mol\cdot g^{-1}$ DM	Units/mg Protein	$nmol\cdot g^{-1}$ DM	$\mu mol\cdot g^{-1}$ DM	$nmol\cdot g^{-1}$ DM
Treatment	MDA [a]	H_2O_2	SOD	ASC	TRE	GSH
420 ppm						
Control	0.0080 b	0.191 a	0.0304 c	0.101 b	0.0892 bc	0.192 b
High Temperature	0.0076 b	0.206 a	0.0306 c	0.123 ab	0.103 abc	0.188 b
Low Temperature	0.0074 b	0.198 a	0.0396 ab	0.106 b	0.0834 c	0.189 b
720 ppm						
Control	0.0078 b	0.183 a	0.0392 ab	0.191 a	0.126 ab	0.177 b
High Temperature	0.0046 b	0.206 a	0.0330 bc	0.150 ab	0.116 abc	0.275 a
Low Temperature	0.0200 a	0.266 a	0.0466 a	0.175 a	0.133 a	0.188 b
Treatment [b,c]	**	NS	**	NS	NS	*
CO_2	NS	NS	*	**	**	NS
Trt *CO_2	**	NS	NS	NS	NS	*

[a] MDA—malondialdehyde; H_2O_2—peroxide; SOD—superoxide dismutase; ASC—ascorbic acid; TRE—trehalose; GSH—glutathione. [b] The standard error of mean was MDA = 0.001976; H_2O_2 = 0.03045; SOD = 0.00272; ASC = 0.02317; TRE = 0.01383; GSH = 0.01776. [c] NS represents non-significant $p > 0.05$; *, ** represent significance levels at $p \leq 0.05$ and $p \leq 0.01$ respectively; within columns, values followed by the same letters are not significantly different.

3. Discussion

Abiotic stress as a result of increased fluctuations of temperature and atmospheric CO_2 affects the productivity of many important crops, including basil [17,22]. In particular, extreme temperature stress during seedling growth reduced the yield of basil. Early season sowing is one of the methods for avoiding yield loss in basil due to high temperatures. However, when producers plant basil earlier in the season, the crop can be exposed to sub-optimal temperatures. Previous research has indicated that elevated CO_2 is beneficial to C3 crop growth and development [6,23]. The connection to whether increasing CO_2 from the ambient concentrations is beneficial in mitigating the adverse impacts of temperature stress in basil is yet to be determined. Hence, we evaluated the physiological and phytochemistry of basil subjected to varying temperature stress and CO_2 concentrations.

The current research indicated that P_n and g_s reduced with low-temperature stress and increased when subjected to high-temperature stress. These observations agree with the results of Ribeiro et al. [16], Kalisz et al. [17], and Balasooriya et al. [24], who revealed a negative impact of chilling temperatures on basil photosynthesis. Basil plants lower the g_s of their leaves to adapt adequately to low-temperature stress. The lower g_s are linked to preventing leaf water loss (wilting), which are directly related to decreases in P_n and C_i concentration. However, in the current study, there were no observations of wilting plants due to decreased g_s. The lower g_s and E and decreased P_n and C_i further complements the results of Saibo et al. [25], who proposed that reduced g_s are an essential factor influencing P_n decrease compared to non-stomatal limitations including reduced -Rubisco activity and -energy consumption. However, C_i/C_a was observed to reduce under low temperature, suggesting that P_n reduction could also be due to decreased C_i/C_a. The accelerated chlorophyll pigment degradation of basil plants observed when elevated CO_2 interacted with low-temperature stress could also be responsible for decreased P_n. The results of this current study match those observed in basil by Gillig et al. [21]. Basil plants subjected to low-temperature stress exhibit lower gas exchange, decreased g_s, and E, thereby explaining a decrease in the metabolic synthesis of metabolites, cell wall formation, and overall growth. Conversely, when basil plants are subjected to high-temperature stress, there is an increase in metabolic processes, gas exchanges, and morphological changes such as plant height.

It is important to remember that fluctuations in temperature have a dramatic effect on plants' metabolic functions that can lead to morphological and biochemical changes that can ultimately decrease crop yields.

Importantly, the adverse effects of the chilling temperature on P_n were ameliorated when interacted with elevated CO_2. The increased CO_2 inside the leaf would increase leaf photosynthesis through increased activity of the rubisco enzyme and reduced photorespiration [6,24]. In contrast, the high-temperature stress was observed to increase the g_s, E, C_i, apparent quantum yield, maximum photosystem II efficiency, and finally, maximum P_n of basil plants. This observation could be linked to the positive impacts of high-temperature treatments detected on the morphological parameters of basil plants [26]. Comparable results were accounted for by Chang et al. [11]. However, elevated CO_2 did not have significant effects on g_s of basil plants in this study. Hence, proposing that the interactive impacts of elevated CO_2 and temperature stress on P_n were not linked with changes in g_s under temperature stress because reduced g_s were already instigated by low-temperature stress. Photosynthetic ETR and F_v'/F_m' also demonstrated a similar pattern with P_n and g_s. The results additionally propose that the raised P_n and g_s under high-temperature stress could be responsible for the increased ETR chain under temperature stress. F_v'/F_m' is sensitive to chilling conditions and therefore decreased F_v'/F_m' can show stress via photo-inhibition. Moreover, the occurrence of photoinhibition in basil due to chilling stress could result to increased production of ROS in basil, which causes oxidative damage in plants. Kalisz et al. [17] attributed the F_v'/F_m' reduction to the presence of photochemically inactive reaction centers and reduced ETR, which was also observed in this present study. On the contrary, elevated CO_2 interacted with temperature stress to remarkably attenuated the damage caused by chilling stress on P_n and promoted F_v'/F_m' by maintaining proper redox balance. Generally, photosynthesis is perhaps the most susceptible parameter to physiological processes, especially when C3 (basil) plants are subjected to low-temperature stress [27]. Thus, when basil plants are exposed to chilling stress, a decrease of P_n, g_s, E, and metabolites such as chlorophylls and carotenoids can have a considerable impact on plant growth and development.

Similar to most of our results, increased growth temperature from 30/22 °C to 38/30 °C under elevated CO_2 produced more chlorophyll content of basil plants. Specifically, Chl a and Chl b increased 18% and 35%, respectively. These results suggest that subjecting basil to heat stress and elevated CO_2 does not cause a drastic loss in Chl content, and this matches those observed in thermophilic plants [28,29]. Zhou et al. [30] reported that in tomato plants, a significant increase of 35% and 31% in Chl a and Chl b was observed, respectively, when the growth temperature increased from 25/20 °C to 38/30 °C. On the other hand, cold temperature stress under elevated CO_2 significantly decreased the concentrations of basil Chl a with no change in the concentration of Chl b. Corroborating the present results, Kalisz et al. [17] and Liu et al. [31] reported a significant reduction in cold-sensitive crops when subjected to chilling stress. Contrary to previous research, it can be conceivably hypothesized that elevated CO_2 does not mitigate the adverse effects of chilling stress on basil chlorophyll concentrations. The data also suggests that reduction in chlorophyll under low-temperature stress may be due to decreased metabolic functions that induce growth and development and production of metabolites to protect the plants. Conversely, under high-temperature stress conditions, growth and metabolic processes increase due to elevated P_n, g_s, and E.

Temperature is an important factor that regulates several compounds in basil that have health benefits to humans, including carotenoids, chlorophylls, phenolics, and epicuticular wax [10]. Previous research by Al-Huqail et al. [18] found that basil under any stress conditions caused a decrease in its carotenoid content. For instance, both cold and heat stress impair the xanthophyll cycle pigments of basil plants in the current study. Likewise, elevated CO_2 caused a decrease in the xanthophyll cycle of basil plants compared to control treatments. Loladze et al. [32] ascribed the negative impacts of CO_2 levels on the xanthophyll cycle of C3 plants to reduce non-photochemical quenching (NPQ) integral photoprotection. Thus, these results may be due to more of the energy produced by increasing CO_2 levels is directed toward photosynthesis and less toward heat dissipation. However, subjecting basil plants to low-temperature stress led to a significant increase in total phenolics content. While under high-temperature treatment, total phenolics of basil decreased considerably. These findings contradict many studies [18,33,34] because increased phenolics are usually associated with thermophilic plant defense's mechanism against heat stress. Moreover, epicuticular wax decreased significantly both under heat and cold stress conditions, which is unexpected because increased wax content is always utilized as a physiological trait for selecting thermophilic plants [35].

Furthermore, the decreased Chl, P_n, g_s, E, and F_v'/F_m' observed in this study could promote a significant proportion of the light energy utilized during photosynthesis to induce excessive accumulation of ROS in basil tissues, thereby impairing the overall growth of the plant. Previous studies have shown that temperature stress increases ROS production (e.g., H_2O_2 and O_2), which causes oxidative stress in plants [36]. However, temperature and elevated CO_2 treatments and their interactions had no significant effects on H_2O_2 content in this study. Thus, suggesting that heat stress did not induce oxidative damage in basil plants, which could further support basil's tolerance to heat stress. Moreover, Al Jaouni et al. [2] demonstrated that elevated CO_2 positively impacts basil antioxidant compounds. Correspondingly, results from this study indicated that increasing the CO_2 concentration from 420 ppm to 720 ppm ameliorated the adverse effects of ROS by increasing TRE and ASC content. An increment in antioxidant enzyme activity has been described to be linked with plant resilience to heat stress [37]. It is important to note that the maximum increase was found in SOD content among the antioxidant enzymes, indicating that when basil was subjected to elevated CO_2, this enzyme had played a critical part in decreasing the damaging consequences of ROS. SOD is considered an enzymatic antioxidant curbing the harmful effects of elevated ROS by catalyzing the dismutation of superoxide radicals to H_2O_2 and O_2 [36]. Hence, the increased contents of SOD in basil plants under elevated CO_2 would reduce the toxic effects of elevated ROS levels. Additionally, GSH is a nonenzymatic antioxidant, which minimizes the damage caused by ROS in plants and protects the photosynthetic apparatus from oxidative damage [27]. Hence, basil's tolerance to heat stress could be linked to the increased GSH antioxidant levels observed in this study when high-temperature stress interreacted with elevated CO_2.

Low-temperature stress frequently induces harm to cell membranes. MDA is a crucial indicator of membrane system injuries and cellular metabolism deterioration [31]. Basil grown at low temperature under elevated CO_2 significantly increased the MDA content, whereas the MDA content decreased under high temperature at elevated CO_2. This result suggests that the combination of low-temperature stress and elevated CO_2 treatments is more damaging than the interaction of heat stress and elevated CO_2. Thus, indicating increased resistance of basil to heat stress due to low levels of MDA observed.

4. Materials and Methods

4.1. Growth Condition and Plant Material

Basil 'Genovese' (Johnny's Selected Seeds, Winslow, ME) seeds were planted in polyvinyl-chloride pots (15.2 cm diameter by 30.5 cm height). The lower part of each pot was loaded up with 500 g gravel while the upper parts were filled with a mixture of sand and soil (3:1 VV) in the soil-plant-atmosphere-research (SPAR) units at the Rodney Foil Plant Science research facility of Mississippi State University, Mississippi State, MS, USA, June-July 2019. The SPAR units can control environmental conditions, including temperature and CO_2 concentration levels, at estimated set points. More information on the SPAR chamber subtleties were earlier portrayed by Reddy et al. [38] and Wijewardana et al. [39].

Six seeds previously selected by size and quality were planted in each pot, and approximately 14 days after sowing (DAS), the plants were thinned to one plant per pot. Throughout the experiment, basil plants were irrigated with full-strength Hoagland's nutrient solution [40] three times daily (7 am, 12 pm, and 5 pm) via an automated computer-controlled drip system.

The experiment was organized in a randomized complete block design within a three by two factorial arrangement with temperature and CO_2 treatments. A total of six SPAR chambers represented three blocks with ten replications. Each SPAR chamber consisted of three rows of pots with ten pots per row in each SPAR chamber. All environmental growing conditions, except for temperature and CO_2, were kept the same throughout the experiment.

4.2. Temperature and CO_2 Treatments

Basil plants were randomly assigned to each chamber consisting of 20/12 (day/night), 30/22, and 38/30 °C in combination with ambient (420 ppm) or elevated (720 ppm) CO_2 concentrations. The day and night time temperatures were respectively initiated at dawn and one hour after nightfall. Table 4 shows the average environmental conditions in which the experiment was conducted. During the period of this experiment, three temperature treatments, 20/12, 30/22, and 38/30 °C, were regarded as low, optimum, and high temperatures, respectively, for basil growth and development.

Table 4. Temperature stress treatments based on the percentage of daily evapotranspiration (ET) imposed at 14 days after sowing, mean day/night temperature, mean day chamber CO_2 concentration, mean day/night vapor pressure deficit (VPD), and mean day/night evapotranspiration (ET) during the experimental period 38 days for each treatment.

Treatments		Measured Temperature (°C)	CO_2 (ppm)	VPD (kPa)	Mean ET (H_2O L·d^{-1})
		Day/night	Day	Day/night	Day/night
Control	30/22 °C, 420 ppm	26.27 ± 0.02	430.47 ± 0.98	1.82 ± 0.01	14.64 ± 1.41
Control + High CO_2	30/22 °C, 720 ppm	26.34 ± 0.01	731.21 ± 1.52	1.98 ± 0.01	12.60 ± 1.27
High Temperature	38/30 °C, 420 ppm	32.16 ± 0.49	434.19 ± 1.21	2.80 ± 0.07	8.74 ± 0.64
Low Temperature	20/12 °C, 420 ppm	19.53 ± 0.56	431.08 ± 0.66	0.89 ± 0.08	8.59 ± 0.47
High Temperature + High CO_2	38/30 °C, 720 ppm	32.09 ± 0.49	728.79 ± 0.83	2.87 ± 0.07	18.41 ± 1.86
Low Temperature + High CO_2	20/12 °C, 720 ppm	19.56 ± 0.57	724.78 ± 0.35	0.95 ± 0.09	6.39 ± 0.37

4.3. Physiology and Gas Exchange Measurements

The OJIP fluorescence readings were taken utilizing a FluorPen FP 100 (Photon Systems Instruments, Drasov, Czech Republic) on the second, most developed basil leaf. The minimal fluorescence (Fo), which was estimated at 50 μs when all PSII reaction centers are open, maximal fluorescence (Fm) when all PSII response focuses are shut, and the steady-state state fluorescence (Fs) were recorded in each plant at 17 DAT.

Analogous to the leaf chlorophyll readings, the photosynthesis and fluorescence parameters of basil leaf subjected to different treatments were recorded between 10 am and 12 pm with LI-6400XT portable photosynthesis system (LiCor Biosciences, Inc., Lincoln, NE, USA) at 17 DAT. These parameters include P_n, E, g_s, internal CO_2 concentration (C_i), electron transport rate (ETR), and the quantum efficiency (F_v'/F_m'). The internal to external CO_2 ratio was calculated by the relationship C_i/C_a. The conditions of the leaf chamber were set at light intensity (PAR) of 1500 μmolm^{-2} s^{-1}, the relative humidity of 50%, the CO_2 concentration of 410 μmol mol^{-1}, and the flow rate through the chamber was regulated to 500 mol s^{-1}. The temperature of the chamber was set at the current temperatures (22, 30, or 38 °C) the readings were taken.

4.4. Carotenoid and Chlorophyll Analysis

Carotenoid and chlorophyll pigments were extracted and analyzed from freeze-dried basil tissues, according to Kopsell et al. [41,42], with few changes as portrayed in Barickman et al. [43].

4.5. Epicuticular Wax Content Determination

The epicuticular leaf waxes were extracted and quantitively analyzed in accordance with the method of Ebercon et al. [44] with minor modifications as described by Singh and Reddy [45].

4.6. Antioxidant and Oxidative Analysis

4.6.1. Malondialdehyde (MDA)

The basil leaf samples were analyzed for MDA content by adapting the procedure used by Heath and Packer [46]. Fresh leaf tissue (500 mg) was homogenized in 0.1% trichloroacetic acid (TCA). The homogenate was centrifuged at $11,320 \times g$ for 5 min, and a 1mL aliquot of the supernatant was treated with 4 mL 0.5% thiobarbituric acid in 20% TCA; the mixture was heated at 958 °C for 30 min and afterward immediately cooled in an ice bath. After centrifugation at $5700 \times g$ for 10 min, the absorbance of the supernatant was recorded at 532 nm. The MDA content was determined by its extinction coefficient of $155 \text{ mM}^{-1} \text{ cm}^{-1}$ and expressed as nmol g^{-1} DW.

4.6.2. Hydrogen Peroxide (H_2O_2)

The content of H_2O_2 was estimated according to the previously reported procedure of Mukherjee and Choudhuri [47]. Fresh leaf tissue (500 mg) was homogenized in 5 mL chilled acetone (80%) and filtered through Whatman filter paper, and 4 mL titanium reagent was added, followed by 5 mL ammonia solution. The mixture was centrifuged at $5030 \times g$, and the supernatant was disposed of. The residue was dissolved with 1 M H_2SO_4, and the absorbance was recorded at 410 nm. The extinction coefficient of H_2O_2 is $0.28 \text{ mmol}^{-1} \text{ cm}^{-1}$. The content of H_2O_2 in samples was acquired from a standard curve using pure H_2O_2 and expressed as µmol g^{-1} DW.

4.6.3. Superoxide Dismutase (SOD)

The SOD activity was estimated using the method of Dhindsa et al. [48] with minor modifications reported by Awasthi et al. [49].

4.6.4. Ascorbic Acid (ASC)

ASC was assessed following the method of Mukherjee and Choudhuri [47]. Details of extraction and analysis were described in Awasthi et al. [49].

4.6.5. Glutathione (GSH)

Basil leaf samples were analyzed for reduced GSH according to the method of Griffith [50], with few changes detailed in Awasthi et al. [49].

4.6.6. Trehalose (TRE)

Trehalose concentration was estimated according to the method of Trevelyan and Harrison [51] and the Anthrone method of Brin [52]. The enzymes associated with TRE metabolism were assayed as per the procedures of Pramanik and Imai [53], with few changes. Trehalose-6-phosphate synthase (TPS) activity was assayed, according to Hottiger et al. [54], which determined the release of UDP from UDP-glucose, involving glucose-6-phosphate. Trehalose-6-phosphate phosphatase (TPP) activity was assayed according to the method of Klutts et al. [55] by measuring the release of inorganic phosphate from trehalose-6-phosphate. Trehalase activity was determined by the activation of phosphorylation using cAMP (cyclic adenosine monophosphate) and assayed by measuring the glucose concentration [56].

5. Data Analysis

The experimental design was a randomized complete block in a factorial arrangement with three temperature treatments, two CO_2 treatments, three-block, and ten replications. Data were analyzed using the PROC GLIMMIX analysis of variance (ANOVA) followed by mean separation. Statistical analysis of the data was performed using SAS (version 9.4; SAS Institute, Cary, NC, USA). The standard errors were based on the pooled error term from the ANOVA table. Duncan's multiple range test ($p \leq 0.05$) was used to differentiate between treatment classifications when F-values were significant for the main effects. Model-based values were reported rather than the unequal standard error from a data-based calculation because pooled errors reflected the statistical testing. Diagnostic tests were conducted to ensure that treatment variances were statistically equal before pooling.

6. Conclusions

The interaction of temperature stress and elevated CO_2 significantly impacted the physiological processes and phytonutrient concentrations of basil plants. Decreasing the basil's growth temperature to 20/12 °C significantly reduced P_n and g_s, with detrimental impacts on the basil plants' growth. Furthermore, low-temperature stress-induced excessive ROS production, which caused harm to the photosynthetic apparatus, as confirmed by reduced F_v'/F_m', low ETR, and altered oxidized and reduced states of PSII and PSI. Additionally, the accelerated chlorophyll and carotenoid pigment degradation of basil plants were observed when elevated CO_2 interacted with low-temperature stress.

Contrarily, elevated CO_2 remarkably ameliorated the damage caused by low-temperature stress to photosynthetic apparatus. Thus, elevated CO_2 promoted leaf C_i/C_a and increased SOD, TRE, and ASC antioxidant levels. Basil, being a thermophilic plant, was observed to increase its chlorophyll and carotenoid concentration, apparent quantum yield, and maximum photosystem II efficiency when subjected to high-temperature stress. Likewise, increased growth temperature for basil under elevated CO_2 produced more antioxidant content. The findings of this study recommend that varying the growth temperature of basil plants would significantly affect the growth and development rates of basil compared to increasing the CO_2 concentrations, which mitigated the constraining effects of temperature stress.

Author Contributions: T.C.B., Conceptualization, Methodology, Validation, Formal Analysis, Investigation, Resources, Data curation, Writing—Original Draft, Writing—Review and Editing, Visualization, Supervision, Project Administration, Funding Acquisition; O.J.O., Formal Analysis, Writing—Original Draft, Writing—Review and Editing; A.S., Methodology, Validation, Investigation; C.H.W., Methodology, Validation, Formal Analysis, Investigation; K.R.R., Conceptualization, Methodology, Validation, Formal Analysis, Investigation, Resources, Data curation, Writing—Review and Editing, Visualization, Supervision, Project Administration, Funding Acquisition; W.G., Conceptualization, Methodology, Validation, Resources, Funding Acquisition. All authors have read and agreed to the published version of the manuscript.

Funding: This material is based on the work supported by the USDA-NIFA Hatch Project under accession number 149,210 and the National Institute of Food and Agriculture, 2019-34263-30552, and MIS 043,050 funded this research.

Institutional Review Board Statement: Not applicable.

Informed Consent Statement: Not applicable.

Data Availability Statement: The data presented in this study are available on request from the corresponding author.

Acknowledgments: We thank David Brand and Thomas Horgan for their technical assistance and graduate students at the Vegetable Physiology Laboratory and the Environmental Plant Physiology Laboratory for helping during data collection.

Conflicts of Interest: The authors declare no conflict of interest.

References

1. Field, C.B.; Barros, V.R.; Dokken, D.J.; Mach, K.J.; Mastrandrea, M.D.; Bilir, T.E.; Chatterjee, M.; Ebi, K.L.; Estrada, Y.O.; Genova, R.C.; et al. *IPCC. Climate Change 2014: Impacts, Adaptation, and Vulnerability*; Part A: Global and Sectoral Aspects. Contribution of Working Group II to the Fifth Assessment Report of the Intergovernmental Panel on Climate Change; Field, C.B., Barros, V.R., Dokken, D.J., Mach, K.J., Mastrandrea, M.D., Bilir, T.E., Chatterjee, M., Ebi, K.L., Estrada, Y.O., Genova, R.C., et al., Eds.; Cambridge University Press: Cambridge, UK; New York, NY, USA, 2014; p. 1132.
2. Al Jaouni, S.; Saleh, A.M.; Wadaan, M.A.M.; Hozzein, W.N.; Selim, S.; AbdElgawad, H. Elevated CO_2 induces a global metabolic change in basil (*Ocimum Basilicum* L.) and peppermint (*Mentha Piperita* L.) and improves their biological activity. *J. Plant Physiol.* **2018**, *224–225*, 121–131. [CrossRef] [PubMed]
3. Prentice, I.C.; Farquhar, G.D.; Fasham, M.J.R.; Goulden, M.L.; Heimann, M.; Jaramillo, V.J.; Kheshgi, H.S.; Le Quéré, C.; Scholes, R.J.; Wallace, D.W.R. The carbon cycle and atmospheric carbon dioxide. In *Climate Change 2001: The Scientific Basis*; Cambridge University Press: Cambridge, UK, 2001; pp. 183–237.
4. Stocker, T.F.; Qin, D.; Plattner, G.-K.; Tignor, M.; Allen, S.K.; Boschung, J.; Nauels, A.; Xia, Y.; Bex, V.; Midgley, P.M. *Climate Change 2013: The Physical Science Basis*; Intergovernmental Panel on Climate Change, Working Group I Contribution to the IPCC Fifth Assessment Report (AR5); Cambridge University Press: New York, NY, USA, 2013; p. 25.
5. USGCRP. *Impacts, Risks, and Adaptation in the United States: Fourth National Climate Assessment, Volume II*; Reidmiller, D.R., Avery, C.W., Easterling, D.R., Kunkel, K.E., Lewis, K.L.M., Maycock, T.K., Stewart, B.C., Eds.; U.S. Global Change Research Program: Washington, DC, USA, 2018; p. 1515. [CrossRef]
6. Reddy, A.R.; Reddy, K.R.; Hodges, H.F. Interactive effects of elevated carbon dioxide and growth temperature on photosynthesis in cotton leaves. *Plant Growth Regul.* **1998**, *26*, 33–40. [CrossRef]
7. Dong, J.; Gruda, N.; Li, X.; Tang, Y.; Zhang, P.; Duan, Z. Sustainable vegetable production under changing climate: The impact of elevated co2 on yield of vegetables and the interactions with environments-A review. *J. Clean. Prod.* **2020**, *253*, 119920. [CrossRef]
8. Solomon, S.; Manning, M.; Marquis, M.; Qin, D. *Climate Change 2007—The Physical Science Basis: Working Group I Contribution to the Fourth Assessment Report of the IPCC*; Cambridge University Press: Cambridge, UK, 2007; Volume 4.
9. Simon, J.E.; Quinn, J.; Murray, R.G. Basil: A Source of Essential Oils. In *Adv. New Crops*; Janick, J., Simon, J.E., Eds.; Timber Press: Portland, OR, USA, 1990; pp. 484–489.
10. Kopsell, D.A.; Kopsell, D.E.; Curran-Celentano, J. Carotenoid and chlorophyll pigments in sweet basil grown in the field and greenhouse. *Hort. Sci.* **2005**, *40*, 1230–1233. [CrossRef]
11. Chang, X.; Alderson, P.; Wright, C. Effect of temperature integration on the growth and volatile oil content of basil (*Ocimum Basilicum* L.). *J. Hortic. Sci. Biotechnol.* **2005**, *80*, 593–598. [CrossRef]
12. Hiltunen, R.; Holm, Y. Essential of oil of *Ocimum*. In *Basil: The Genus Ocimum*; Hiltunen, R., Holm, Y., Eds.; Harwood Academic Publishers: Amsterdam, The Netherlands, 1999; pp. 113–135.
13. Kumar, B.; Gupta, E.; Yadav, R.; Singh, S.C.; Lal, R.K. Temperature effects on seed germination potential of holy basil (*Ocimum tenuiflorum*). *Seed Technol.* **2014**, *36*, 75–79.
14. Mortensen, L.M. The effect of air temperature on growth of eight herb species. *Am. J. Plant Sci.* **2014**, *5*, 1542–1546. [CrossRef]
15. Taiz, L.; Zeiger, E.; Møller, I.M.; Murphy, A. *Plant Physiology and Development*, 6th ed.; Sinauer Associates: Sunderland, MA, USA, 2015; ISBN 978-1-60535-255-8.
16. Ribeiro, P.; Simon, J.E. Breeding sweet basil for chilling tolerance. In *Issues in New Crops and New Uses*; Janick, J., Whipkey, A., Eds.; ASHS Press: Alexandria, VA, USA, 2007; pp. 302–305.
17. Kalisz, A.; Jezdinský, A.; Pokluda, R.; Sękara, A.; Grabowska, A.; Gil, J. Impacts of chilling on photosynthesis and chlorophyll pigment content in juvenile basil cultivars. *Hortic. Environ. Biotechnol.* **2016**, *57*, 330–339. [CrossRef]
18. Al-Huqail, A.; El-Dakak, R.M.; Sanad, M.N.; Badr, R.H.; Ibrahim, M.M.; Soliman, D.; Khan, F. Effects of climate temperature and water stress on plant growth and accumulation of antioxidant compounds in sweet basil (*Ocimum basilicum* L.) leafy vegetable. *Scientifica* **2020**, *3808909*. [CrossRef]
19. Singh, H.; Poudel, M.R.; Dunn, B.L.; Fontanier, C.; Kakani, G. Effect of greenhouse CO_2 supplementation on yield and mineral element concentrations of leafy greens grown using nutrient film technique. *Agronomy* **2020**, *10*, 323. [CrossRef]
20. Watanabe, C.K.; Sato, S.; Yanagisawa, S.; Uesono, Y.; Terashima, I.; Noguchi, K. Effects of elevated CO_2 on levels of primary metabolites and transcripts of genes encoding respiratory enzymes and their diurnal patterns in *Arabidopsis thaliana*: Possible relationships with respiratory rates. *Plant Cell Physiol.* **2014**, *55*, 341–357. [CrossRef] [PubMed]
21. Gillig, S.; Heinemann, R.; Hurd, G.; Pittore, K.; Powell, D. Response of basil (*Ocimum basilicum*) to increased CO_2 levels. In *E&ES359 Global Climate Change, Johan Varekamp*; Wesleyan University: Middletown, CT, USA, 2008.
22. Walters, K.J.; Currey, C.J. Growth and development of basil species in response to temperature. *HortScience* **2019**, *54*, 1915–1920. [CrossRef]
23. Brand, D.; Wijewardana, C.; Gao, W.; Reddy, K.R. Interactive effects of carbon dioxide, low temperature, and ultraviolet-b radiation on cotton seedling root and shoot morphology and growth. *Front. Earth Sci.* **2016**, *10*, 607–620. [CrossRef]
24. Balasooriya, H.N.; Dassanayake, K.B.; Seneweera, S.; Ajlouni, S. Interaction of elevated carbon dioxide and temperature on strawberry (*Fragaria × Ananassa*) growth and fruit yield. *Int. J. Biol. Biomol. Agric. Food Biotechnol. Eng. World Acad. Sci. Eng. Technol. Int. Sci. Index* **2018**, *12*, 279–287. [CrossRef]

25. Saibo, N.J.M.; Lourenço, T.; Oliveira, M.M. Transcription factors and regulation of photosynthetic and related metabolism under environmental stresses. *Ann. Bot.* **2009**, *103*, 609–623. [CrossRef] [PubMed]

26. Barickman, T.C.; Olorunwa, O.J.; Sehgal, A.; Walne, C.H.; Reddy, K.R.; Gao, W. Interactive impacts of temperature and elevated CO_2 on Basil (*Ocimum Basilicum* L.) root and shoot morphology and growth. *Horticulturae* **2021**, *7*, 112. [CrossRef]

27. Yuan, L.; Yuan, Y.; Liu, S.; Wang, J.; Zhu, S.; Chen, G.; Hou, J.; Wang, C. Influence of high temperature on photosynthesis, antioxidative capacity of chloroplast, and carbon assimilation among heat-tolerant and heat-susceptible genotypes of nonheading chinese cabbage. *HortScience* **2017**, *52*, 1464–1470. [CrossRef]

28. Camejo, D.; Rodríguez, P.; Angeles Morales, M.; Miguel Dell'Amico, J.; Torrecillas, A.; Alarcón, J.J. High temperature effects on photosynthetic activity of two tomato cultivars with different heat susceptibility. *J. Plant Physiol.* **2005**, *162*, 281–289. [CrossRef]

29. Wahid, A.; Gelani, S.; Ashraf, M.; Foolad, M.R. Heat tolerance in plants: An overview. *Environ. Exp. Bot.* **2007**, *61*, 199–223. [CrossRef]

30. Zhou, R.; Yu, X.; Li, X.; Mendanha dos Santos, T.; Rosenqvist, E.; Ottosen, C.-O. Combined high light and heat stress induced complex response in tomato with better leaf cooling after heat priming. *Plant Physiol. Biochem.* **2020**, *151*, 1–9. [CrossRef]

31. Liu, W.; Yu, K.; He, T.; Li, F.; Zhang, D.; Liu, J. The low temperature induced physiological responses of avena nuda l., a cold-tolerant plant species. *Sci. World J.* **2013**, *2013*, 658793. [CrossRef]

32. Loladze, I.; Nolan, J.M.; Ziska, L.H.; Knobbe, A.R. Rising atmospheric CO_2 lowers concentrations of plant carotenoids essential to human health: A Meta-Analysis. *Mol. Nutr. Food Res.* **2019**, *63*, 1801047. [CrossRef]

33. Shamloo, M.; Babawale, E.A.; Furtado, A.; Henry, R.J.; Eck, P.K.; Jones, P.J.H. Effects of genotype and temperature on accumulation of plant secondary metabolites in canadian and australian wheat grown under controlled environments. *Sci. Rep.* **2017**, *7*, 9133. [CrossRef]

34. Sublett, W.L.; Barickman, T.C.; Sams, C.E. Effects of elevated temperature and potassium on biomass and quality of dark red 'lollo rosso' lettuce. *Horticulturae* **2018**, *4*, 11. [CrossRef]

35. Jumrani, K.; Bhatia, V.S. Interactive effect of temperature and water stress on physiological and biochemical processes in soybean. *Physiol. Mol. Biol. Plants* **2019**, *25*, 667–681. [CrossRef]

36. Gill, S.S.; Tuteja, N. Reactive oxygen species and antioxidant machinery in abiotic stress tolerance in crop plants. *Plant Physiol. Biochem.* **2010**, *48*, 909–930. [CrossRef] [PubMed]

37. Duan, M.; Feng, H.-L.; Wang, L.-Y.; Li, D.; Meng, Q.-W. Overexpression of thylakoidal ascorbate peroxidase shows enhanced resistance to chilling stress in tomato. *J. Plant Physiol.* **2012**, *169*, 867–877. [CrossRef] [PubMed]

38. Reddy, K.; Read, J.J.; McKinion, J.M. Soil-Plant-Atmosphere-Research (SPAR) facility: A tool for plant research and modeling. *Biotronics* **2001**, *30*, 27–50.

39. Wijewardana, C.; Hock, M.; Henry, B.; Reddy, K.R. Screening corn hybrids for cold tolerance using morphological traits for early-season seeding. *Crop Sci.* **2015**, *55*, 851–867. [CrossRef]

40. Hoagland, D.R.; Arnon, D.I. *The Water-Culture Method for Growing Plants without Soil*, 2nd ed.; Circular 347; California Agricultural Experiment Station: Berkeley, CA, USA, 1950; p. 347.

41. Kopsell, D.A.; Kopsell, D.E.; Lefsrud, M.G.; Curran-Celentano, J.; Dukach, L.E. Variation in lutein, β-carotene, and chlorophyll concentrations among brassica oleracea cultigens and seasons. *HortScience* **2004**, *39*, 361–364. [CrossRef]

42. Kopsell, D.A.; Kopsell, D.E.; Curran-Celentano, J. carotenoid pigments in kale are influenced by nitrogen concentration and form. *J. Sci. Food Agric.* **2007**, *87*, 900–907. [CrossRef]

43. Barickman, T.C.; Kopsell, D.A.; Sams, C.E. Abscisic acid impacts tomato carotenoids, soluble sugars, and organic acids. *HortScience* **2016**, *51*, 370–376. [CrossRef]

44. Ebercon, A.; Blum, A.; Jordan, W.R. A rapid colorimetric method for epicuticular wax contest of sorghum leaves. *Crop Sci.* **1977**, *17*, 179–180. [CrossRef]

45. Singh, S.K.; Reddy, K.R. Regulation of photosynthesis, fluorescence, stomatal conductance and water-use efficiency of cowpea (*Vigna Unguiculata* [L.] *Walp.*) under drought. *J. Photochem. Photobiol. B Biol.* **2011**, *105*, 40–50. [CrossRef]

46. Heath, R.L.; Packer, L. Photoperoxidation in isolated chloroplasts: I. Kinetics and stoichiometry of fatty acid peroxidation. *Arch. Biochem. Biophs.* **1968**, *125*, 189–198. [CrossRef]

47. Mukherjee, S.P.; Choudhuri, M.A. Implications of water stress-induced changes in the levels of endogenous ascorbic acid and hydrogen peroxide in vigna seedlings. *Physiol. Plant.* **1983**, *58*, 166–170. [CrossRef]

48. Dhindsa, R.S.; Plumb-Dhindsa, P.; Thorpe, T.A. Leaf senescence: Correlated with increased levels of membrane permeability and lipid peroxidation, and decreased levels of superoxide dismutase and catalase. *J. Exp. Bot.* **1981**, *32*, 93–101. [CrossRef]

49. Awasthi, R.; Gaur, P.; Turner, N.C.; Vadez, V.; Siddique, K.H.M.; Nayyar, H. Effects of individual and combined heat and drought stress during seed filling on the oxidative metabolism and yield of chickpea (*Cicer Arietinum*) genotypes differing in heat and drought tolerance. *Crop Past. Sci.* **2017**, *68*, 823–841. [CrossRef]

50. Griffith, O.W. Determination of glutathione and glutathione disulfide using glutathione reductase and 2-vinylpyridine. *Anal. Biochem.* **1980**, *106*, 207–212. [CrossRef]

51. Trevelyan, W.E.; Harrison, J.S. Studies on yeast metabolism. Fractionation and microdetermination of cell carbohydrates. *Biochem. J.* **1952**, *50*, 298–303. [CrossRef]

52. Brin, M. [89] Transketolase: Clinical aspects. In *Methods in Enzymology*; Academic Press: Cambridge, MA, USA, 1966; Volume 9, pp. 506–514. ISBN 0076-6879.

53. Pramanik, H.R.; Imai, R. Functional identification of a trehalose 6-phosphate phosphatase gene that is involved in transient induction of trehalose biosynthesis during chilling stress in rice. *Plant Mol. Biol.* **2005**, *58*, 751–762. [CrossRef] [PubMed]
54. Hottiger, T.; Boller, T.; Wiemken, A. Rapid changes of heat and desiccation tolerance correlated with changes of trehalose content in saccharomyces cerevisiae cells subjected to temperature shifts. *FEBS Lett.* **1987**, *220*, 113–115. [CrossRef]
55. Klutts, S.; Pastuszak, I.; Edavana, V.K.; Thampi, P.; Pan, Y.-T.; Abraham, E.C.; Carroll, J.D.; Elbein, A.D. Purification, cloning, expression, and properties of mycobacterial trehalose-phosphate phosphatase. *J. Biol. Chem.* **2003**, *278*, 2093–2100. [CrossRef] [PubMed]
56. Einig, W.; Hampp, R. Carbon partitioning in Norway spruce: Amounts of Fructose 2, 6-bisphosphate and of intermediates of starch/sucrose synthesis in relation to needle age and degree of needle loss. *Trees* **1990**, *4*, 9–15. [CrossRef]

Article

Climate Change, Crop Yields, and Grain Quality of C_3 Cereals: A Meta-Analysis of [CO_2], Temperature, and Drought Effects

Sinda Ben Mariem [1], David Soba [1], Bangwei Zhou [2], Irakli Loladze [3], Fermín Morales [1] and Iker Aranjuelo [1,*]

[1] Instituto de Agrobiotecnología (IdAB), CSIC-Gobierno de Navarra, Avda. de Pamplona 123, 31192 Mutilva, Spain; sinda.ben@csic.es (S.B.M.); david.soba@unavarra.es (D.S.); fermin.morales@csic.es (F.M.)

[2] Key Laboratory of Vegetation Ecology, Institute of Grassland Science, Northeast Normal University, Ministry of Education, Changchun 130024, China; zhoubw599@nenu.edu.cn

[3] Bryan Medical Center, Bryan College of Health Sciences, Lincoln, NE 68506, USA; loladze@gmail.com

* Correspondence: iker.aranjuelo@csic.es; Tel.: +34-948168015

Abstract: Cereal yield and grain quality may be impaired by environmental factors associated with climate change. Major factors, including elevated CO_2 concentration ([CO_2]), elevated temperature, and drought stress, have been identified as affecting C_3 crop production and quality. A meta-analysis of existing literature was performed to study the impact of these three environmental factors on the yield and nutritional traits of C_3 cereals. Elevated [CO_2] stimulates grain production (through larger grain numbers) and starch accumulation but negatively affects nutritional traits such as protein and mineral content. In contrast to [CO_2], increased temperature and drought cause significant grain yield loss, with stronger effects observed from the latter. Elevated temperature decreases grain yield by decreasing the thousand grain weight (TGW). Nutritional quality is also negatively influenced by the changing climate, which will impact human health. Similar to drought, heat stress decreases starch content but increases grain protein and mineral concentrations. Despite the positive effect of elevated [CO_2], increases to grain yield seem to be counterbalanced by heat and drought stress. Regarding grain nutritional value and within the three environmental factors, the increase in [CO_2] is possibly the more detrimental to face because it will affect cereal quality independently of the region.

Keywords: cereals; yield and quality; high [CO_2]; predicted future climate; high temperature; grain quality traits; drought stress

Citation: Ben Mariem, S.; Soba, D.; Zhou, B.; Loladze, I.; Morales, F.; Aranjuelo, I. Climate Change, Crop Yields, and Grain Quality of C_3 Cereals: A Meta-Analysis of [CO_2], Temperature, and Drought Effects. *Plants* **2021**, *10*, 1052. https://doi.org/10.3390/plants10061052

Academic Editor: Othmane Merah

Received: 31 March 2021
Accepted: 21 May 2021
Published: 24 May 2021

Publisher's Note: MDPI stays neutral with regard to jurisdictional claims in published maps and institutional affiliations.

1. Introduction

Food security is threatened by the impacts of climate change on agriculture and by increasing the world population [1,2]. Actually, climate change has already slowed global agricultural productivity growth, and in a recent study, Ortiz-Bobea et al. [3] found that anthropogenic climate change (ACC) has reduced global agricultural total factor productivity since 1961 by about 21%, with a greater impact for warm regions such as Africa (-34%) than for cooler regions such as Europe and Central Asia (-7.1%). Over the next few decades, climate change is expected to affect more the world's supply of cereal grains, impacting their quantity and quality due to the complex effects of elevated atmospheric [CO_2] and changing temperature and rainfall patterns on crops [4]. Cereals contribute to a substantial part of the world's plant-derived food production and comprise a majority of the crops harvested. In fact, FAO statistics show that in 2016, sugar cane had the highest production globally, followed by corn, wheat, and rice [5]. Adding to that, according to the Foreign Agricultural Service/USDA, preliminary world production in 2018 of maize, wheat, and rice was estimated at around 1076, 763, and 495 million tons, respectively [6]. Further, their nutritional quality has a significant impact on human well-being and health, especially in the developing world [7]. Thus, one of the major challenges that plant breeders are facing currently is to increase cereal grain production while taking into consideration an adequate grain nutrient content.

Numerous effects of elevated atmospheric [CO_2] on plants have been documented through a photosynthesis-mediated CO_2 fertilization effect, including increased carbon (C) assimilation, growth, yield, and C content [8,9]. Thus, elevated [CO_2] could enhance the concentration of photosynthesis-derived carbohydrates in grains, starch being the major component [10–12]. Since grains are predominantly composed of carbohydrates (mostly in the form of starch), it has been suggested that increases in starch concentrations can cause a dilution effect onother nutrients, including proteins, lipids, vitamins, and minerals. In addition, adjustments in the photosynthetic apparatus and later on the redistribution from senescing leaves to grains must be considered as the key mechanisms. Due to the different biochemistry of C_3 and C_4 photosynthesis, the positive effect of elevated [CO_2] on photosynthesis is more pronounced in C_3 crops such as wheat and rice but less notable in C_4 crops such as maize [13]. Rising [CO_2] is likely to lead to "globally imbalanced plant stoichiometry (relative to pre-industrial times)" [14], which in turn would "intensify the already acute problem of micronutrient malnutrition" [14], particularly regarding minerals such asFe, Zn, and I, as well asprotein (or N) [12,15,16]. Elevated [CO_2] has been reported to decrease mineral concentrations in barley grains (−6.9%), rice grains (−7.2%), and wheat grains (−7.6%), and increasing the ratio of non-structural carbohydrates (TNC) to protein by 6–47% in grains and tubers [9]. For the grain crops barley, rice, and wheat, the reduction in protein mediated by elevated [CO_2] was reported to be 15, 10, and 10%, respectively [15]. In their meta-analysis of the impact of elevated [CO_2] on wheat grains, Broberg et al. [12] found a significant reduction in the concentration of the majority of minerals (Ca, Cd, Cu, Fe, Mg, Mn, P, S, and Zn), while B and Na were not significantly affected, and K was significantly increased (<2%). These meta-analytic results are in line with those from individual wheat FACE experiments [17–21]. Two minerals, Fe and Zn, are already deficient in the diets of hundreds of millions of people, and CO_2-induced reductions in Fe and Zn have been reported in the edible parts of major crops [9,22,23] and are projected to have negative effects on human nutrition [24,25]. Furthermore, emerging evidence points to elevated [CO_2] affecting nutrients beyond protein and minerals that are essential to humans, such as vitamins and carotenoids [26,27]. The decrease in mineral concentrations is notable in C_3 plants but less so in C_4 plants [9,23] and is consistent with differences in physiology; the simulation of carbohydrate production by elevated [CO_2] is stronger in C_3 plants, while reduced transpiration is present in both C_3 and C_4 plants.

During the last two decades, the air temperature has increased by 0.85 °C [28]. In fact, annual average minimum temperatures in Spain have increased over the last century by 1.5 °C, and by 0.6 °C during the last 25 years [29]. The most probable outcome of climate ensemble model projections foresees increasesof1.8 to 4.0 °C by the end of the 21st century (2090–2099) relative to the period 1980–1999. These numbers originate from the best estimate of greenhouse gas time series deduced from the six marker scenarios alone [30]. Heat stress is a major constraint to sustainable cereal production, with reductions in grain yield being associated with high temperatures during the reproductive or grain-filling stages in wheat [31,32] and rice [33–35]. High-temperature impacts on grain filling can vary enormously, depending on timing (days after anthesis) and duration. Both chronic moderately high temperatures (25–35 °C) and heat shocks (>35 °C) during the grain-filling phase are frequently associated with an increase in grain protein concentration in wheat [31,36,37] and rice [35,38]. Indeed, high temperature primarily impacts the accumulation of starch in wheat grain, with accumulation beginning earlier than under cooler temperatures, the duration of its accumulation also being reduced, and the result is a greater concentration of protein in the grain. Further, the duration of protein accumulation is reduced, while the rate of protein accumulation is substantially increased. In addition, leaves senesce before the heads mature, suggesting that high temperatures might enhance N remobilization from leaves and stems [39,40]. Moreover, the timing and duration of heat stress during grain filling have been shown to be important sources of variation in dough properties in wheat [41]. Grain protein and mineral composition are quality characteristics that can change due to high temperature, and they respond to changes in enzymes involved in starch and protein synthesis. Yang et al. [42] observed that the activity of glutamate synthase was enhanced by heat stress, while sucrose phosphate synthase, sucrose synthase, and soluble starch synthase were significantly decreased during grain filling. However, Monjardino et al. [43] found that protein concentration was negatively affected by heat stress during the early stage of endosperm development. They found that among the protein fractions, zeins are the most affected by heat stress. In fact, zein accumulation was repressed under high temperature rather than being degradedin the early developmental stages. In rice, elevated temperature also alters grain protein and mineral nutrient composition [35,44]. Ferreira et al. [36] showed that the total quantity of N per grain in wheat is generally little affected by the growing temperature but, due to the above-mentioned lower grain yield, the percentage of N on a dry weight basis rises under higher temperatures. Similar increases in the percentage of dry weight have been reported for wheat [45,46].

Increasing greenhouse gas emissions may also lead to rainfall reductions in the coming decades, which will increase the frequency and intensity of drought in the Mediterranean basin [47–49]. Climate change projections for the Mediterranean region indicate a precipitation decrease of 25–30% for the last decades of the 21st century [50]. Adding to that, the seasonality of rainfall is much more important. In fact, the expected shortage in Mediterranean rainfall should impact summer precipitation much more than winter precipitation. Mediterranean crop growth, however, is mainly driven by winter rain. Moreover, drought is considered one of the most important factors limiting crop yields around the world. Wheat crop responses to water scarcity depend on several factors, including plant development status, duration, and intensity of the stress and genetic variables [51]. Although rainfall during winter has been traditionally abundant and coincides with the lowest evapotranspiration rates, the occurrence of drought in winter during the early stages of the crop cycle has been recently reported [52].

This can further constrain wheat growth and thus final grain yield, mostly through a decrease in ear density and the number of kernels per unit crop area [53,54]. Grain yield reductions mediated by drought have been widely reported in wheat [55,56], and depending on the genotype, the reductions may reach up to 50%. The TGW is also reduced significantly, above 30% in droughted wheat [51,56,57]. Drought stress leads to reduced photosynthetic area and acceleration of leaf senescence during late grain filling in cereals, resulting in a shorter grain-filling period. In wheat, this smaller photosynthetic area and accelerated leaf senescence limit the amount of assimilates translocated to the grain, which implies reductions in grain yield [51]. Grain composition is also affected. Drought stress affects starch accumulation [51,58] more severely than N accumulation during grain filling, putatively influencing the conversion of sucrose into starch [59]. This tends to increase the grain protein concentration (expressed as % protein) in wheat [35,55,57] and rice [60]. In some cases, the opposite effect has been observed in wheat [61,62], possibly related to differences in stress levels and plant development status [63]. Knowledge of the effects of drought stress on grain mineral composition is scarce [7,64]. Crusciol et al. [60] explained the increase in rice grain N, Ca, Mg, Fe, and Zn concentration under rainfed conditions being due to a dilution effect because productivity was higher in irrigated than rainfed systems.

All the above-mentioned changes in grain composition linked to the changing environmental conditions are expected to have important implications for the nutritional quality of foods. During the last decade, different meta-analyses have characterized elevated $[CO_2]$ effects on crop yield and quality traits. However, comparatively little attention has been given to how other target environmental parameters such as temperature and drought will affect crop yield, and especially grain nutritional characteristics. Considering the economic and social importance of cereal crops and the impact of climate change not only on grain production but also on the nutritional value, this meta-analysis aims to provide an overview of the effects and interactions of multiple climate stressors, specifically high [CO$_2$], drought and elevated temperatures, on the productivity and grain quality of C$_3$ cereals.

2. Results

2.1. [CO$_2$], Temperature and Drought Stress Effects on Grain Yield Components

The overall effect of elevated [CO$_2$] on C$_3$ crops resulted insignificant increases in grain yield and thousand grain weight (TGW) of30.10% and 7.41%, respectively (Figure 1). Nevertheless, a contrasting drastic loss in grain yield and TGW was observed under high temperatures and drought stress. Results presented in Figure 1 indicate that the heat and drought stress effects were similarfor TGW and recorded −20.17% and −20.29% reductions, respectively, but the negative effect of drought on cereal grain yield was larger than the effect of elevated temperatures (−70.53% vs. −24.85%).

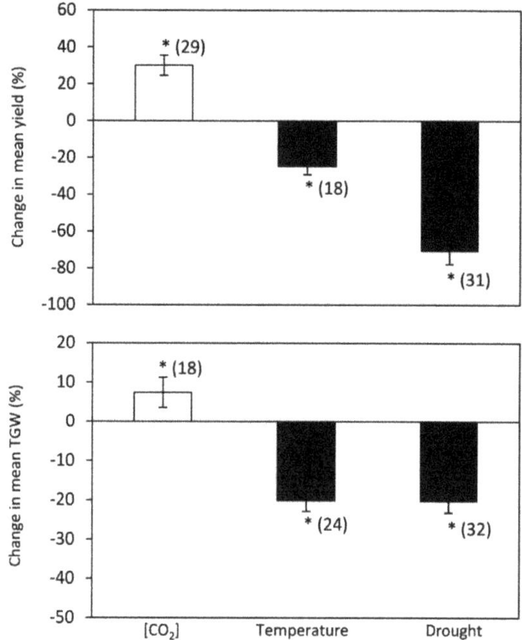

Figure 1. Change (%) in the mean of yield and TGW of plants grown under elevated [CO$_2$], high temperature, and drought stress relative to the control. Data within parentheses indicate the number of observations. Error bars indicate the standard error of the mean. * indicates a statistically significant difference at $p < 0.05$.

2.2. [CO$_2$], Temperature and Drought Stress Effects on Grain Quality

2.2.1. Starch

In cereals grown under high [CO$_2$], there was a significant increase in grain starch concentration (5.65%), whereas there was a significant decrease (−9.91%) under elevated temperature (Figure 2). Regarding water availability, there was no significant change as responses were sprayed in a broad range of both positive and negative changes (Figure 2).

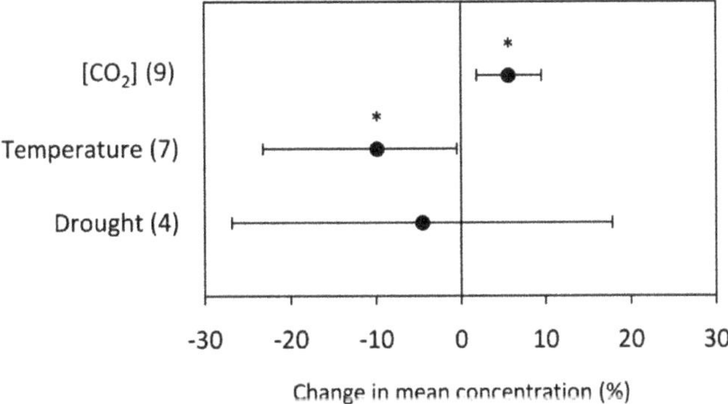

Figure 2. Change (%) in the mean concentration of grain starch of plants grown under elevated [CO_2], high temperature, and drought stress relative to the control. Data within parentheses indicate the number of observations. Error bars indicate 95% CI. * indicates a statistically significant difference at $p < 0.05$.

2.2.2. Total Protein

Grain total protein concentration was negatively affected (−8.90%) by high [CO_2]. However, it was significantly increased by temperature and drought (10.40% and 12.44%, respectively, as shown in Figure 3. Among the proteins that were studied, the gluten, gliadin, and glutenin concentrations were analyzed under elevated [CO_2]. The grain gluten, gliadin, and glutenin concentrations presented in Figure 4 reveal a significant decrease in the gluten and gliadin concentrations (−11.54% and −7.41%, respectively). In contrast, rising [CO_2] decreased the glutenin concentration, but it was not significant. Regarding the effects of drought and heat stress on these proteins, it was not possible to generate statistically powerful results due to the low amount of data (less than three repetitions).

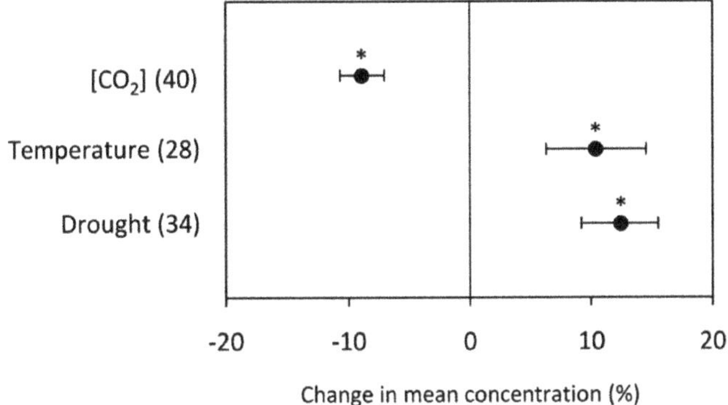

Figure 3. Change (%) in the mean concentration of grain total protein of plants grown under elevated [CO_2], high temperature, and drought stress relative to the control. Data within parentheses indicate the number of observations. Error bars indicate 95% CI. * indicates a statistically significant difference at $p < 0.05$.

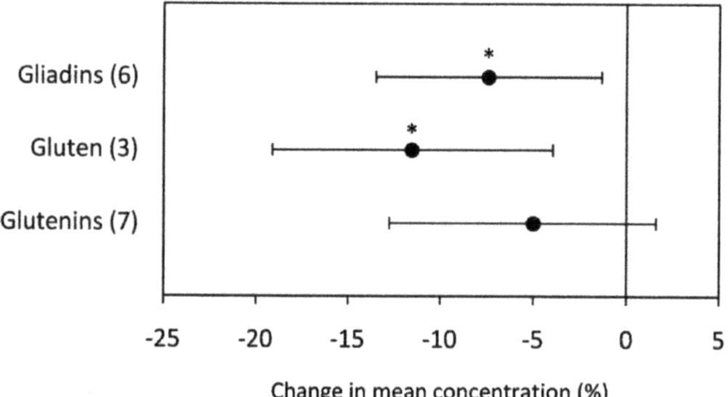

Figure 4. Change (%) in the mean concentration of grain gliadins, gluten, and glutenins of plants grown under elevated [CO_2] relative to ambient level. Data within parentheses indicate the number of observations. Error bars indicate 95% CI. * indicates a statistically significant difference at $p < 0.05$.

2.2.3. Mineral Composition

The results presented in Figure 5 show an overall decrease in micro-macronutrients in C_3 grains under elevated [CO_2]. Across all the data, the mean change ranged between −4.70% (recorded for P) and −39.41% (recorded for Mo). The changes in B and Se were not significant. Among all the measured elements, only Na concentration increased significantly (52.05%) under high [CO_2]. Heat stress had no significant effect on any of the grain mineral concentrations, and this could be due to data scarcity and small sample sizes leading to high data variability. Slight increases in Mg and N of1.91% and 6.31%, respectively, were recorded, whereas the Ca, Fe, Mn, and Zn concentrations were reduced. Regarding water scarcity, the data analysis showed distinct effects between minerals (Figure 5). In fact, drought stress induced an accumulation of Ca and N in grains and recorded a significant increase by 19.92% and 9.56%, respectively. However, no significant increase was obtained regarding Fe, Mg, P, and Zn concentrations. Under low water availability, S and K concentrations declined, but not significantly, by −10.43% and −7.59%, respectively.

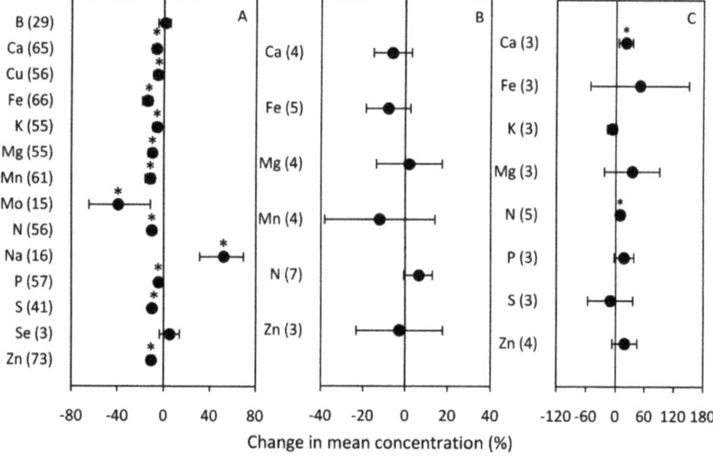

Figure 5. Change (%) in the mean concentration of grain minerals of plants grown under (**A**) elevated [CO_2]; (**B**) high temperature; (**C**) drought stress relative to the control. Data within parenthesis indicate the number of observations. Error bars indicate 95% CI. * indicates statistically significant difference at $p < 0.05$.

3. Discussion

3.1. [CO₂], Temperature and Drought Stress Effects on Grain Yield Components

Current scientific knowledge indicates that grain yield and quality will face serious challenges under the projected future climate. In line with previous papers [65–67], our meta-analysis shows that the predicted elevated [CO_2] will increase crop grain production [8,17,68]. However, as noted by studies conducted over recent decades, it is essential to consider that the [CO_2]-derived "fertilization" effect might decline or be eliminated when combined with stressful growth conditions, such as drought and temperature stress [69–71]. Moreover, in cereals such as wheat, increased grain yields have been associated with increases in the numbers of tillers and grains per spike rather than spike number or grain size [72,73]. The results of the current study have also revealed an association with an increase in the number of grains rather than their weight (larger increase in grain yield than TGW).

Both high temperature and drought negatively affected crop yield. Data analysis showed that yield was more markedly affected under drought than under heat stress conditions. Lower yields in stressed plants can be associated with (i) a shortened duration of the grain-filling period and/or (ii) a lowered photosynthetic rate during grain-filling. Dixit et al. [67] applied a crop simulation model to assess the impact of climate change on wheat production and found a loss of 15% in wheat grain yield in stressed plants, which was associated with a reduction in the number of days to reach grain maturity. Indeed, Mitchell et al. [74] attributed the direct negative effect of rising temperature on wheat yield to the temperature-dependent shortening of the phenological stages. Such decreases in the duration of grain filling would imply a shorter time available for accumulating resources for grain formation [31,46]. The time and duration of heat stress could cause different physiological responses in the plant, therefore, affect crop production. Many studies have reported that heat stress applied prior to anthesis negatively affects the grain yield of wheat due to many reasons [31,75]. High temperature accelerates leaf senescence and reduces post heading duration [75]. Adding to that, heat stress significantly reduces seed germination and negatively affects microspores and pollen cells, leading to non-functional florets or abortion of fertile florets and resulting in male sterility [76]. In fact, the decline in grain yields under high day temperatures was primarily caused by a reduction in the seed set percentage. Meanwhile, under high night temperature, the combination of decreased spikelet number per panicle, grain weight, and biomass production in addition to decreased seed set percentage contributed to the grain yield loss [77]. Altenbach et al. [51] reported that high temperature during anthesis promoted both grain shrinkage and a decrease in weight. Additionally, under heat stress conditions, plants tend to have a shorter grain-filling period, which reduces grain size and thousand kernel weight, while under drought conditions, plants tend to produce fewer grains per spikelet (and/or fewer tillers) [78]. This finding matches the TGW analyses stated above. In fact, heat stress and water scarcity showed similar effects on TGW in the current data analysis, suggesting that under drought conditions, the drastic decline in C_3 cereal yields is instead linked to a decrease in grain number produced per plant.

3.2. Effects of [CO₂], Temperature and Drought Stress on Grain Quality

Another major consideration is the effect of climate change on grain quality. While crop breeding is already much more focused on yield traits, comparatively little attention has been given to grain quality traits. This is a matter of great concern because, as described in more detail below, environmental stress will affect the relative abundance of starch, protein, and minerals [9,79,80].

3.2.1. Starch

Starch is the most abundant end-product of cereal growth and development, representing around 70% of the dry weight (w/w) of grains [81]. Rising [CO_2] increases photosynthetic rates in C_3 plants; increased carbohydrate translocation from the source (leaves and stems) to the sink (grains) is expected to increase the starch content in grains [82]. Indeed, the current data analysis has shown that growth under elevated [CO_2] has a significant positive effect on the grain starch concentration, which contrasts with the non-significant results reported by Högy and Fangmeier [83] and Broberg et al. [12]. Fangmeier et al. [84] reported that elevated [CO_2] significantly increased starch only for plants under high levels of N fertilizer.

Despite no significant effect due to drought, we revealed an overall decrease in grain starch concentration under drought stress. Worch et al. [85] observed that changes in endosperm starch content positively correlated with grain yield and concluded that grain starch content is one of the leading causes of reduced yield in crops subjected to drought conditions. This can be due to water deficit compromising both production of photoassimilates (source of carbon skeletons for the synthesis of starch) and the activity of enzymes involved in starch biosynthesis in the endosperm. Thus, the lower starch content observed in grains of genotypes subjected to water deficit could be correlated with the availability of reducing sugars [86].

Elevated temperature also negatively affected the starch concentration in grains. It has been reported that the reduction in starch concentration under high-temperature conditions is due to two factors; (i) shortening of the grain-filling period, which may reduce the duration of starch accumulation [51], and (ii) impairment of starch metabolism. While data for grains of plants exposed to high temperatures are scarce, Hawker and Jenner [87] and Keeling et al. [88] reported the inhibition of starch metabolism by high temperature (generally around 30 °C), possibly due to thermal denaturation negatively affecting the activity of starch synthase.

3.2.2. Total Protein

Elevated [CO_2] has been documented to reduce grain protein (or N) content in edible parts of crops [14–16]. In line with these earlier studies, the current meta-analysis showed that elevated [CO_2] significantly decreased grain protein concentrations. This reduction has been associated with increased photosynthesis and accumulation of grain carbohydrates, leading to reductions in the amount of grain protein (due to a dilution effect) [17,35]. However, Goufo et al. [89] reported decreases in protein without associated increases in starch in grains of rice exposed to elevated [CO_2]. Decreased protein concentrations in cereal grains under elevated [CO_2] might be a consequence of reduced leaf protein concentrations in photosynthetic tissues, leading to decreased seed protein [84,90]. The suppression of nitrate assimilation by elevated [CO_2] could be another contributor [91]. Our study also showed that there was a change in protein composition in grains of plants grown at elevated [CO_2]. In line with the results of Wieser et al. [92] and Högy et al. [17], gluten, gliadins, and glutenin concentrations decreased under increasing [CO_2]. Differences in the amounts and proportions of gluten protein fractions and types have significant effects on dough mixing and rheological characteristics. One of the most important characteristics for baking quality is bread volume, which has been strongly correlated with crude protein, total gluten proteins, and glutenin macropolymers [4,93]. Consequently, a reduction in bread quality can be expected due to the higher sensitivity of gluten fractions to elevated [CO_2].

Grain protein content is sensitive to environmental conditions and controlled by a number of factors, particularly the duration and rate of grain filling and the availability of assimilates, which are negatively affected in crops subjected to stressful growth conditions [94,95]. In contrast to elevated [CO_2], we found that high temperatures increased the grain protein concentration by 10.4%, which could be attributed to greater remobilization of shoot-derived protein. The grain protein concentration is expressed as a percentage of grain dry mass, which alongside the lower size and weight of the affected grains (also detected in our meta-analysis), would contribute to them having lower carbohydrate levels and consequently higher grain protein [96]. We note that the increase in grain protein concentration (10.4%) is almost the same as the decrease in grain starch concentration (-9.9%), suggesting that starch depletion increases the relative content of total protein.

Drought affects plant phenology and physiology. Water scarcity has been previously described as reducing photosynthetic rates, shortening the grain-filling period [11,97], and accelerating leaf senescence after anthesis. We detected significant increases in grain total protein associated with low water availability. Bhullar and Jenner [98] reported that during the grain-filling period, drought stress hinders the conversion of sucrose into starch but has a milder effect on protein biosynthesis. Our findings did not corroborate Bhullar and co-workers' conclusions. As mentioned before, the fact that the grain starch concentration was not significantly affected by drought would discard the lower carbohydrate level as a factor that induces increased grain protein content. Singh et al. [7] observed that together with lower rates of carbohydrate accumulation in the grain of plants subjected to drought, the increase in flour protein was mainly due to higher rates of grain N accumulation. The present meta-analysis supports this assertion because grain N concentration was affected by drought. Adding to that, the increased grain protein concentration under drought could be explained by the shortened maturation time common to stress conditions, which tends to favor protein over starch accumulation in cereal grains [99]. Drought, among other stresses, accelerates the translocation of senescence-inducing resources (including amino acids) from leaves to seeds during grain filling. Several studies have demonstrated that the contribution of reserve mobilization to the final grain yield is higher under stressful conditions than relatively well-irrigated conditions [100–102].

3.2.3. Mineral Composition

The present study showed that elevated $[CO_2]$ leads to an impoverishment of macro/microelements in grains. Moreover, there is a variation among minerals in the magnitude of the reductions, and this supports previous results [17,18,20,26,103]. In fact, only the Na concentration was significantly increased, with surprisingly few studies having investigated this element in relation to the effect of $[CO_2]$, and so there is little background information to explain this trend. Basically, most studies have focused on the main minerals that affect human health, such as Fe, Zn, P, K, and Ca, and have underlined a common decline in these minerals under rising $[CO_2]$. With respect to our results, the concentrations of Zn, Fe, S, Ca, Mg, P, Mn, K, and Mo were significantly decreased. Such reductions have been associated with increased production of spikes and grains that translates into a grain nutrient-dilution effect, diminishing the nutritional value. Furthermore, by reducing transpiration (linked to stomatal closure due to long-term exposure to elevated $[CO_2]$), high $[CO_2]$ can reduce the mass flow in the soil toward roots, which diminishes the availability of mobile minerals in the rhizosphere [14]. While carbohydrate dilution should lower all other nutrients in plant tissues evenly [104], other effects of elevated $[CO_2]$ on plant physiology are not evenly distributed among the minerals. For example, reduction in transpiration and elevated biosynthesis affect some minerals more than others. This means that the stoichiometry of plants exposed to elevated $[CO_2]$ should "differ not only in C:(other elements) ratios but also in the ratios among other elements (e.g., C:N, N:P, and P:Zn should be different)" [14]. Indeed, Loladze's meta-analysis of over 7500 pairs of observations from studies of elevated $[CO_2]$ published over 30 years (1984–2014) showed a significant reduction in foliar Mg (and N, P, K, Ca, S, Fe, Zn, and Cu) but not the Mn content in C_3 plants, and underlying biochemical mechanisms responsible for the increased Mn:Mg ratio have been proposed [105].

Changes in the elemental composition in grains are also detected under heat and drought stress. Previous studies suggested that both stress factors tend to increase mineral concentrations (including Fe, N, S, Zn, K, and P). However, the low number of reports means that there is relatively large uncertainty about the magnitude of the increase. The observed increase in grain protein and N concentrations (and the concomitant decrease in starch) under elevated temperature means that there is more N per unit of starch [106]. In addition, Fe and Zn tend to increase under drought. Although water plays a significant role in mineral uptake and later mobilization within the plant, with these processes decreasing during water stress, our meta-analysis agrees with Ge et al. [107] reporting that soil drought stress improved transport mechanisms and/or routes for some minerals, such as Fe and Zn, leading to increased grain concentrations of these elements. Moreover, according to other studies [107,108], the increase in the levels of Fe and Zn may be related to the more efficient remobilization of these nutrients from leaves to grains. However, according to other authors [109], the increase in Fe and Zn concentrations is linked to sink strength at the single grain level. More specifically, Miller et al. [109] observed in maize how the mineral content in drought-sensitive genotypes (which produced lower numbers of grains than the tolerant ones) was higher than in fully watered plants. According to this explanation, the increase in nutrients in the grains may be related to the number of grains formed, with each grain being a specific sink [86]. Furthermore, as we mentioned above, heat and drought cause a decrease in the number and size of cereal grains, which suggests that there might be a concentration effect due to the smaller grains [110].

4. Materials and Methods

4.1. Data Search and Selection Criteria

To find relevant studies related to the issue of the current meta-analysis, literature searches of primary research in published peer-reviewed journal sources were conducted from Google, Web of Science, and Scopus in June 2017. To search the literature, the following keywords were used: grain yield, cereal, high [CO_2], elevated temperature, drought stress, climate change, and C_3 grain quality. More than 150 papers were found, but 78 articles were selected according to the following criteria: (i) the article studies the effect of at least one climate parameter, including [CO_2], temperature, and drought, (ii) the article contains at least one response variable from the following list: grain yield, thousand grain weight (TGW), starch, total protein, gluten, glutenins, gliadins, and a set of minerals (Al, N, B, Ca, Cd, Co, Cr, Cu, Fe, K, Mo, Mg, Mn, Na, Ni, P, Pb, S, Se, Si, and Zn). The most abundant C_3 species that are reported in the literature are wheat, rice, and barley. All papers included in this meta-analysis were published between 1990 and 2019 (Table A1). The study is based on comparing plants grown at elevated [CO_2] (550–900 ppm) using Open Top Chamber (OTC) facilities or in the field using Free-Air-CO_2-Enrichment (FACE) systems with those grown at ambient [CO_2] (currently at ca. 400 ppm). Studies comparing different ranges of temperature, from ambient (10–25 °C) to elevated temperature (28–37 °C), and two levels of irrigation (limited irrigation or well-watered) are also included in the current report. Response means of plants grown under the different environmental conditions stated previously were taken from tables. The time of occurrence of stress during the crop cycle and the duration of stress applied differ among the studies, as indicated in Table S1. Most studies reported that treatments were maintained until the end of the experiments, when the plants reached maturity.

4.2. Data Analysis

All the data described above were organized in an Excel datasheet pairwise (control and experimental value) for each experimental factor ([CO_2], temperature, water) (Table S1). The datasheet was loaded into and analyzed in RStudio v1.1.456 [111]. For the effect size metric, we used the natural log of the response ratio, $lnR = ln(HF/LF)$, where LF and HF are reported mean nutrient concentrations at low and high treatment, respectively, with the treatment being any of the three climate factors considered in this study (CO_2, temperature, or water). The log response ratio eliminates asymmetry between percentage decreases limited to 100% and unlimited percentage increases; it is a standard approach for analyzing elevated [CO_2] and other ecological studies [112]. After performing statistical analyses, all the results were back-transformed to regular percentage changes using the formula: $(exp(lnR) - 1)*100\%$. For estimating the 95% confidence intervals for the mean effect size, a non-parametric test, namely bootstrapping with 999 replacements, was used for sample sizes of seven or more (i.e., when seven or more independent studies reported any given nutrient concentration at low and high treatments) [27]. The advantage of this approach is that it does not require the distribution of effect sizes to be normal. However, for the confidence intervals to be accurate, they can be applied only for sufficiently large sample sizes (>7). For sample sizes <7, we had a choice of discarding the data completely, which would result in the loss of potentially valuable information, or making a normality assumption and applying a parametric method. We chose the latter for sample sizes of 3 to 7. No confidence intervals were derived for sample sizes of two or less. In all cases, unweighted methods were used, with each study having equal weight.

5. Conclusions and Perspectives

This study highlights that while current and near-future environmental conditions will severely affect cereal yield, the nutritional value of cereal grain will also be affected.

It seems that within the three factors related to climate change investigated, the rise in atmospheric $[CO_2]$ is possibly the one more detrimental and difficult to face because elevated $[CO_2]$ will impact grain quality traits all over the world while the impacts of the increase in temperature and the decrease in water availability will be localized or easy to counterbalance. In fact, although the increase in $[CO_2]$ might promote yield enhancement and starch accumulation through higher rates of photosynthesis, the grains of these plants will have lower concentrations of total proteins and minerals, leading to reduced baking quality and deficient nutritional value. On the other hand, even if both high temperature and drought severely decrease crop yields, the available data shows that grain quality will be differentially affected. Heat stress will negatively affect grain starch concentration due to depleted starch biosynthesis metabolism and shortening of the grain-filling period, but it might increase total proteins and N concentration. Regarding water availability effects, grain yield could be conditioned by the final starch concentration of affected plants. Adding to the increase in the Fe and Zn concentrations, we found that total protein concentrations are significantly increased, which is probably due to a dilution effect on starch and the accelerated reserve remobilization from source to sink to compensate for the nutrient uptake deficit that results from low soil water content. According to numerous climatic models, precipitation patterns are expected to change in the future with more frequent drought events in semiarid and arid regions but, it is also predicted that in other regions, precipitation will likely increase. Therefore, while drought and elevated temperature can be potentially mitigated (by increasing irrigation, planting crops at higher altitudes within a given latitude, or displaced to cooler and wet latitudes within a country), the effect of rising $[CO_2]$ is present at all latitudes and will act independently of where crops will be established. Hence, $[CO_2]$-induced reductions in grain quality would be much more challenging to mitigate.

Our study highlights the fact that within the context of the present and near-future environments, it is crucial to increase crop yield through the development of stress-adapted cultivars. While the current breeding programs and agricultural incentives are almost exclusively yield-based, breeding for improved cereal quality can meaningfully improve the nutritional status of humanity. For this purpose, a better understanding of how environmental growth conditions (such as elevated temperature, drought, etc.) affect grain yield and nutritional parameters of cereals will help developing more nutrient-dense crops. Adding to that, exploring genetic diversity and variability of major crops is needed to discover genotypes more resilient to ongoing climate change.

Supplementary Materials: The following is available online at https://www.mdpi.com/article/10.3390/plants10061052/s1, Table S1: The collected data and information used in the meta-analysis.

Author Contributions: Conceptualization, I.A.; methodology, I.A.; software, I.L.; validation, I.A., F.M., and B.Z.; investigation, S.B.M. and D.S.; writing—original draft preparation, S.B.M.; writing—review and editing, S.B.M., I.L., and I.A. All authors have read and agreed to the published version of the manuscript.

Funding: This work was supported by European Interest Group (EIG) CONCERT-Japan (IRUEC), the Spanish Ministry of Science and Innovation (Spanish MINECO projects PCIN-2017-007 and PID2019-110445RB-I00). Sinda Ben Mariem had a PhD grant from the Navarra Government.

Institutional Review Board Statement: Not applicable.

Informed Consent Statement: Not applicable.

Data Availability Statement: Data is contained within the article or supplementary material.

Conflicts of Interest: The authors declare no conflict of interest.

Appendix A

Table A1. List of papers and species used for the data analysis of each environmental factor.

Papers	Species
High [CO$_2$]	
[92]	*Triticum durum* L.
[113]	*Triticum aestivum* L.
[17]	*Triticum aestivum* L.
[114]	*Triticum aestivum* L.
[115]	*Triticumaestivum* L./*Hordeumvulgare*
[26]	*Oryza sativa* L.
[116]	*Oryza sativa* L.
[73]	*Triticum aestivum* L.
[22]	*Triticumdurum* L./*Oryza sativa* L.
[117]	*Triticum aestivum* L.
[20]	*Triticum aestivum* L.
[118]	*Triticum aestivum* L.
[119]	*Triticum aestivum* L.
[120]	*Triticum aestivum* L.
[121]	*Triticum aestivum* L.
[122]	*Triticum aestivum* L.
[84]	*Triticum aestivum* L.
[123]	*Triticum aestivum* L.
[124]	*Triticum aestivum* L.
[125]	*Triticum aestivum* L.
[126]	*Triticum aestivum* L.
[127]	*Triticum aestivum* L.
[74]	*Triticum aestivum* L.
[128]	*Triticum aestivum* L.
[129]	*Triticum durum* L.
[38]	*Oryza sativa* L.
[80]	*Triticum durum* L.
[18]	*Triticum durum* L.
[19]	*Triticum aestivum* L.
[130]	*Triticum aestivum* L.
[131]	*Oryza sativa* L.
[132]	*Triticum aestivum* L.
[133]	*Triticum aestivum* L.
[134]	*Triticum durum* L.
[35]	*Oryza sativa* L.
[135]	*Oryza sativa* L.
Drought stress	
[136]	*Triticum aestivum* L.
[57]	*Triticum durum* L.
[137]	*Triticum aestivum* L.
[138]	*Triticum aestivum* L.
[139]	*Triticum durum* L.
[140]	*Triticum durum* L.
[141]	*Triticum aestivum* L.

Table A1. *Cont.*

Papers	Species
[55]	*Triticum durum* L.
[62]	*Triticum aestivum* L.
[80]	*Triticum durum* L.
[60]	*Oryza sativa* L.
[61]	*Triticum aestivum* L.
[129]	*Triticum durum* L.
[126]	*Triticum aestivum* L.
[56]	*Triticum aestivum* L.
[142]	*Triticum aestivum* L.
[37]	*Triticumaestivum* L., *Triticumdurum* L.
[143]	*Triticum aestivum* L.
[144]	*Triticum aestivum* L.
[145]	*Triticum aestivum* L.
[51]	*Triticum aestivum* L.
Heat stress	
[36]	*Triticum durum* L.
[51]	*Triticum aestivum* L.
[32]	*Triticum aestivum* L.
[146]	*Triticumaestivum* L., *Triticumdurum* L.
[59,147]	*Triticum aestivum* L.
[45]	*Triticum aestivum* L.
[148]	*Triticumaestivum* L., *Triticumdurum* L.
[31]	*Triticum durum* L.
[145]	*Triticum durum* L.
[39]	*Triticum aestivum* L.
[74]	*Triticum aestivum* L.
[149]	*Triticum aestivum* L.
[46]	*Triticum aestivum* L.
[150]	*Triticumaestivum* L., *Triticumdurum* L.
[151]	*Triticum aestivum* L.
[37]	*Triticumaestivum* L., *Triticumdurum* L.
[152]	*Triticum aestivum* L.
[153]	*Triticum aestivum* L.
[41]	*Triticum aestivum* L.
[154]	*Triticum durum* L.

References

1. Food and Agriculture Organization of the United Nations (FAO). *The Future of Food and Agriculture—Alternative Pathways to 2050. Summary Version*; FAO: Rome, Italy, 2018; 60p.
2. United Nations. United Nations Population Division (UNPD). Available online: http://www.un.org/en/development/desa/population/ (accessed on 12 March 2019).
3. Ortiz-Bobea, A.; Ault, T.R.; Carillo, C.M.; Chambers, R.G.; Lobell, D.B. Anthropogenic climate change has slowed global agricultural productivity growth. *Nat. Clim. Chang.* **2021**, *11*, 306–312. [CrossRef]
4. Nuttall, J.G.; O'Leary, G.J.; Panozzo, J.F.; Walker, C.K.; Barlow, K.M.; Fitzgerald, G.J. Models of grain quality in wheat—A review. *Field Crop. Res.* **2017**, *202*, 136–145. [CrossRef]
5. Food and Agriculture Organization of the United Nations (FAO). *World Food and Agriculture. Statistical Pocketbook*; FAO: Rome, Italy, 2018; 254p.
6. Foreign Agriculture Service/United States Department of Agriculture. *World Agricultural Production*; Circular Series WAP; USDA: Washington, DC, USA, 2019.
7. Singh, S.; Gupta, A.K.; Kaur, N. Influence of drought and sowing time on protein composition, antinutrients, and mineral contents of wheat. *Sci. World J.* **2012**, *2012*, 1–9. [CrossRef]
8. Ainsworth, E.A.; Long, S.P. What have we learned from 15 years of free-air CO_2 enrichment (FACE)? A meta-analytic review of the responses of photosynthesis, canopy properties and plant production to rising CO_2. *New Phytol.* **2005**, *165*, 351–372. [CrossRef]
9. Loladze, I. Hidden shift of the ionome of plants exposed to elevated CO_2 depletes minerals at the base of human nutrition. *eLife* **2014**, *3*, e02245. [CrossRef]

10. Zhang, G.; Sakai, H.; Tokida, T.; Usui, Y.; Zhu, C.; Nakamura, H.; Yoshimoto, M.; Fukuoka, M.; Kobayashi, K.; Hasegawa, T. The effects of free-air CO_2 enrichment (FACE) on carbon and nitrogen accumulation in grains of rice. *J. Exp. Bot.* **2013**, *64*, 3179–3188. [CrossRef]

11. Zhang, X.; Cai, J.; Wollenweber, B.; Liu, F.; Dai, T.; Cao, W.; Jiang, D. Multiple heat and drought events affect grain yield and accumulations of high molecular weight glutenin subunits and glutenin macropolymers in wheat. *J. Cereal Sci.* **2013**, *57*, 134–140. [CrossRef]

12. Broberg, M.; Högy, P.; Pleijel, H. CO_2-Induced Changes in Wheat Grain Composition: Meta-Analysis and Response Functions. *Agronomy* **2017**, *7*, 32. [CrossRef]

13. Ghannoum, O.; Caemmerer, S.V.; Ziska, L.H.; Conroy, J.P. The growth response of C_4 plants to rising atmospheric CO_2 partial pressure: A reassessment. *Plant Cell Environ.* **2000**, *23*, 931–942. [CrossRef]

14. Loladze, I. Rising atmospheric CO_2 and human nutrition: Toward globally imbalanced plant stoichiometry? *Trends Ecol. Evol.* **2002**, *17*, 457–461. [CrossRef]

15. Taub, D.R.; Miller, B.; Allen, H. Effects of elevated CO_2 on the protein concentration of food crops: A meta-analysis. *Glob. Chang. Biol.* **2008**, *14*, 565–575. [CrossRef]

16. Medek, D.E.; Schwartz, J.; Myers, S.S. Estimated effects of future atmospheric CO_2 concentrations on protein intake and the risk of protein deficiency by country and region. *Environ. Health Perspect.* **2017**, *125*, 1–8. [CrossRef]

17. Högy, P.; Wieser, H.; Köhler, P.; Schwadorf, K.; Breuer, J.; Franzaring, J.; Muntifering, R.; Fangmeier, A. Effects of elevated CO_2 on grain yield and quality of wheat: Results from a 3-year free-air CO_2 enrichment experiment. *Plant Biol.* **2009**, *11* (Suppl.1), 60–69. [CrossRef]

18. Fernando, N.; Panozzo, J.; Tausz, M.; Norton, R.M.; Fitzgerald, G.J.; Seneweera, S. Rising atmospheric CO_2 concentration affects mineral nutrient and protein concentration of wheat grain. *Food Chem.* **2012**, *133*, 1307–1311. [CrossRef]

19. Fernando, N.; Panozzo, J.; Tausz, M.; Norton, R.M.; Fitzgerald, G.J.; Myers, S.; Nicolas, M.E.; Seneweera, S. Intraspecific variation of wheat grain quality in response to elevated $[CO_2]$ at two sowing times under rain-fed and irrigation treatments. *J. Cereal Sci.* **2014**, *59*, 137–144. [CrossRef]

20. Fernando, N.; Panozzo, J.; Tausz, M.; Norton, R.M.; Neumann, N.; Fitzgerald, G.J.; Seneweera, S. Elevated CO_2 alters grain quality of two bread wheat cultivars grown under different environmental conditions. *Agric. Ecosyst. Environ.* **2014**, *185*, 24–33. [CrossRef]

21. Fernando, N.; Panozzo, J.; Tausz, M.; Norton, R.M.; Fitzgerald, G.J.; Myers, S.; Walter, C.; Stangoules, J.; Seneweera, S. Wheat grain quality under increasing atmospheric CO_2 concentrations in a semi-arid cropping system. *J. Cereal Sci.* **2012**, *56*, 684–690. [CrossRef]

22. Dietterich, L.H.; Zanobetti, A.; Kloog, I.; Huybers, P.; Leakey, A.D.B.; Bloom, A.J.; Carlisle, E.; Fernando, N.; Fitzgerald, G.; Hasegawa, T.; et al. Impacts of elevated atmospheric CO_2 on nutrient content of important food crops. *Sci. Data* **2015**, *2*, 1–8. [CrossRef] [PubMed]

23. Myers, S.S.; Zanobetti, A.; Kloog, I.; Huybers, P.; Leakey, A.D.; Bloom, A.J.; Carlisle, E.; Dietterich, L.H.; Fitzgerald, G.; Hasegawa, T.; et al. Rising CO_2 threatens human nutrition. *Nature* **2014**, *510*, 139–142. [CrossRef]

24. Weyant, C.; Brandeau, M.L.; Burke, M.; Lobell, D.B.; Bendavid, E.; Basu, S. Anticipated burden and mitigation of carbon-dioxide-induced nutritional deficiencies and related diseases: A simulation modeling study. *PLoS Med.* **2018**, *15*, e1002586. [CrossRef]

25. Beach, R.H.; Sulser, T.B.; Crimmins, A.; Cenacchi, N.; Cole, J.; Fukagawa, N.K.; Mason-D'Croz, D.; Myers, S.; Sarofin, M.C.; Smith, M.; et al. Combining the effects of increased atmospheric carbon dioxide on protein, iron, zinc availability and projected climate change on global diets: A modelling study. *Lancet Planet. Health* **2019**, *3*, e307–e317. [CrossRef]

26. Zhu, C.; Kobayashi, K.; Loladze, I.; Zhu, J.; Jiang, Q.; Xu, X.; Liu, G.; Seneweera, S.; Ebi, K.L.; Drewnowski, A.; et al. Carbon dioxide (CO_2) levels this century will alter the protein, micronutrients and vitamin content of rice grains with potential health consequences for the poorest rice-dependent countries. *Sci. Adv.* **2018**, *4*, 1–9. [CrossRef] [PubMed]

27. Loladze, I.; Nolan, J.M.; Ziska, L.H.; Knobbe, A.R. Rising atmospheric CO_2 lowers concentrations of plant carotenoids essential to human health: A meta-analysis. *Mol. Nutr. Food Res.* **2019**, *63*, 1–9. [CrossRef]

28. Intergovernmental Panel on Climate Change (IPCC). Climate Change 2014: Synthesis Report. In *Contribution of Working Groups I, II and III to the Fifth Assessment Report of the Intergovernmental Panel on Climate Change*; Core Writing Team, Pachauri, R.K., Meyer, L.A., Eds.; IPCC: Geneva, Switzerland, 2014; 151p.

29. Lopez-Bustins, J.A.; Pla, E.; Nadal, M.; De Herralde, F.; Savé, R. Global change and viticulture in the Mediterranean region: A case of study in north-eastern Spain. *Span. J. Agric. Res.* **2014**, *12*, 78–88. [CrossRef]

30. Meehl, G.A.; Stocker, T.F.; Collins, W.D.; Friedlingstein, P.; Gaye, A.T.; Gregory, J.M.; Kitoh, A.; Knutti, R.; Murphy, J.M.; Noda, A.; et al. Global Climate Projections. In *Climate Change 2007: The Physical Science Basis. Contribution of Working Group I to the Fourth Assessment Report of the Intergovernmental Panel on Climate Change*; Solomon, S., Qin, D., Manning, M., Chen, Z., Marquis, M., Averyt, K.B., Tignor, M., Miller, H.L., Eds.; Cambridge University Press: Cambridge, UK, 2007; pp. 747–845.

31. Lizana, X.C.; Calderini, D.F. Yield and grain quality of wheat in response to increased temperatures at key periods for grain number and grain weight determination: Considerations for the climatic change scenarios of Chile. *J. Agric. Sci.* **2013**, *151*, 209–221. [CrossRef]

32. Castro, M.; Peterson, G.J.; Rizza, M.D.; Dellavalle, P.D.; Vázquez, D.; Ibáñez, V.; Ross, A. Influence of heat stress on wheat grain characteristics and protein molecular weight distribution. *Wheat Prod. Stress Environ.* **2007**, *12*, 365–371. [CrossRef]

33. Usui, Y.; Sakai, H.; Tokida, T.; Nakamura, H.; Nakagawa, H.; Hasegawa, T. Heat-tolerant rice cultivars retain grain appearance quality under free-air CO_2 enrichment. *Rice* **2014**, *7*, 1–9. [CrossRef] [PubMed]

34. Bahuguna, R.N.; Jha, J.; Pal, M.; Shah, D.; Lawas, L.M.; Khetarpal, S.; Jagadish, S.V.K. Physiological and biochemical characterization of NERICA-L-44: A novel source of heat tolerance at the vegetative and reproductive stages in rice. *Physiol. Plant.* **2015**, *154*, 543–559. [CrossRef]

35. Chaturvedi, A.K.; Bahuguna, R.N.; Pal, M.; Shah, D.; Maurya, S.; Jagadish, K.S.V. Elevated CO_2 and heat stress interactions affect grain yield, quality and mineral nutrient composition in rice under field conditions. *Field Crop. Res.* **2017**, *206*, 149–157. [CrossRef]

36. Ferreira, M.S.L.; Martre, P.; Mangavel, C.; Girousse, C.; Rosa, N.N.; Samson, M.F.; Morel, M.H. Physicochemical control of durum wheat grain filling and glutenin polymer assembly under different temperature regimes. *J. Cereal Sci.* **2012**, *56*, 58–66. [CrossRef]

37. Guzmán, C.; Autrique, J.E.; Mondal, S.; Singh, R.P.; Govindan, V.; Morales-Dorantes, A.; Posadas-Romano, G.; Crossa, J.; Ammar, K.; Peña, R.J. Response to drought and heat stress on wheat quality, with special emphasis on bread-making quality, in durum wheat. *Field Crop. Res.* **2016**, *186*, 157–165. [CrossRef]

38. Jing, L.; Wang, J.; Shen, S.; Wang, Y.; Zhu, J.; Wang, Y.; Yang, L. The impact of elevated CO_2 and temperature on grain quality of rice grown under open-air field conditions. *J. Sci. Food Agric.* **2016**, *96*, 3658–3667. [CrossRef] [PubMed]

39. Dupont, F.M.; Hurkman, W.J.; Vensel, W.H.; Tanaka, C.; Kothari, K.M.; Chung, O.K.; Altenbach, S.B. Protein accumulation and composition in wheat grains: Effects of mineral nutrients and high temperature. *Eur. J. Agron.* **2006**, *25*, 96–107. [CrossRef]

40. Savill, G.P.; Michalski, A.; Powers, S.J.; Wan, Y.; Tosi, P.; Buchner, P.; Hawkesford, M.J. Temperature and nitrogen supply interact to determine protein distribution gradients in the wheat grain endosperm. *J. Exp. Bot.* **2018**, *69*, 3117–3126. [CrossRef]

41. Wardlaw, I.F. Interaction Between Drought and Chronic High Temperature During Kernel Filling in Wheat in a Controlled Environment. *Ann. Bot.* **2002**, *90*, 469–476. [CrossRef]

42. Yang, H.; Gu, X.; Ding, M.; Lu, W.; Lu, D. Heat stress during grain filling affects activities of enzymes involved in grain protein and starch synthesis in waxy maize. *Sci. Rep.* **2018**, *8*, 1–9. [CrossRef] [PubMed]

43. Monjardino, P.; Smith, A.G.; Jones, R.J. Heat stress effects on protein accumulation of maize endosperm. *Crop. Sci.* **2005**, *45*, 1203–1210. [CrossRef]

44. Ziska, L.H.; Namuco, O.; Moya, T.; Quilang, J. Growth and yield response of field-grown tropical rice to increasing carbon dioxide and air temperature. *Agron. J.* **1997**, *89*, 45–53. [CrossRef]

45. Randall, P.J.; Moss, H.J. Some effects of temperature regime during grain filling on wheat quality. *Aust. J. Agric. Res.* **1990**, *41*, 603–617. [CrossRef]

46. Gooding, M.J.; Ellis, R.H.; Shewry, P.R.; Schofield, J.D. Effects of restricted water availability and increased temperature on the grain filling, drying and quality of winter wheat. *J. Cereal Sci.* **2003**, *37*, 295–309. [CrossRef]

47. Habash, D.Z.; Kehel, Z.; Nachit, M. Genomic approaches for designing durum wheat ready for climate change with a focus on drought. *J. Exp. Bot.* **2009**, *60*, 2805–2815. [CrossRef]

48. Intergovernmental Panel on Climate Change (IPCC). Summary for Policymakers. In *Climate Change 2013: The Physical Science Basis. Contribution of Working Group I to the Fifth Assessment Report of the Intergovernmental Panel on Climate Change*; Stocker, T.F., Qin, D., Plattner, G.-K., Tignor, M., Allen, S.K., Boschung, J., Nauels, A., Xia, Y., Bex, V., Midgley, P.M., Eds.; Cambridge University Press: Cambridge, UK; New York, NY, USA, 2013.

49. McKersie, B. Planning for food security in a changing climate. *J. Exp. Bot.* **2015**, *66*, 3435–3450. [CrossRef]

50. Giorgi, F.; Lionello, P. Climate change projections for the Mediterranean region. *Glob. Planet. Chang.* **2008**, *63*, 90–104. [CrossRef]

51. Altenbach, S.B.; DuPont, F.M.; Kothari, K.M.; Chan, R.; Johnson, E.L.; Lieu, D. Temperature, water and fertilizer influence the timing of key events during grain development in a US spring wheat. *J. Cereal Sci.* **2003**, *37*, 9–20. [CrossRef]

52. Russo, A.C.; Gouveia, C.M.; Trigo, R.M.; Liberato, M.L.R.; DaCamara, C. The influence of circulation weather patterns at different spatial scales on drought variability in the Iberian Peninsula. *Front. Environ. Sci.* **2015**, *3*, 1–15. [CrossRef]

53. Araus, J.L.; Slafer, G.A.; Royo, C.; Serret, M.D. Breeding for yield potential and stress adaptation in cereals. *Crit. Rev. Plant Sci.* **2008**, *27*, 377–412. [CrossRef]

54. Rebolledo, M.C.; Luquet, D.; Courtois, B.; Henry, A.; Soulié, J.C.; Rouan, L.; Dingkuhn, M. Can early vigour occur in combination with drought tolerance and efficient water use in rice genotypes? *Funct. Plant Biol.* **2013**, *40*, 582–594. [CrossRef] [PubMed]

55. Kiliç, H.; Yagbasanlar, T. The Effect of drought stress on grain yield, yield components and some quality traits of durum wheat (*Triticum turgidum* ssp. durum) Cultivars. *Not. Bot. Hort. Agrobot. Cluj-Napoca* **2010**, *38*, 164–170. [CrossRef]

56. Balla, K.; Rakszegi, M.; Li, Z.; Békés, F.; Bencze, S.; Veisz, O. Quality of winter wheat in relation to heat and drought shock after anthesis. *Czech J. Food Sci.* **2011**, *29*, 117–128. [CrossRef]

57. Houshmand, S.; Arzani, A.; Mirmohammadi-Maibody, S.A.M. Effects of salinity and drought stress on grain quality of durum wheat. *Commun. Soil Sci. Plant Anal.* **2014**, *45*, 297–308. [CrossRef]

58. Sharma, S.; Carena, M.J. Grain quality in Maize (*Zea mays* L.): Breeding implications for short-season drought environments. *Euphytica* **2016**, *212*, 247–260. [CrossRef]

59. Panozzo, J.F.; Eagles, H.A. Cultivar and environmental effects on quality characters in wheat. I. *Aust. J. Agric. Res.* **1998**, *49*, 757–766. [CrossRef]

60. Crusciol, C.A.C.; Arf, O.; Soratto, R.P.; Mateus, G.P. Grain quality of upland rice cultivars in response to cropping systems in the Brazilian tropical savanna. *Sci. Agric.* **2008**, *65*, 468–473. [CrossRef]
61. Singh, S.; Gupta, A.K.; Gupta, S.K.; Kaur, N. Effect of sowing time on protein quality and starch pasting characteristics in wheat (*Triticum aestivum* L.) genotypes grown under irrigated and rain-fed conditions. *Food Chem.* **2010**, *122*, 559–565. [CrossRef]
62. Abd El-Kareem, T.H.A.; El-Saidy, A.E.A. Evaluation of yield and grain quality of some bread wheat genotypes under normal irrigation and drought stress conditions in calcareous soils. *J. Biol. Sci.* **2011**, *11*, 156–164. [CrossRef]
63. Flagella, Z.; Giuliani, M.M.; Giuzio, L.; Volpi, Z.; Masci, S. Influence of water deficit on durum wheat storage protein composition and technological quality. *Eur. J. Agron.* **2010**, *33*, 197–207. [CrossRef]
64. Peleg, Z.; Saranga, Y.; Yazici, A.; Fahima, T.; Ozturk, L.; Cakmak, I. Grain zinc, iron and protein concentrations and zinc-efficiency in wild emmer wheat under contrasting irrigation regimes. *Plant Soil* **2008**, *306*, 57–67. [CrossRef]
65. Wilcox, J.; Makowski, D. A meta-analysis of the predicted effects of climate change on wheat yields using simulation studies. *Field Crop. Res.* **2014**, *156*, 180–190. [CrossRef]
66. Knox, J.; Daccache, A.; Hess, T.; Haro, D. Meta-analysis of climateimpacts and uncertainty on crop yields in Europe. *Environ. Res. Lett.* **2016**, *11*, 1–10. [CrossRef]
67. Dixit, P.N.; Telleria, R.; Al Khatib, A.N.; Allouzi, S.F. Decadal analysis of impact of future climate on wheat production in dry Mediterranean environment: A case of Jordan. *Sci. Total. Environ.* **2018**, *610*, 219–233. [CrossRef]
68. Kimball, B.A.; Morris, C.F.; Pinter, P.J.; Wall, G.W.; Hunsaker, D.J.; Adamsen, F.J.; Lamorte, R.L.; Leavitt, S.W.; Thompson, T.L.; Matthias, A.D.; et al. Elevated CO_2, drought, and soil nitrogen effects on wheat grain quality. *New Phytol.* **2001**, *150*, 295–303. [CrossRef]
69. Aranjuelo, I.; Cabrera-Bosquet, L.; Morcuende, R.; Avice, J.-C.; Nogués, S.; Araus, J.L.; Martínez-Carrasco, R.; Pérez, P. Does ear C sink strength contribute to overcoming photosynthetic acclimation of wheat plants exposed to elevated CO_2? *J. Exp. Bot.* **2011**, *62*, 3957–3969. [CrossRef] [PubMed]
70. Aranjuelo, I.; Sanz-Sáez, A.; Jauregui, I.; Irigoyen, J.J.; Araus, J.L.; Sánchez-Díaz, M.; Erice, G. Harvest index, a parameter conditioning responsiveness of wheat plants to elevated CO_2. *J. Exp. Bot.* **2013**, *64*, 1879–1892. [CrossRef] [PubMed]
71. Lobell, D.B.; Schlenker, W.; Costa-Roberts, J. Climate trends and global crop production since 1980. *Science* **2011**, *333*, 616–620. [CrossRef] [PubMed]
72. Bourgault, M.; Dreccer, M.F.; James, A.T.; Chapman, S.C. Genotypic variability in the response to elevated CO_2 of wheat lines differing in adaptive traits. *Funct. Plant Biol.* **2013**, *40*, 172–184. [CrossRef] [PubMed]
73. Pleijel, H.; Högy, P. CO_2 dose-response functions for wheat grain, protein and mineral yield based on FACE and open-top chamber experiments. *Environ. Pollut.* **2015**, *198*, 70–77. [CrossRef] [PubMed]
74. Mitchell, R.A.C.; Lawlor, D.W.; Mitchell, V.J.; Gibbard, C.L.; White, E.M.; Porter, J.R. Effects of elevated CO_2 concentration and increased temperature on winter wheat: Test of ARCWHEAT1 simulation model. *Plant Cell Environ.* **1995**, *18*, 736–748. [CrossRef]
75. Talukder, S.K.; Babar, M.A.; Vijaylakshmi, K.; Poland, J.; Prasad, P.V.V.; Bowden, R.; Fritz, A. Mapping QTL for the traits associated with heat tolerance in wheat (*Triticum aestivum* L.). *BMC Genet.* **2014**, *15*, 97. [CrossRef]
76. Akter, N.; Islam, M.R. Heat stress effects and management in wheat. A review. *Agron. Sustain. Dev.* **2017**, *37*, 1–17. [CrossRef]
77. Xiong, D.; Ling, X.; Huang, J.; Peng, S. Meta-analysis and dose-response analysis of high temperature effects on rice yield and quality. *Environ. Exp. Bot.* **2017**, *141*, 1–9. [CrossRef]
78. Li, Y.; Wu, Y.; Hernandez-Espinosa, N.; Peña, R.J. The influence of drought and heat stress on the expression of end-use quality parameters of common wheat. *J. Cereal Sci.* **2013**, *57*, 73–78. [CrossRef]
79. Ziska, L.H.; Bunce, J.A.; Shimono, H.; Gealy, D.R.; Baker, J.T.; Newton, P.C.D.; Matthew, P.R.; Jagadish, K.S.V.; Zhu, C.; Howden, M.; et al. Security and climate change: On the potential to adapt global crop production by active selection to rising atmospheric carbon dioxide. *Proc. R. Soc. B Biol. Sci.* **2012**, *279*, 4097–4105. [CrossRef] [PubMed]
80. Goicoechea, N.; Bettoni, M.M.; Fuertes-Mendizábal, T.; González-Murua, C.; Aranjuelo, I. Durum wheat quality traits affected by mycorrhizal inoculation, water availability and atmospheric CO_2 concentration. *Crop Pasture Sci.* **2016**, *67*, 147–155. [CrossRef]
81. Jung, K.H.; An, G.; Ronald, P.C. Towards a better bowl of rice: Assigning function to tens of thousands of rice genes. *Nature* **2008**, *9*, 91–101. [CrossRef] [PubMed]
82. Thitisaksakul, M.; Jiménez, R.C.; Arias, M.C.; Beckles, D.M. Effects of environmental factors on cereal starch biosynthesis and composition. *J. Cereal Sci.* **2012**, *56*, 67–80. [CrossRef]
83. Högy, P.; Fangmeier, A. Effects of elevated atmospheric CO_2 on grain quality of wheat. *J. Cereal Sci.* **2008**, *48*, 580–591. [CrossRef]
84. Fangmeier, A.; De Temmerman, L.; Mortensen, L.; Kemp, K.; Burke, J.; Mitchell, R.; Van Oijen, M.; Weigel, H.J. Effects on nutrients and on grain quality in spring wheat crops grown under elevated CO_2 concentrations and stress conditions in the European, multiple-site experiment "ESPACE-wheat". *Eur. J. Agron.* **1999**, *10*, 215–229. [CrossRef]
85. Worch, S.; Rajesh, K.; Harshavardhan, V.T.; Pietsch, C.; Korzun, V.; Kuntze, L.; Börner, A.; Wobus, U.; Röder, M.S.; Sreenivasulu, N. Haplotyping, linkage mapping and expression analysis of barley genes regulated by terminal drought stress influencing seed quality. *BMC Plant Biol.* **2011**, *11*, 1–14. [CrossRef]
86. Avila, R.G.; Da Silva, E.M.; Magalhães, P.C.; De Alvarenga, A.A.; De Oliveira Lavinsky, A. Drought changes yield and organic and mineral composition of grains of four maize genotypes. *Acad. J. Agric. Res.* **2017**, *5*, 243–250. [CrossRef]
87. Hawker, J.S.; Jenner, C.F. High temperature affects the activity of enzymes in the committed pathway of starch synthesis in developing wheat endosperm. *Aust. J. Plant Physiol.* **1993**, *20*, 197–209. [CrossRef]

88. Keeling, P.L.; Bacon, P.J.; Holt, D.C. Elevated temperature reduced starch deposition in wheat endosperm by reducing the activity of soluble starch synthase. *Planta* **1993**, *191*, 342–348. [CrossRef]

89. Goufo, P.; Falco, V.; Brites, C.M.; Wessel, D.F.; Kratz, S.; Rosa, E.A.S.; Carranca, C.; Trindade, H. Effect of Elevated Carbon Dioxide Concentration on Rice Quality: Nutritive Value, Color, Milling and Cooking/Eating Qualities. *Cereal Chem.* **2014**, *91*, 513–521. [CrossRef]

90. Fangmeier, A.; Chrost, B.; Ho, P.; Krupinska, K. CO_2 enrichment enhances flag leaf senescence in barley due to greater grain nitrogen sink capacity. *Environ. Exp. Bot.* **2000**, *44*, 151–164. [CrossRef]

91. Bloom, A.J.; Burger, M.; Kimball, B.A.; Pinter, P.J. Nitrate assimilation is inhibited by elevated CO_2 in field-grown wheat. *Nat. Clim. Chang.* **2014**, *4*, 477–480. [CrossRef]

92. Wieser, H.; Manderscheid, R.; Erbs, M.; Weigel, H.J. Effects of elevated atmospheric CO_2 concentrations on the quantitative protein composition of wheat grain. *J. Agric. Food Chem.* **2008**, *56*, 6531–6535. [CrossRef] [PubMed]

93. Weegels, P.L.; Hamer, R.J.; Schofield, J.D. Critical review: Functional properties of wheat glutenin. *J. Cereal Sci.* **1996**, *23*, 1–17. [CrossRef]

94. Yang, J.; Zhang, J. Grain filling of cereals under soil drying. *New Phytol.* **2006**, *169*, 223–236. [CrossRef]

95. Brdar, M.D.; Kraljević-Balalić, M.M.; Kobiljski, B.D. The parameters of grain filling and yield components in common wheat (*Triticum aestivum* L.) and durum wheat (*Triticum turgidum* L. var. durum). *Cent. Eur. J. Biol.* **2008**, *3*, 75–82. [CrossRef]

96. Barnabás, B.; Jäger, K.; Fehér, A. The effect of drought and heat stress on reproductive processes in cereals. *Plant Cell Environ.* **2008**, *31*, 11–38. [CrossRef]

97. Gallé, A.; Csiszár, J.; Secenji, M.; Guóth, A.; Cseuz, L.; Tari, I.; Györgyey, J.; Erdei, L. Drought response strategies during grain filling in wheat. Glutathione transferase activity and expression pattern in flag leaves. *J. Plant Physiol.* **2009**, *170*, 1389–1399. [CrossRef]

98. Bhullar, S.S.; Jenner, C.F. Effects of temperature on conversion of sucrose to starch in the developing wheat endosperm. *Aust. J. Plant. Physiol.* **1986**, *13*, 605–615. [CrossRef]

99. Wang, Y.; Frei, M. Stressed food—The impact of abiotic environmental stresses on crop quality. *Agric. Ecosyst. Environ.* **2011**, *141*, 271–286. [CrossRef]

100. Blum, A. Improving wheat grain filling under stress by stem reserve mobilization. *Euphytica* **1998**, *100*, 77–83. [CrossRef]

101. Yang, J.; Zhang, J.; Wang, Z.; Zhu, Q.; Liu, L. Involvement of abscisic acid and cytokinins int the senescence and remobilization of carbon reserves in wheat subjected to water stress during grain filling. *Plant Cell. Environ.* **2003**, *26*, 1621–1631. [CrossRef]

102. Srivastava, A.; Srivastava, P.; Sharma, A.; Sarlach, R.S.; Bains, N.S. Effect of stem reserve mobilization on grain filling under drought stress conditions in recombinant inbred population of wheat. *J. Appl. Nat. Sci.* **2017**, *9*, 1–5. [CrossRef]

103. Houshmandfar, A.; Fitzgerald, G.J.; Tausz, M. Elevated CO_2 decreases both transpiration flow and concentrations of Ca and Mg in the xylem sap of wheat. *J. Plant Physiol.* **2015**, *174*, 157–160. [CrossRef] [PubMed]

104. Gifford, R.M.; Barrett, D.J.; Lutze, J.L. The effects of elevated $[CO_2]$ on the C:N and C:P mass ratios of plant tissues. *Plant Soil* **2000**, *224*, 1–14. [CrossRef]

105. Bloom, A.J.; Lancaster, K.M. Manganese binding to Rubisco could drive a photorespiratory pathway that increases the energy efficiency of photosynthesis. *Nat. Plants* **2018**, *4*, 414–422. [CrossRef]

106. Stone, P.J.; Grast, P.W.; Nicolas, M.E. The influence of recovery temperature on the effects of a brief heat shock on wheat. III. grain protein composition and dough properties. *J. Cereal Sci.* **1997**, *25*, 129–141. [CrossRef]

107. Ge, T.D.; Sui, F.G.; Nie, S.; Sun, N.B.; Xiao, H.; Tong, C.L. Differential responses of yield and selected nutritional compositions to drought stress in summer maize grains. *J. Plant. Nut.* **2010**, *33*, 1811–1818. [CrossRef]

108. Farahani, S.M.; Mazaheri, D.; Chaichi, M.; Afshari, R.T.; Savaghebi, G. Effect of seed vigour on stress tolerance of barley (*Hordeum vulgare*) seed at germination stage. *Seed Sci. Technol.* **2010**, *38*, 494–507. [CrossRef]

109. Miller, R.O.; Jacobsen, J.S.; Skogley, E.O. Aerial accumulation and partitioning of nutrients by hard red spring wheat. *Commun. Soil Sci. Plant Anal.* **1994**, *24*, 2389–2407. [CrossRef]

110. Velu, G.; Guzman, C.; Mondal, S.; Autrique, J.E.; Huerta, J.; Singh, R.P. Effect of drought and elevated temperature on grain zinc and iron concentrations in CIMMYT spring wheat. *J. Cereal Sci.* **2016**, *69*, 182–186. [CrossRef]

111. RStudio Team. *RStudio: Integrated Development Environment for R*; RStudio Inc.: Boston, MA, USA, 2016.

112. Hedges, L.V.; Gurevitch, J.; Curtis, P.S. The meta-analysis of response ratios in experimental ecology. *Ecology* **1999**, *80*, 1150–1156. [CrossRef]

113. Högy, P.; Keck, M.; Niehaus, K.; Franzaring, J.; Fangmeier, A. Effects of atmospheric CO_2 enrichment on biomass, yield and low molecular weight metabolites in wheat grain. *J. Cereal Sci.* **2010**, *52*, 215–220. [CrossRef]

114. Högy, P.; Brunnbauer, M.; Koehler, P.; Schwadrof, K.; Breuer, J.; Franzaring, J.; Zhunusbayeva, D.; Fangmeier, A. Grain quality characteristics of spring wheat (*Triticum aestivum*) as affected by free-air CO_2 enrichment. *Environ. Exp. Bot.* **2013**, *88*, 11–18. [CrossRef]

115. Erbs, M.; Manderscheid, R.; Jansen, G.; Seddig, S.; Pacholski, A.; Weigel, H.J. Effects of free-air CO_2 enrichment and nitrogen supply on grain quality parameters and elemental composition of wheat and barley grown in a crop rotation. *Agric. Ecosyst. Environ.* **2010**, *136*, 59–68. [CrossRef]

116. Usui, Y.; Sakai, H.; Tokida, T.; Nakamura, H.; Nakagawa, H.; Hasegawa, T. Rice grain yield and quality responses to free-air CO_2 enrichment combined with soil and water warming. *Glob. Chang. Biol.* **2016**, *22*, 1256–1270. [CrossRef]

117. Panozzo, J.F.; Walker, C.K.; Partington, D.L.; Neumann, N.C.; Tausz, M.; Seneweera, S.; Fitzgerald, G.J. Elevated carbon dioxide changes grain protein concentration and composition and compromises baking quality. A FACE study. *J. Cereal Sci.* **2014**, *60*, 461–470. [CrossRef]
118. Bencze, S.; Veisz, O.; Bedő, Z. Effects of high atmospheric CO_2 on the morphological and heading characteristics of winter wheat. *Cereal Res. Comm.* **2004**, *32*, 233–240. [CrossRef]
119. Blumenthal, C.; Rawson, H.M.; McKenzie, E.; Gras, P.W.; Barlow, E.W.R.; Wrigley, C.W. Changes in wheat grain quality due to doubling the level of atmospheric CO_2. *Cereal Chem.* **1996**, *73*, 762–766.
120. Conroy, J.; Seneweera, S.; Basra, A.; Rogers, G.; Nissen-Wooller, B. Influence of rising atmospheric CO_2 concentrations and temperature on growth, yield and grain quality of cereal crops. *Aust. J. Plant. Physiol.* **1994**, *21*, 741–758. [CrossRef]
121. De la Puente, L.S.; Perez, P.P.; Carrasco, R.M.; Morcuende, R.M.; Del Molino, L.M.M. Action of elevated CO_2 and high temperature on the mineral chemical composition of two varieties of wheat. *Agrochimica* **2000**, *44*, 221–230.
122. Fangmeier, A.; Grüters, U.; Högy, P.; Vermehren, B.; Jäger, H.J. Effects of elevated CO_2, nitrogen supply and tropospheric ozone on spring wheat—II. Nutrients (N, P, K, S, Ca, Mg, Fe, Mn, Zn). *Environ. Pollut.* **1997**, *96*, 43–59. [CrossRef]
123. Fangmeier, A.; Grüters, U.; Vermehren, B.; Jager, H.J. Responses of some cereal cultivars to CO_2 enrichment and tropospheric ozone at different levels of nitrogen supply. *J. Appl. Bot. Food Qual.* **1996**, *70*, 12–18.
124. Wroblewitz, S.; Hüther, L.; Manderscheid, R.; Weigel, H.J.; Wätzig, H.; Dänicke, S. Effect of Rising atmospheric carbon dioxide concentration on the protein composition of cereal grain. *J. Agric. Food Chem.* **2014**, *62*, 6616–6625. [CrossRef]
125. Weigel, H.J.; Manderscheid, R. CO_2 enrichment effects on forage and grain nitrogen content of pasture and cereal plants. *J. Crop Improv.* **2005**, *13*, 73–89. [CrossRef]
126. Wu, D.X.; Wang, G.X.; Bai, Y.F.; Liao, J.X. Effects of elevated CO_2 concentration on growth, water use, yield and grain quality of wheat under two soil water levels. *Agric. Ecosyst. Environ.* **2004**, *104*, 493–507. [CrossRef]
127. Rogers, G.S.; Gras, P.W.; Batey, I.L.; Milham, P.J.; Payne, L.; Conroy, J.P. The influence of atmospheric CO_2 concentration on the protein, starch and mixing properties of wheat flour. *Aust. J. Plant Physiol.* **1998**, *25*, 387–393. [CrossRef]
128. Manderscheid, R.; Bender, J.; Jäger, H.J.; Weigel, H.J. Effects of season long CO_2 enrichment on cereals. II. Nutrient concentrations and grain quality. *Agric. Ecosyst. Environ.* **1995**, *54*, 175–185. [CrossRef]
129. Erice, G.; Sanz-Sáez, A.; González-Torralba, J.; Mendez-Espinoza, A.M.; Urretavizcaya, I.; Nieto, M.T.; Serret, M.D.; Araus, J.L.; Irigoyen, J.J.; Aranjuelo, I. Impact of elevated CO_2 and drought on yield and quality traits of a historical (*Blanqueta*) and a modern (*Sula*) durum wheat. *J. Cereal Sci.* **2019**, *87*, 194–201. [CrossRef]
130. Carlisle, E.; Myers, S.; Raboy, V.; Bloom, A. The Effects of Inorganic Nitrogen form and CO_2 Concentration on Wheat Yield and Nutrient Accumulation and Distribution. *Front. Plant Sci.* **2012**, *3*, 1–13. [CrossRef]
131. Guo, H.; Zhu, J.; Zhou, H.; Sun, Y.; Yin, Y.; Pei, D.; Ji, R.; Wu, J.; Wang, X. Elevated CO_2 levels affects the concentrations of copper and cadmium in crops grown in soil contaminated with heavy metals under fully open-air field conditions. *Environ. Sci. Technol.* **2011**, *45*, 6997–7003. [CrossRef]
132. Ma, H.; Zhu, J.; Xie, Z.; Liu, G.; Zeng, Q.; Han, Y. Responses of rice and winter wheat to free-air CO_2 enrichment (China FACE) at rice/wheat rotation system. *Plant Soil* **2007**, *294*, 137–146. [CrossRef]
133. Pleijel, H.; Danielsson, H. Yield dilution of grain Zn in wheat grown in open-top chamber experiments with elevated CO_2 and O_3 exposure. *J. Cereal Sci.* **2009**, *50*, 278–282. [CrossRef]
134. Beleggia, R.; Fragasso, M.; Miglietta, F.; Cattivelli, L.; Menga, V.; Nigro, F.; Pecchioni, N.; Fares, C. Mineral composition of durum wheat grain and pasta under increasing atmospheric CO_2 concentrations. *Food Chem.* **2018**, *242*, 53–61. [CrossRef]
135. Sakai, H.; Tokida, T.; Usui, Y.; Nakamura, H.; Hasegawa, T. Yield responses to elevated CO_2 concentration among Japanese rice cultivars released since 1882. *Plant Prod. Sci.* **2019**, *22*, 352–366. [CrossRef]
136. Chang-Xing, Z.; Ming-Rong, H.; Zhen-Lin, W.; Yue-Fu, W.; Qi, L. Effects of different water availability at post-anthesis stage on grain nutrition and quality in strong-gluten winter wheat. *Comptes Rendus Biol.* **2009**, *332*, 759–764. [CrossRef]
137. Eivazi, A.; Habibi, F. Sensitivity wheat genotypes for grain yield and quality traits to drought stress. *Int. J. Agron. Plant Prod.* **2012**, *3*, 738–747.
138. Baric, M.; Keresa, S.; Sarcevic, H.; HabusJercic, I.; Horvat, D.; Drezner, G. Influence of drought during the grain filling period to the yield and quality of winter wheat (*T. aestivum* L.). In Proceedings of the 3rd International Congress "Flour-Bread 05" and 5th Croatian Congress of Cereal Technologists, Opatija, Croatia, 26–29 October 2005; pp. 19–24.
139. Houshmand, S.; Arzani, A.; Maibody, S.A.M. Influences of drought and salt stress on grain quality of durum wheat. Genetic Variation for Plant Breeding. In Proceedings of the 17th Eucarpia General Congress, Tulln, Austria, 8–11 September 2004; pp. 383–386.
140. Arzani, A. Grain quality of durum wheat germplasm as affected by heat and drought stress at grain filling period. *Wheat Inf. Serv.* **2002**, *94*, 9–13.
141. Saint Pierre, C.; Peterson, C.J.; Ross, A.S.; Ohm, J.B.; Verhoeven, M.C.; Larson, M.; Hoefer, B. White wheat grain quality changes with genotype, nitrogen fertilization, and water stress. *Agron. J.* **2008**, *100*, 414–420. [CrossRef]
142. Souza, E.J.; Martin, J.M.; Guttieri, M.J.; O'Brien, K.M.; Habernicht, D.K.; Lanning, S.P.; McLean, R.; Carlson, G.R.; Talbert, L.E. Influence of genotype, environment, and nitrogen management on spring wheat quality. *Crop Sci.* **2004**, *44*, 425–432. [CrossRef]
143. Mkhabela, M.; Bullock, P.; Gervais, M.; Finlay, G.; Sapirstein, H. Assessing indicators of agricultural drought impacts on spring wheat yield and quality on the Canadian prairies. *Agric. For. Meteorol.* **2010**, *150*, 399–410. [CrossRef]

144. Guttieri, M.J.; Stark, J.C.; O'Brien, K.; Souza, E. Relative sensitivity of spring wheat grain yield and quality parameters to moisture deficit. *Crop. Sci.* **2001**, *41*, 327–335. [CrossRef]
145. Li, Y.F.; Wu, Y.; Hernandez-Espinosa, N.; Peña, R.J. Heat and drought stress on durum wheat: Responses of genotypes, yield, and quality parameters. *J. Cereal Sci.* **2013**, *57*, 398–404. [CrossRef]
146. Dias, A.S.; Bagulho, A.S.; Lidon, F.C. Ultrastructure and biochemical traits of bread and durum wheat grains under heat stress. *Braz. J. Plant Physiol.* **2008**, *20*, 323–333. [CrossRef]
147. Panozzo, J.F.; Eagles, H.A. Cultivar and environmental effects on quality characters in wheat. II. Protein. *Aust. J. Agric. Res.* **2000**, *51*, 629–636. [CrossRef]
148. Corbellini, M.; Canevar, M.G.; Mazza, L.; Ciaffi, M.; Lafiandra, D.; Borghi, B. Effect of the Duration and Intensity of Heat Shock During Grain Filling on Dry Matter and Protein Accumulation, Technological Quality and Protein Composition in Bread and Durum Wheat. *Aust. J. Plant Physiol.* **1997**, *24*, 245–260. [CrossRef]
149. Daniel, C.; Triboi, E. Effects of temperature and nitrogen nutrition on the grain composition of winter wheat: Effects on gliadin content and composition. *J. Cereal Sci.* **2000**, *32*, 45–56. [CrossRef]
150. Labuschagne, M.T.; Elago, O.; Koen, E. The influence of temperature extremes on some quality and starch characteristics in bread, biscuit and durum wheat. *J. Cereal Sci.* **2009**, *49*, 184–189. [CrossRef]
151. Spiertz, J.H.J.; Hamer, R.J.; Xu, H.; Primo-Martin, C.; Don, C.; van der Putten, P.E.L. Heat stress in wheat (*Triticum aestivum* L.): Effects on grain growth and quality traits. *Eur. J. Agron.* **2006**, *25*, 89–95. [CrossRef]
152. Viswanathan, C.; Khanna-Chopra, R. Effect of heat stress on grain growth, starch synthesis and protein synthesis in grains of wheat (*Triticum aestivum* L.) varieties differing in grain weight stability. *J. Agron. Crop. Sci.* **2001**, *186*, 1–7. [CrossRef]
153. Wrigley, C.; Blumenthal, C.; Gras, P.; Barlow, E. Temperature variation during grain filling and changes in wheat-grain quality. *Aust. J. Plant Physiol.* **1994**, *21*, 875–885. [CrossRef]
154. De Leonardis, A.M.; Fragasso, M.; Beleggia, R.; Ficco, D.B.M.; De Vita, P.; Mastrangelo, A.M. Effects of heat stress on metabolite accumulation and composition, and nutritional properties of durum wheat grain. *Int. J. Mol. Sci.* **2015**, *16*, 30382–30404. [CrossRef]

Review

Effects of Elevated CO$_2$ and Heat on Wheat Grain Quality

Xizi Wang and Fulai Liu *

Department of Plant and Environmental Sciences, Faculty of Science, University of Copenhagen,
Højbakkegård Allé 13, DK-2630 Tåstrup, Denmark; xiwa@plen.ku.dk
* Correspondence: fl@plen.ku.dk; Tel.: +45-3533-3392

Abstract: Wheat is one of the most important staple foods in temperate regions and is in increasing demand in urbanizing and industrializing countries such as China. Enhancing yield potential to meet the population explosion around the world and maintaining grain quality in wheat plants under climate change are crucial for food security and human nutrition. Global warming resulting from greenhouse effect has led to more frequent occurrence of extreme climatic events. Elevated atmospheric CO$_2$ concentration (eCO$_2$) along with rising temperature has a huge impact on ecosystems, agriculture and human health. There are numerous studies investigating the eCO$_2$ and heatwaves effects on wheat growth and productivity, and the mechanisms behind. This review outlines the state-of-the-art knowledge regarding the effects of eCO$_2$ and heat stress, individually and combined, on grain yield and grain quality in wheat crop. Strategies to enhance the resilience of wheat to future warmer and CO$_2$-enriched environment are discussed.

Keywords: elevated CO$_2$; grain quality; heat stress

Citation: Wang, X.; Liu, F. Effects of Elevated CO$_2$ and Heat on Wheat Grain Quality. *Plants* **2021**, *10*, 1027. https://doi.org/10.3390/plants10051027

Academic Editor: James Bunce

Received: 30 April 2021
Accepted: 18 May 2021
Published: 20 May 2021

Publisher's Note: MDPI stays neutral with regard to jurisdictional claims in published maps and institutional affiliations.

1. Introduction

Global atmospheric concentration of carbon dioxide (CO$_2$) is expected to reach levels of 420 ppm (RCP2.6) to 1300 ppm (RCP8.5) by the end of this century (IPCC, 2013) with a concomitant rise in mean global temperature of about 2 °C by 2050 (IPCC, 2014). Extreme climatic events (ECEs), such as heat waves and droughts, which often affect plant growth and pose a growing threat to natural and agricultural ecosystems, are predicted to increase in frequency and severity in many cropping areas [1,2]. Wheat is one of the world's major food crops with an average annual global production of over 750 million tons from 2015 to 2019 (http://faostat.fao.org/ (accessed on 14 May 2021)). It is considered an important source of starch and energy. Wheat provides significant amounts of important nutrients, including proteins and mineral elements, as well as other components that are beneficial to human health, such as vitamins (especially vitamin B), phytochemicals and dietary fibers [3]. The effect of elevated atmospheric CO$_2$ concentration (eCO$_2$) on wheat grain yield and grain quality has been well studied [4–6]. A general increase in grain yield and a reduction of grain quality of plants grown under eCO$_2$, especially the decrease of nitrogen (N) concentration and thus also protein contents, have often been reported, leading to the conclusion that eCO$_2$ potentially exacerbates the prevalence of "hidden hunger" for human nutrition [7,8].

Another environmental factor, heat waves, limits wheat yields globally [9]. Air temperatures have increased since the beginning of the century and global temperature is predicted to increase 1.5–5.5 °C in the next 50 to 75 years [10]. Such increases in the temperature can lead to heat stress, a severe threat to wheat production, particularly when it occurs during reproductive and grain-filling phases [11]. Previous studies showed that exposure to temperatures above the optimum temperature for wheat at anthesis and grain filling stage (12 to 20 °C) can significantly reduce grain yield of more than 20% [12,13]. All the physiological processes of wheat plants are sensitive to temperature and can be damaged by heat permanently. Heat stress during anthesis can increase floret abortion [14] and temperatures over 30 °C during floret formation may lead to complete sterility [15]. Heat stress during reproductive stage can lead to pollen sterility, tissue dehydration, lower CO_2 assimilation, increased photorespiration and reduced time to capture resources due to accelerated growth and senescence, consequently reducing the yield [11,16]. Besides the effects on grain yield, heat stress also has great impact on wheat grain quality. It was reported that heat stress could alter the ratio of gliadin to glutenin, which may lead to weaker dough properties and reduce baking quality [17]. However, there was a great diversity of the dough weakening effect under heat stress in other studies [18,19] because wheat quality mainly depends on genotype (G), environment (E) and their interactive effects (G × E) [20]. Several studies carried out at the CIMMYT showed diverse responses of different cultivars in quality traits under drought and heat stressed conditions. Hernandez-Espinosa et al. [21] reported that grain morphology (grain density and size), protein content and flour yield were strongly affected by the environment, while the traits related to gluten quality (gluten strength and extensibility, and bread loaf volume) were mainly determined by genotype, but the environmental and G × E effects were also important, especially for gluten extensibility. Li et al. [22] tested 15 quality parameters and found similar results that environmental factors have large effect on grain yield, while grain hardness and gluten quality-related traits are mainly controlled by genotype. Moreover, they found that drought and heat stress showed contrasting effects on dough rheological properties, where heat stress decreased dough tenacity (increased extensibility), slightly reduced dough strength and increased bread loaf volume while drought stress is the opposite. These results suggest that wheat quality is determined by many inter-related factors.

Although the impact of rising CO_2 concentration and elevated temperature on plant growth and development has been well studied, the combined effects of heat stress with eCO_2 on wheat crop performance remain unclear. Stomatal closure is induced by eCO_2, which leads to the reduction of transpirational canopy cooling; however, higher temperature results in higher water vapor deficits in the air, causing an increased transpiration rate thereby offsetting the effect of stomatal closure induced by eCO_2 on crop water relations [23]. On the other hand, reduced stomatal conductance under eCO_2 will slow transpiration rate thereby alleviating the impact of water stress, which is often concurrent with heat stress [24]. However, plants' ability to sustain leaf gas exchange is dependent on genotype and stress intensity [25]. More detailed studies are required to assess the balance of these processes and how they acclimate to different environmental factors, because under eCO_2 growth environment plants' response to these abiotic stresses can be much more complicated. Therefore, this review summarizes current knowledge regarding the effects of eCO_2 and heat stress on grain yield and quality in wheat and aims to (i) introduce wheat quality traits that are sensitive to abiotic growth conditions, (ii) discuss the impacts of eCO_2 and heat stress on wheat grain yield and quality and the underlying mechanisms.

2. Wheat Quality and Grain Protein

Wheat is the most important staple food in the world due to its wide adaptation to diverse growth conditions and its unique property of gluten protein fraction in the grain. Wheat has evolved itself from emmer wheat into the cultivated species today by both nature and anthropogenic processes since primitive times (ca. 3000–4000 BC). There are two major wheat species nowadays utilized for food production. The first one is common bread wheat (*Triticum aestevum* L.), a hexaploid wheat that is mainly processed into baking products and the other one is tetraploid durum wheat (*T. turgidum* L. var. durum), which is used to make coarse flour (semolina) for pasta making. Wheat species can be classified by grain hardness, color (red, white and amber) and growing season (spring or winter wheat) and within each class, wheat grain can be evaluated by different grading factors [26]. Wheat grain quality is a combination of many specific parameters. The processing and end-use quality is determined by the multiple phenotypic characteristics of grain, flour, dough and final products [27]. Grain protein content (GPC) is considered as one of the most important components affecting the baking quality [28]. Besides, concentration of mineral nutrients, grain hardness (milling properties), grain size, starch content and minor grain constituents such as lipids and soluble proteins also play important roles in determining end-use properties [10,29].

Baking quality mainly depends on the protein composition and concentrations of wheat grains. Wheat grain proteins are divided into three categories according to their solubility in different solvents: (1) water-soluble non-prolamins, albumins and globulins (ALGL), which play important roles in grain metabolism, development and response to environmental factors [30]; (2) gliadins (GLIA), which are soluble in 70% ethanol at room temperature and (3) glutenins (GLUT), which are soluble in dissolving media [31,32]. Gliadins and glutenins are storage proteins, which decide the baking quality, and they are collectively called gluten proteins. Gliadins are mixtures of single polypeptides and are classified into four subgroups: α-, β-, γ- and ω-GLIA, which can be separated by gel electrophoresis at low pH and by reverse phase high performance liquid chromatography (RP-HPLC) [33], whereas glutenins comprise the subunits that are aggregated together by disulphide bonds which are high (HMW-GS) and low molecular weight glutenin subunits (LMW-GS). Gliadins and glutenins are responsible for different biophysical properties where gliadins decide the dough viscosity while glutenins determine the dough elasticity and strength [32,34].

The concentrations and composition of gluten proteins are dependent on both the genotypic (variety) and the environmental factors (climate, fertilization, soil, etc.) [35], especially the availability of N fertilization [3,36]. Wieser and Seilmeier [31] reported that different N fertilization strongly influenced the quantities of gluten proteins where the effect on gliadins was more pronounced than that on glutenins, whereas albumins and globulins were barely affected. Moreover, the proportions of gluten proteins were changed significantly, where the hydrophilic proteins (ω-GLIA, HMW-GS) were increased by higher N supply and hydrophobic proteins (γ-GLIA, LMW-GS) were decreased. Daniel and Triboi [33] showed similar results that the percentage of proteins and gliadins in the flour increased with the increase of N supply and the proportion of ω-GLIA in total gliadin increased with N while the α- and β-GLIA decreased with N fertilization. In durum wheat, heat and/or drought stress during cultivation affect the grain quality attributes. According to Li et al. [37], drought tends to enhance gluten strength through increased lactic acid retention capacity (LARC) and mixograph peak time (MPT), while heat stress tends to decrease these parameters. Guzmán et al. [38] reported that the concentration of micronutrients (Fe and Zn) and flour yellowness (processing and pasta-making quality) in durum wheat were improved by drought but reduced by heat stress except Zn content, which increased under severe heat stress due to "concentration effect" induced by smaller grain. Moreover, heat stress was found to reduce gluten strength (alveograph energy, W) leading to a weakening effect on grain quality in durum wheat. However, in bread wheat, Hernandez-Espinosa et al. [21] did not detect the absolute weakening dough effect under severe heat stressed condition; both drought and heat stress led to higher gluten strength due to the higher protein content level. Joshi et al. [39] found that Fe and Zn concentrations in wheat grains varied across locations and years and are influenced by higher temperature and soil availability of Zn content in 30–60 cm soil depth. However, the effects of growing conditions are only based on quantitative effects, the influences on the structures, quantities and proportions of flour protein groups and the gluten protein types are largely dependent on the variety [31].

3. The Effects of eCO$_2$ on Wheat Plants and the Mechanisms Behind

Elevated CO_2 enhances plant growth (productivity and total biomass) through promoting net CO_2 assimilation rate (A) and improving water use efficiency (WUE) due to reduced stomatal conductance (g_s) and transpiration in C$_3$ plants (e.g., rice and wheat) and therefore leading to a higher yield [40,41]. Wheat grain yield is mainly derived from photosynthates of leaf, stem and ear during the grain filling stage. The translocation of storage of carbohydrates in stem only contributes 5–10% to final grain weight [42], while flag provides major source of photoassimilate to grain during the period of grain development [43].

According to Wang et al. [4], a meta-analysis of the effects of eCO_2 on wheat physiology and yield showed that eCO_2 (450–800 ppm) significantly increased A by 33% and decreased g_s by 23%, Rubisco total activity by 26% and Rubisco content by 14%, hence increasing grain yield by 24%. However, the yield increase in free-air CO_2 enrichment (FACE) experiments was 44% less than those obtained in non-FACE (i.e., enclosure) facilities. While Broberg et al. [41] reported that wheat grain yield increased by 26% under eCO_2 (605 ppm) on an average, mainly through the increase in grain number. Similar results were found in rice where eCO_2 (627 ppm), on an average, increased rice yields by 23% through increasing grain mass, panicle and grain number, while FACE experiments showed only a 12% increase in rice yield [44]. In addition, Wang et al. [45] reported that eCO_2 enhanced rice yield by 20% on average, however, the yield responses to eCO_2 were smaller under FACE conditions (+16%) compared with other methods including greenhouses (+37%), growth chambers (+24%), and open-top chambers (+20%).

Moreover, the type of cultivars also plays an important role in determining the physiological and yield responses of crops to eCO_2. Lv et al. [46] reported that eCO_2 enhanced rice yields by 13.5%, 22.6% and 32.8% for japonica, indica and hybrid cultivars, respectively, in FACE experiments. In wheat, eCO_2 had significantly different effects on A between the two types of wheat cultivars (i.e., spring and winter wheat) with a higher increase (71%) in spring wheat cultivars and only 23% increase in winter wheat cultivars [47]. In addition to grain crops, eCO_2 was also reported to increase the yield of vegetables, ornamentals, non-agricultural herbaceous species and woody species by 33% on average with a doubling atmospheric CO_2 concentration [48]. According to a meta-analysis conducted by Dong et al. [49], eCO_2 (827 ppm) generally enhanced yield of vegetables by 34% mainly through the increasing number of organs and vegetable mass. However, other environmental factors could modulate the eCO_2 effects on yield responses including temperature, light, water availability and nutrient supply. For example, the eCO_2 yield stimulation in wheat plants was stronger in the regions with low agronomic productivity [41] and the enhancement in A of wheat and rice plants was greater under sufficient nutrient compared to that under lower nutrient supply [4,44]. Therefore, optimizing growth environments are required to maximize the eCO_2-induced yield benefit in agricultural systems, such as additional light (PAR) and high N availability [50,51].

Although eCO_2 can enhance the productivity and yield of agricultural crops, it on the other hand alters and decreases the plant quality, depressing the concentrations of macronutrients (i.e., carbohydrates, protein, and fat) and micronutrients (i.e., minerals, vitamins and phytonutrients) [7,8,52]. Dong et al. [53] reported that eCO_2 increased the concentrations of fructose, glucose and total soluble sugar in the edible part of vegetables but decreased the concentrations of protein, N, magnesium (Mg), iron (Fe) and zinc (Zn) by 9.5%, 18.0%, 9.2%, 16.0% and 9.4%, respectively. A meta-analysis conducted by Taub et al. [54] on several major food crops showed that eCO_2 (540–958 ppm) reduced cereal grain protein concentrations by 10–15% on an average in wheat, barley and rice and reduced tuber protein concentration in potato by 14%. Moreover, for soybean, there was a smaller but statistically significant decrease of protein concentration of 1.4% compared with that grown under ambient CO_2 concentration (aCO_2). In rice, the nutritive value of grains was also negatively affected by eCO_2 under FACE conditions through a decrease in protein by 6% and copper (Cu) content by 20% [55].

It is well known that eCO_2 has direct effect on C and N metabolism in wheat, resulting in changes in the chemical composition in plants [8,56]. In a FACE experiment, Högy et al. [5] reported that eCO_2 leads to an overall reduction in grain protein concentration by 7.4% in spring wheat cultivars, particularly the N- and glutamine-rich gliadin fraction, thus lowering the gluten concentration and reducing baking quality. Along with the lowered protein concentration in grain, the composition of amino acids and their concentrations were also modified under eCO_2, and the size distribution was significantly shifted towards smaller grains [5]. Blandino et al. [57] also suggested a 7% decrease of protein content in four winter wheat cultivars, accompanied by a reduction in dough strength of plants grown under FACE conditions. A three-year field trial in Australia demonstrated that eCO_2 consistently decreased baking quality and grain protein content in wheat, and protein composition changed towards a greater glutenin/gliadin ratio in all years [58]. Moreover, eCO_2 could also reduce concentration of minerals in wheat grain. For instance, a significant decrease in concentrations of Zn and Fe in wheat grain has been reported [6]. Similar reduction in grain S [59], Ca, Fe and Zn [60] concentrations was documented for wheat grown under eCO_2.

There are many different hypotheses explaining these changes of grain quality traits (summarized in Figure 1) and the most frequently mentioned one is the dilution effect, which includes two aspects, biomass dilution and functional dilution [61]. Biomass dilution mainly results from the accumulation of non-structural carbohydrates (NSC) (i.e., soluble sugars, starch, fructans, etc.) which are initial long-term C-storage products of enhanced photosynthesis induced by eCO_2 and therefore reduces the concentration of other constituents [62]. It means that proteins and minerals are diluted by the increased photosynthetic assimilation of carbon, which poses a great threat to future food security because the calories in the food may be sufficient but undernourishment with essential mineral nutrients. For example, eCO_2 greatly increased the ratio of C to N by increasing the starch and total NSC concentrations and decreasing grain N concentration in wheat [63]. However, Taub et al. [54] found that the magnitude of the negative eCO_2 effect on wheat grains was smaller under high soil N conditions than under low soil N with the decrease of grain protein concentrations by 9.8% and 16.4%, respectively, suggesting that the dilution effect may be compensated by a higher nutrient supply. Besides, a relative increase in the synthesis of C-based secondary compounds that are low in N (e.g., lignins, tannins or other polyphenolics) may lead to dilution as well [64], but there was no consistent response [65–67].

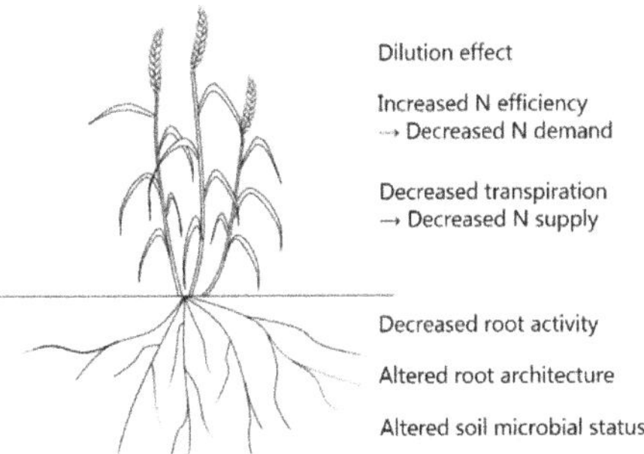

Figure 1. Possible mechanisms explaining the effects of eCO_2 on wheat grain quality.

Functional dilution refers to a decrease in dry mass concentration of N (Nm) due to increased shoot specific activity, which means that N concentration declines due to the accumulation of additional photosynthates by shoots [61]. This theory bases on the functional-balance model:

$$Nm \propto \frac{\text{root mass} \times \text{rate (absorption)}}{\text{leaf mass} \times \text{rate (photosynthesis)}}, \tag{1}$$

where the tissue N concentrations are dependent on the relative activities of roots and shoots and the partitioning of photosynthates is controlled by the relative rates of root absorption of soil nutrients and leaf photosynthesis [68,69]. Functional dilution seems to be pervasive because increased photosynthetic rates for plants grown under eCO_2

is frequently observed if we assume the shoot specific activity equal to photosynthetic rate [70].

Within either biomass dilution or functional dilution, all other mineral elements except C, H and O that are assimilated through photosynthesis should be diluted to a similar ratio. However, there are many heterogeneous responses of different mineral concentrations for crops under eCO_2. For instance, the decrease in Zn concentrations in rice grains under eCO_2 was significantly different from those in Cu, Ca, B and P [6] indicating that dilution is not the only mechanism responsible for decreasing nutrient concentrations in plants under eCO_2 [54,61,71]. Li et al. [72] reported that both the concentrations of K, Ca and Mg in wheat organs and the total accumulations of these elements in the plants were significantly decreased by eCO_2. Moreover, they also found decreases of the concentration of these minerals in the xylem sap, suggesting that the reduced mineral concentrations were not only because of dilution effect but also due to a reduced nutrient acquisition by roots. Within the functional balance concept mentioned above, decreased specific root activity could also play a role in the reduction of nutrient concentrations in plants through a decreased root uptake rate under eCO_2, but the effect is larger for plants grown in soil than in hydroponics [73]. It is reported that the average mean decrease of specific root uptake rate of N (i.e., uptake per unit root mass or length) for plants grown in solid media is 16.4% [74–76]. While the results for root uptake kinetics in solution were quite different and variable [69], but the overall trend was to increase rather than decrease the specific N uptake [77,78]. Therefore, there are some elements influencing the root uptake that are not present in hydroponics. This indicates that the soil microbiomes, rhizosphere conditions and/or root architecture may play a role in nutrient uptake as well.

When it comes to the acquisition of nutrients, there are two aspects: uptake and demand by the plants. On the one hand, eCO_2 could affect the ability of soil-root system to supply N, which refers to source effect. On the other hand, eCO_2 could also increase source use efficiencies, which allows plants to sustain growth in a lower N concentration leading to a lower demand of nutrients (demand effect) [61]. Besides the root specific activity mentioned above, it is widely recognized that the eCO_2-induced decrease in plant mineral uptake associated with the reduced mass flow or diffusion of mineral ions from the soil solution to the root surface due to lower transpiration rate which results from a reduced stomatal conductance under eCO_2. There is evidence that eCO_2 enhances the root growth, which may enable the plants to acquire more nutrients; while a number of studies have also indicated that eCO_2 depresses root hydraulic conductance [79] probably via down-regulating genes encoding aquaporin hereby reducing the mass flow. In addition, it has been proposed that the effect of transpiration rate on mineral uptake is more pronounced with those primarily transported via mass flow (e.g., N, Ca and Mg) than those via diffusion (e.g., P, K and most micronutrients) [80]. This may lead to a shift in the stoichiometry, further affecting nutritive value of the grain [7]. However, to date these possible effects of eCO_2 on root water and mineral uptake have not been fully illustrated.

4. The Effects of Heat Stress on Wheat Grain Quality

Wheat is a temperate crop adapted to temperatures below 30 °C and the threshold temperature during post-anthesis stage is 26 °C [18]. Heat stress (over 35 °C) has a huge impact on both grain yield and grain quality during anthesis and grain-filling phase; however, most of the studies focused on the heat stress effects on grain yield components [18,81,82], only a few studies investigated the impact of heat stress on grain quality traits [17,18,35,83,84]. Increased temperature affects grain development due to the limitation of assimilate supply, grain-filling duration and rate, and starch biosynthesis and deposition [11]. The effects of heat stress on wheat grain quality are summarized in Table 1. The response of wheat to heat stress varied between genotypes and the degree of heat-caused damage depends on the intensity, duration and frequency of heat stress [83,85].

Table 1. The effects of heat stress on different grain quality traits.

Traits	Impacts	References
Gliadin/glutenin ratio	+/−/ns	[17,18]
Grain protein content (GPC)	+	[21,35]
Starch content	−	[86–88]
Dough strength (W)	+/−/ns	[21,37,38]
Gluten extensibility (L)	+	[19,22]
Dough tenacity (P)	−	[22]
Bread loaf volume (LV)	+	[22,38]
Micronutrient concentration (Zn, Fe)	+/−	[38,39]

The "+" sign indicates an increase and the "−" sign indicates a decrease in the performance of the trait of interest.

Although extremely high temperatures have more detrimental impact on grain quality, within moderate or chronic temperature range (15–35 °C), there are also marked changes in grain quality. For example, dough strength (measured by resistance to extension of a dough piece in the Babender Extensograph) was increased with rise in daily average temperature up to about 30 °C; however, when the temperature was above this threshold value to max. 36 °C, even applied for only 3 days, it tended to decrease dough strength. This occurred independent of the timing of the stress, but the degree of reduction varied with different growth stages [89]. In addition, the protein content per grain increases without a change for starch when the temperature rises between 15 and 21 °C during grain-filling stage; however, when temperature reaches 30 °C, the deposition of both protein and starch reduces but with more pronounced decrease in starch than protein [86]. These discrepancies indicate that different growth stages have different threshold and sensitivity to high temperatures. However, the overall higher temperatures generally reduce wheat grain yield through producing smaller grains, reducing grain number and lowering kernel weight [15,90,91], and alter grain quality by increasing GPC but decreasing starch deposition and functional properties of wheat flour [18,35,87], although different wheat genotypes respond differently under heat stress [84,92,93]. For example, Castro et al. [84] evaluated 14 spring wheat genotypes to characterize their response to high temperatures and detected a significant genotype × treatment interaction, suggesting that varieties possess a thermos tolerant response could be used as genetic sources for breeding heat tolerance wheat cultivars.

GPC is the most important characteristic determining wheat grain quality and is in essence determined by the relative rates and durations of protein and starch synthesis [88]. Starch accounting for 65–75% of wheat grain dry weight and over 80% of endosperm weight, which is a decisive factor of grain yield and flour quality [94]. During grain development, starch is deposited into three types of granules differing in size and their formative period in amyloplasts. Large lenticular A-type granules with diameters greater than 15.9 μm are synthesized early during endosperm development; spherical B-type granules with diameters between 5.3 μm and 15.9 μm are produced during mid-development and smaller C-type granules with diameters less than 5.3 μm are initiated late in development [95]. It is reported that high temperatures applied post anthesis reduced the duration of starch accumulation and starch content and modified the size distribution of starch granules with less B-type granules produced in the grain under higher temperatures [88,96]. For the functional properties, starch is composed of two classes of glucose polymers: straight-chained amylose, which is an almost linear α-1,4 glucan molecule comprising 25–30% of grain starch, and highly branched amylopectin, which constitutes 70–75% grain starch [97,98]. Under heat stress during grain-filling stage, the amylose to amylopectin ratio increases, leading to a reduction in dough elasticity [99]. Extension of α-1,4 glucan chains is catalyzed by starch synthases, which are sensitive to heat stress [100], indicating that high temperatures decrease metabolism and enzyme activities involved in starch biosynthesis and reducing the rate of conversion of sucrose to starch [101]. The decreasing starch deposition affects protein concentration by allowing more N per unit of starch [91], leading to smaller grain size, which as a result causing a decrease in milling quality [83].

Under heat stress, the reduction of dough properties of wheat is mainly associated with the reduction of glutenins in gluten proteins especially the HMW-GS [102] that accounts for the genotypic variance in wheat quality [35]. Although there is a general increase in grain protein proportion relative to starch content under high temperatures, due to less temperature sensitivity of N accumulation than starch deposition [88], the protein composition alters towards a poorer flour quality by the following reasons summarized by Blumenthal et al. [102]. First, heat stress decreases synthesis of glutenins therefore leading to the reduction in glutenin/gliadin ratio with gliadin synthesis being maintained or increased [103]. This is explained by the molecular mechanism that there are heat-shock elements (HSE) in the upstream of coding regions of gliadin but not for glutenin [104,105]. However, the effect also depends on genotypes. Stone and Nicolas [18] reported variety difference in glutenin/gliadin ratio in response to heat stress where only one variety (Oxley) showed a decrease in this ratio and Sun 9E-16 increased in this component while the other three varieties had no significant response to high temperature. This suggested that wheat varieties vary in their response of gliadin synthesis to heat stress. The second hypothesis is that heat shocks lower the degree of polymerization of glutenin subunits and reduce the large sized glutenin polymers by altering the formation of disulphide bonds between glutenin peptides therefore weakening the dough properties [106]. In addition, heat-shock proteins (HSP) play an important role in determining the dough-protein function. On one hand, HSP per se is involved in guiding the formation, folding and polymerization of peptides in the grain and can disaggregate and hydrolyze deformed proteins under stress conditions. Therefore, under heat stress, HSP may break off the glutenin synthesis and polymerization thus influencing dough structure [99]. On the other hand, the synthesis of HSP and their prevailing presence in the mature grain could result in the heat-related loss of dough quality. However, according to Blumenthal et al. [102], although the concentration of HSP 70 in mature grain increased under a few days' heat treatment, there was no strong correlation between the amount of HSP 70 and the loss of dough strength. Moreover, they failed to find the presence of HSE in the upstream coding region of the genes for HMW-GS, suggesting that the weakening of dough properties under heat stress may be more relevant to the degree of polymerization of glutenin molecules and the roles of HSP family during grain development.

Besides GPC, heat stress also affects mineral nutrition of wheat grains with a general decline in the concentration of micronutrients, especially Fe, Zn and Mn [38,107], but the results strongly depend on varieties and meteorological factors. For example, Dias et al. [108] reported that under heat stress, Fe concentration in the stems and leaves decreased in bread wheat but increased in durum wheat, and Mn concentration increased significantly in shoot during grain-filling stage for all genotypes except Golia, which is a less heat tolerant genotype. In addition, Kumar et al. [107] reported that under heat stress, the accumulation of Fe and Mn in flag leaf and spike diminished significantly during booting and grain-filling stages, while some of the varieties showed increased accumulation of micronutrients in either flag leaf or spike. Narendra et al. [109] found a high Zn or Fe content in some heat-tolerant varieties under heat stress, suggesting that these genotypes can be used for future breeding to cope with the problem of malnutrition. Furthermore, it is reported that the proportions of gliadins and polymeric protein were less affected by heat stress in the grain with high Zn concentrations, indicating that grain Zn nutrition may interact with grain-filling temperatures and alter protein composition under heat stress [110].

5. The Interactive Effects of eCO$_2$ and Heat Stress on Wheat Plants

So far, there are many studies of plant response to single eCO$_2$ or high temperature stress, but the research on the combined effect of these two factors on wheat plants is quite limited, especially on grain quality traits. Kadam et al. [111] summarized that the eCO$_2$ ×

high temperature interaction strongly depends on the growth and developmental stages of plants. During vegetative stage, heat stress decreases net CO_2 assimilation rates in wheat grown under aCO_2, but eCO_2 moderates the negative effect of heat stress on canopy photosynthesis. This is because increased temperature will decrease the ratio of solubility of CO_2 and of O_2 in water and reduce the specificity of Rubisco for CO_2 relative to O_2 [112], leading to a preference to oxygenation rather than carboxylation in C_3 photosynthesis. Rising CO_2 concentration can inhibit photorespiration and increase ribulose bisphosphate (RuBp) regeneration capacity thus increasing the net photosynthesis. However, the response varies between species. For example, in rice, leaf photosynthesis increased up to 63%, while in sorghum, the increase was not significant [111]. In wheat, Abdelhakim et al. [113] tested the physiological responses of different spring wheat genotypes of heat-tolerance to $eCO_2 \times$ heat interaction and found that under eCO_2, all genotypes showed higher photosynthetic rate and maintained maximum quantum efficiency of PSII under heat stress compared to aCO_2.

Despite the advantage brought by eCO_2 on the vegetative tissue, during the reproductive phase (especially anthesis and grain-filling stage), temperature is the main factor determining the wheat grain yield and quality when exposed to both eCO_2 and heat stress [111]. This is due to the irreversible damage of high temperatures to the anabolic and metabolic processes in wheat flowers and grains, which has been mentioned in other studies only focusing on the heat stress effects [11,25]. Chavan et al. [81] reported that although eCO_2 alleviates the negative impact of heat stress on photosynthesis in wheat, grain yield was reduced equally under aCO_2 and eCO_2 by heat stress due to grain abortion and shortened grain filling duration. Moreover, Macabuhay et al. [114] found an increase in stem water-soluble carbohydrates (WSC) under eCO_2 in both control and heat stressed wheat plants; however, there is limitation of WSC translocation to grains under heat stress, which decreases WSC remobilization. Therefore, the overriding impact of heat stress on crop production seems to limit the potential benefits provided by carbon fertilization of eCO_2. However, there are some positive responses of wheat in the combined situation (reviewed by Kadam et al. [111]) such as increase in grain yield, number of ears and harvest index. While for thousand grain weight, number of spikelets and grain number, the results were inconsistent, indicating a partial ameliorative effect of eCO_2 on grain yield when combined with heat stress. Furthermore, varietal difference, seasonal conditions and different experimental treatment applied also contribute to the inconsistency of the plant response under such situation. Besides, the interaction of weeds such as little seed canary grass (*Phalaris minor*) and common lambsquarters (*Chenopodium murale*) in wheat field could accelerate the yield loss under eCO_2 and thermal scenario due to the similar photosynthetic pathway and nutritional level in weeds and main crop, and the greater response of weeds than crop to eCO_2 condition [115].

Although eCO_2 cannot buffer the negative impact of heat stress on reproductive stage, the reduction of grain N concentration under eCO_2 could be slightly alleviated in heat-treated plants because N accumulation is less sensitive to temperature than C in wheat grain resulting in less carbohydrates relative to N [114]. However, overall, the grain quality seems to decrease in the case of eCO_2 and high temperature combined situation mainly due to the reduction of storage protein under both conditions [33,116], especially the synthesis of GLUT polymers leading to a diminished dough functionality and baking quality. The effect of eCO_2 and heat stress interactions on mineral nutrient composition of wheat grains was rarely investigated. However, a study in rice under field conditions showed a significant decrease in grain mineral content, including Ca, Mg, Cu, Fe, Mn and Zn under combined eCO_2 and heat stress situation across different cultivars. Moreover, some of the micronutrients (Cu, Fe and Zn) were reduced more prominently as compared to eCO_2 alone [117], suggesting that heat stress may further exacerbate the negative effect of eCO_2 on grain mineral nutrition.

6. Conclusions

Although eCO_2 increases wheat grain yield, its negative impact on grain quality poses a great threat to human nutrition. This is mainly due to the dilution effect by higher accumulation of photosynthetic assimilates of C and the decreased root N uptake associated with lower transpiration rate and mass flow under eCO_2. Heat stress generally counteracts the positive effect of eCO_2 on yield components and may aggravate the negative effect of eCO_2 on grain quality because wheat is quite sensitive to high temperature stress especially during anthesis and grain-filling stage, which leads to permanent and irreversible damage during flower and grain development. However, grain quality is strongly dependent on variety and environment, and different quality attributes show diverse responses to abiotic stresses. Since eCO_2 cannot protect wheat plants from high temperature stress, other environmental factors should be taken into consideration, such as enhancing nutrient fertilizing and improving soil water content, etc. Selecting and breeding new genotypes of heat tolerance could be another way to deal with climate change and the increasing demand for food. Moreover, the contemporary crop models for evaluating the effects of different environmental conditions on wheat quality can provide new insights into adaptation strategies to cope with the impacts of climate change on global crop production and grain quality.

Author Contributions: Conceptualization, X.W. and F.L.; writing—original draft preparation, X.W.; writing—review and editing, F.L.; supervision, F.L. All authors have read and agreed to the published version of the manuscript.

Funding: This research received no external funding.

Institutional Review Board Statement: Not applicable.

Informed Consent Statement: Not applicable.

Data Availability Statement: The article does not contain original data.

Conflicts of Interest: The authors declare no conflict of interest.

References

1. Mirza, M.M.Q. Climate change and extreme weather events: Can developing countries adapt? *Clim. Policy* **2003**, *3*, 233–248. [CrossRef]
2. Ray, D.K.; West, P.C.; Clark, M.; Gerber, J.S.; Prishchepov, A.V.; Chatterjee, S. Climate change has likely already affected global food production. *PLoS ONE* **2019**, *14*, e0217148. [CrossRef]
3. Shewry, P.R.; Hey, S.J. The contribution of wheat to human diet and health. *Food Energy Secur.* **2015**, *4*, 178–202. [CrossRef] [PubMed]
4. Wang, L.; Feng, Z.; Schjoerring, J.K. Effects of elevated atmospheric CO_2 on physiology and yield of wheat (*Triticum aestivum* L.): A meta-analytic test of current hypotheses. *Agric. Ecosyst. Environ.* **2013**, *178*, 57–63. [CrossRef]
5. Högy, P.; Wieser, H.; Köhler, P.; Schwadorf, K.; Breuer, J.; Franzaring, J.; Muntifering, R.; Fangmeier, A. Effects of elevated CO_2 on grain yield and quality of wheat: Results from a 3-year free-air CO_2 enrichment experiment. *Plant Biol.* **2009**, *11*, 60–69. [CrossRef] [PubMed]
6. Myers, S.S.; Zanobetti, A.; Kloog, I.; Huybers, P.; Leakey, A.D.B.; Bloom, A.J.; Carlisle, E.; Dietterich, L.H.; Fitzgerald, G.; Hasegawa, T.; et al. Increasing CO_2 threatens human nutrition. *Nature* **2014**, *510*, 139–142. [CrossRef] [PubMed]
7. Loladze, I. Rising atmospheric CO_2 and human nutrition: Toward globally imbalanced plant stoichiometry? *Trends Ecol. Evol.* **2002**, *17*, 457–461. [CrossRef]
8. Loladze, I. Hidden shift of the ionome of plants exposed to elevated CO_2 depletes minerals at the base of human nutrition. *Elife* **2014**, *2014*, 1–29.
9. Cossani, C.M.; Reynolds, M.P. Physiological Traits for Improving Heat Tolerance in Wheat 1 [W]. *Plant Physiol.* **2020**, *160*, 1710–1718. [CrossRef]
10. Melaku, T.A. Environmental Impact on Processing Quality of Wheat Grain. *Int. J. Food Sci. Nutr. Diet.* **2019**, 1–8. [CrossRef]
11. Farooq, M.; Bramley, H.; Palta, J.A.; Siddique, K.H.M.M. Heat stress in wheat during reproductive and grain-filling phases. *Crit. Rev. Plant Sci.* **2011**, *30*, 491–507. [CrossRef]
12. McDonald, G.K.; Sutton, B.G.; Ellison, F.W. The effect of time of sowing on the grain yield of irrigated wheat in the Namoi Valley, New South Wales. *Aust. J. Agric. Res.* **1983**, *34*, 229–240. [CrossRef]
13. Mullarkey, M.; Jones, P. Isolation and analysis of thermotolerant mutants of wheat. *J. Exp. Bot.* **2000**, *51*, 139–146. [CrossRef] [PubMed]

14. Wardlaw, I.F.; Wrigley, C.W. Heat Tolerance in Temperate Cereals: An Overview. *Funct. Plant Biol.* **1994**, *21*, 695. [CrossRef]
15. Saini, H.S.; Aspinall, D. Abnormal sporogenesis in wheat (*Triticum aestivum* L.) induced by short periods of high temperature. *Ann. Bot.* **1982**, *49*, 835–846. [CrossRef]
16. Kaše, M.; Čatský, J. Maintenance and growth components of dark respiration rate in leaves of C3 and C4 plants as affected by leaf temperature. *Biol. Plant.* **1984**, *26*, 461–470. [CrossRef]
17. Blumenthal, C.S.; Barlow, E.W.R.; Wrigley, C.W. Growth environment and wheat quality: The effect of heat stress on dough properties and gluten proteins. *J. Cereal Sci.* **1993**, *18*, 3–21. [CrossRef]
18. Stone, P.; Nicolas, M. Wheat Cultivars Vary Widely in Their Responses of Grain Yield and Quality to Short Periods of Post-Anthesis Heat Stress. *Funct. Plant Biol.* **1994**, *21*, 887. [CrossRef]
19. Fleitas, M.C.; Mondal, S.; Gerard, G.S.; Hernández-Espinosa, N.; Singh, R.P.; Crossa, J.; Guzmán, C. Identification of CIMMYT spring bread wheat germplasm maintaining superior grain yield and quality under heat-stress. *J. Cereal Sci.* **2020**, *93*, 102981. [CrossRef]
20. Williams, R.M.; O'Brien, L.; Eagles, H.A.; Solah, V.A.; Jayasena, V. The influences of genotype, environment, and genotype $\frac{3}{4}$ environment interaction on wheat quality. *Aust. J. Agric. Res.* **2008**, *59*, 95–111. [CrossRef]
21. Hernández-Espinosa, N.; Mondal, S.; Autrique, E.; Gonzalez-Santoyo, H.; Crossa, J.; Huerta-Espino, J.; Singh, R.P.; Guzmán, C. Milling, processing and end-use quality traits of CIMMYT spring bread wheat germplasm under drought and heat stress. *Field Crop. Res.* **2018**, *215*, 104–112. [CrossRef]
22. Li, Y.; Wu, Y.; Hernandez-Espinosa, N.; Peña, R.J. The influence of drought and heat stress on the expression of end-use quality parameters of common wheat. *J. Cereal Sci.* **2013**, *57*, 73–78. [CrossRef]
23. Kimball, B.A.; Pinter, P.J.; Garcia, R.L.; Lamorte, R.L.; Wall, G.W.; Hunsaker, D.J.; Wechsung, G.; Wechsung, F.; Kartschall, T. Productivity and water use of wheat under free-air CO_2 enrichment. *Glob. Chang. Biol.* **1995**, *1*, 429–442. [CrossRef]
24. Morison, J.I.L. Response of plants to CO_2 under water limited conditions. *Vegetatio* **1993**, *104*, 193–209. [CrossRef]
25. Wahid, A.; Gelani, S.; Ashraf, M.; Foolad, M.R. Heat tolerance in plants: An overview. *Environ. Exp. Bot.* **2007**, *61*, 199–223. [CrossRef]
26. Peña, R.J. Wheat for bread and other foods. In *Bread Wheat Improvement and Production*; Food and Agriculture Organization of the United Nations: Rome, Italy, 2002; pp. 483–542.
27. Battenfield, S.D.; Guzmán, C.; Gaynor, R.C.; Singh, R.P.; Peña, R.J.; Dreisigacker, S.; Fritz, A.K.; Poland, J.A. Genomic Selection for Processing and End-Use Quality Traits in the CIMMYT Spring Bread Wheat Breeding Program. *Plant Genome* **2016**, *9*. [CrossRef]
28. Zhao, L.; Zhang, K.P.; Liu, B.; Deng, Z.Y.; Qu, H.L.; Tian, J.C. A comparison of grain protein content QTLs and flour protein content QTLs across environments in cultivated wheat. *Euphytica* **2010**, *174*, 325–335. [CrossRef]
29. Coles, G.D.; Hartunian-Sowa, S.M.; Jamieson, P.D.; Hay, A.J.; Atwell, W.A.; Fulcher, R.G. Environmentally-induced variation in starch and non-starch polysaccharide content in wheat. *J. Cereal Sci.* **1997**, *26*, 47–54. [CrossRef]
30. Arena, S.; D'Ambrosio, C.; Vitale, M.; Mazzeo, F.; Mamone, G.; Di Stasio, L.; Maccaferri, M.; Curci, P.L.; Sonnante, G.; Zambrano, N.; et al. Differential representation of albumins and globulins during grain development in durum wheat and its possible functional consequences. *J. Proteom.* **2017**, *162*, 86–98. [CrossRef] [PubMed]
31. Wieser, H.; Seilmeier, W. The influence of nitrogen fertilisation on quantities and proportions of different protein types in wheat flour. *J. Sci. Food Agric.* **1998**, *76*, 49–55. [CrossRef]
32. Payne, P.I.; Holt, L.M.; Jackson, E.A.; Law, C.N.; Damania, A.B.; Trans, P.; Lond, R.S. Wheat storage proteins: Their genetics and their potential for manipulation by plant breeding. *Philos. Trans. R. Soc. Lond. B Biol. Sci.* **1984**, *304*, 359–371.
33. Daniel, C.; Triboi, E. Effects of temperature and nitrogen nutrition on the grain composition of winter wheat: Effects on gliadin content and composition. *J. Cereal Sci.* **2000**, *32*, 45–56. [CrossRef]
34. Dier, M.; Hüther, L.; Schulze, W.X.; Erbs, M.; Köhler, P.; Weigel, H.J.; Manderscheid, R.; Zörb, C. Elevated Atmospheric CO_2 Concentration Has Limited Effect on Wheat Grain Quality Regardless of Nitrogen Supply. *J. Agric. Food Chem.* **2020**, *68*, 3711–3721. [CrossRef] [PubMed]
35. Uhlen, A.K.; Hafskjold, R.; Kalhovd, A.H.; Sahlström, S.; Longva, Å.; Magnus, E.M. Effects of cultivar and temperature during grain filling on wheat protein content, composition, and dough mixing properties. *Cereal Chem.* **1998**, *75*, 460–465. [CrossRef]
36. Kunkulberga, D.; Linina, A.; Ruza, A. Effect of nitrogen fertilization on protein content and rheological properties of winter wheat wholemeal. *Foodbalt* **2019**, *2019*, 88–92.
37. Li, Y.; Wu, Y.; Hernandez-espinosa, N.; Peña, R.J. Heat and drought stress on durum wheat: Responses of genotypes, yield, and quality parameters. *J. Cereal Sci.* **2013**, *57*, 398–404. [CrossRef]
38. Guzmán, C.; Autrique, J.E.; Mondal, S.; Singh, R.P.; Govindan, V.; Morales-dorantes, A.; Posadas-Romano, G.; Crossa, J.; Ammar, K.; Javier, R.; et al. Response to drought and heat stress on wheat quality, with special emphasis on bread-making quality, in durum wheat. *Field Crop. Res.* **2016**, *186*, 157–165. [CrossRef]
39. Joshi, A.K.; Crossa, J.; Arun, B.; Chand, R.; Trethowan, R.; Vargas, M.; Ortiz-monasterio, I. Genotype x environment interaction for zinc and iron concentration of wheat grain in eastern Gangetic plains of India. *Field Crop. Res.* **2010**, *116*, 268–277. [CrossRef]
40. Long, S.P.; Ainsworth, E.A.; Rogers, A.; Ort, D.R. Rising Atmospheric Carbon Dioxide: Plants FACE the future. *Annu. Rev. Plant Biol.* **2004**, *55*, 591–628. [CrossRef]
41. Broberg, M.C.; Högy, P.; Feng, Z.; Pleijel, H. Effects of elevated CO_2 on wheat yield: Non-linear response and relation to site productivity. *Agronomy* **2019**, *9*, 243. [CrossRef]

42. Wardlaw, I.F.; Porter, H.K. The redistribution of stem sugars in wheat during grain development. *Aust. J. Biol. Sci.* **1967**, *20*, 309–318. [CrossRef]

43. Evans, L.T.; Rawson, H.M. Photosynthesis and respiration by the flag leaf and components of the ear during grain development in wheat. *Aust. J. Biol. Sci.* **1970**, *23*, 245–254. [CrossRef]

44. Ainsworth, E.A. Rice production in a changing climate: A meta-analysis of responses to elevated carbon dioxide and elevated ozone concentration. *Glob. Chang. Biol.* **2008**, *14*, 1642–1650. [CrossRef]

45. Wang, J.; Wang, C.; Chen, N.; Xiong, Z.; Wolfe, D.; Zou, J. Response of rice production to elevated [CO_2] and its interaction with rising temperature or nitrogen supply: A meta-analysis. *Clim. Chang.* **2015**, *130*, 529–543. [CrossRef]

46. Lv, C.; Huang, Y.; Sun, W.; Yu, L.; Zhu, J. Response of rice yield and yield components to elevated [CO_2]: A synthesis of updated data from FACE experiments. *Eur. J. Agron.* **2020**, *112*, 125961. [CrossRef]

47. Barnes, J.D.; Ollerenshaw, J.H.; Whitfield, C.P. Effects of elevated CO_2 and/or O_3 on growth, development and physiology of wheat (*Triticum aestivum* L.). *Glob. Chang. Biol.* **1995**, *1*, 129–142. [CrossRef]

48. Kimball, B.A. Carbon Dioxide and Agricultural Yield: An Assemblage and Analysis of 430 Prior Observations 1. *Agron. J.* **1983**, *75*, 779–788. [CrossRef]

49. Dong, J.; Gruda, N.; Li, X.; Tang, Y.; Zhang, P.; Duan, Z. Sustainable vegetable production under changing climate: The impact of elevated CO_2 on yield of vegetables and the interactions with environments-A review. *J. Clean. Prod.* **2020**, *253*, 119920. [CrossRef]

50. Fierro, A.; Tremblay, N.; Gosselin, A. Supplemental carbon dioxide and light improved tomato and pepper seedling growth and yield. *HortScience* **1994**, *29*, 152–154. [CrossRef]

51. Demmers-Derks, H.; Mitchell, R.A.C.; Mitchell, V.J.; Lawlor, D.W. Response of sugar beet (*Beta vulgaris* L.) yield and biochemical composition to elevated CO_2 and temperature at two nitrogen applications. *Plant Cell Environ.* **1998**, *21*, 829–836. [CrossRef]

52. Chumley, H.; Hewlings, S. The effects of elevated atmospheric carbon dioxide [CO_2] on micronutrient concentration, specifically iron (Fe) and zinc (Zn) in rice; a systematic review. *J. Plant Nutr.* **2020**, *43*, 1571–1578. [CrossRef]

53. Dong, J.; Gruda, N.; Lam, S.K.; Li, X.; Duan, Z. Effects of elevated CO_2 on nutritional quality of vegetables: A review. *Front. Plant Sci.* **2018**, *9*, 1–11. [CrossRef] [PubMed]

54. Taub, D.R.; Miller, B.; Allen, H. Effects of elevated CO_2 on the protein concentration of food crops: A meta-analysis. *Glob. Chang. Biol.* **2008**, *14*, 565–575. [CrossRef]

55. Yang, L.; Wang, Y.; Dong, G.; Gu, H.; Huang, J.; Zhu, J.; Yang, H.; Liu, G.; Han, Y. The impact of free-air CO_2 enrichment (FACE) and nitrogen supply on grain quality of rice. *Field Crop. Res.* **2007**, *102*, 128–140. [CrossRef]

56. Li, X.; Jiang, D.; Liu, F. Dynamics of amino acid carbon and nitrogen and relationship with grain protein in wheat under elevated CO_2 and soil warming. *Environ. Exp. Bot.* **2016**, *132*, 121–129. [CrossRef]

57. Blandino, M.; Badeck, F.W.; Giordano, D.; Marti, A.; Rizza, F.; Scarpino, V.; Vaccino, P. Elevated CO_2 Impact on Common Wheat (*Triticum aestivum* L.) Yield, Wholemeal Quality, and Sanitary Risk. *J. Agric. Food Chem.* **2020**, *68*, 10574–10585. [CrossRef]

58. Panozzo, J.F.; Walker, C.K.; Partington, D.L.; Neumann, N.C.; Tausz, M.; Seneweera, S.; Fitzgerald, G.J. Elevated carbon dioxide changes grain protein concentration and composition and compromises baking quality. A FACE study. *J. Cereal Sci.* **2014**, *60*, 461–470. [CrossRef]

59. Erbs, M.; Manderscheid, R.; Jansen, G.; Seddig, S.; Pacholski, A.; Weigel, H.J. Effects of free-air CO_2 enrichment and nitrogen supply on grain quality parameters and elemental composition of wheat and barley grown in a crop rotation. *Agric. Ecosyst. Environ.* **2010**, *136*, 59–68. [CrossRef]

60. Fernando, N.; Panozzo, J.; Tausz, M.; Norton, R.; Fitzgerald, G.; Seneweera, S. Rising atmospheric CO_2 concentration affects mineral nutrient and protein concentration of wheat grain. *Food Chem.* **2012**, *133*, 1307–1311. [CrossRef]

61. Taub, D.R.; Wang, X. Why are nitrogen concentrations in plant tissues lower under elevated CO_2? A critical examination of the hypotheses. *J. Integr. Plant Biol.* **2008**, *50*, 1365–1374. [CrossRef]

62. Kuehny, J.S.; Peet, M.M.; Nelson, P.V.; Willits, D.H. Nutrient dilution by starch in CO_2-enriched chrysanthemum. *J. Exp. Bot.* **1991**, *42*, 711–716. [CrossRef]

63. Pandey, V.; Sharma, M.; Deeba, F.; Maurya, V.K.; Gupta, S.K.; Singh, S.P.; Mishra, A.; Nautiyal, C.S. Impact of Elevated CO_2 on Wheat Growth and Yield under Free Air CO_2 Enrichment Impact of Elevated CO_2 on Wheat Growth and Yield under Free Air CO_2 Enrichment. *Am. J. Clim. Chang.* **2017**, *6*, 573–596. [CrossRef]

64. Gifford, R.M.; Barrett, D.J.; Lutze, J.L. The effects of elevated [CO_2] on the C:N and C:P mass ratios of plant tissues. *Plant Soil* **2000**, *224*, 1–14. [CrossRef]

65. Henning, F.P.; Wood, C.W.; Rogers, H.H.; Runion, G.B.; Prior, S.A. Composition and decomposition of soybean and sorghum tissues grown under elevated atmospheric carbon dioxide. *J. Environ. Qual.* **1996**, *25*, 822–827. [CrossRef]

66. Chu, C.C.; Field, C.B.; Mooney, H.A.; Chu, C.C.; Field, C.B.; Mooney, H.A. Effects of CO_2 and nutrient enrichment on tissue quality of two California annuals. *Oecologia* **1996**, *107*, 433–440. [CrossRef]

67. Hattenschwiler, S.; Schweingruber, F.H.; Korner, C. Tree ring responses to elevated CO_2 and increased N deposition in Picea abies. *Plant Cell Environ.* **1996**, *19*, 1369–1378. [CrossRef]

68. Davidson, R.L. Effect of root/leaf temperature differentials on root/shoot ratios in some pasture grasses and clover. *Ann. Bot.* **1969**, *33*, 561–569. [CrossRef]

69. BassiriRad, H.; Gutschick, V.P.; Lussenhop, J. Root system adjustments: Regulation of plant nutrient uptake and growth responses to elevated CO_2. *Oecologia* **2001**, *126*, 305–320. [CrossRef]

70. Ellsworth, D.S.; Reich, P.B.; Naumburg, E.S.; Koch, G.W.; Kubiske, M.E.; Smith, S.D. Photosynthesis, carboxylation and leaf nitrogen responses of 16 species to elevated pCO$_2$ across four free-air CO$_2$ enrichment experiments in forest, grassland and desert. *Glob. Chang. Biol.* **2004**, *10*, 2121–2138. [CrossRef]

71. Runion, G.B.; Entry, J.A.; Prior, S.A.; Mitchell, R.J.; Rogers, H.H. Tissue chemistry and carbon allocation in seedlings of Pinus palustris subjected to elevated atmospheric CO$_2$ and water stress. *Tree Physiol.* **1999**, *19*, 329–335. [CrossRef]

72. Li, X.; Jiang, D.; Liu, F. Soil warming enhances the hidden shift of elemental stoichiometry by elevated CO$_2$ in wheat. *Sci. Rep.* **2016**, *6*, 23313. [CrossRef]

73. Poorter, H.; Nagel, O. The role of biomass allocation in the growth response of plants to different levels of light, CO$_2$, nutrients and water: A quantitative review. *Funct. Plant Biol.* **2000**, *27*, 1191. [CrossRef]

74. Bazzaz, F.A. Ecological Society of America effects of c02 and temperature on growth and. *Ecology* **2015**, *73*, 1244–1259.

75. Israel, D.W.; Rufty, T.W., Jr.; Cure, J.D. Nitrogen and phosphorus nutritional interactions in a CO$_2$ enriched environment. *J. Plant Nutr.* **1990**, *13*, 1419–1433. [CrossRef]

76. Zerihun, A.; Gutschick, V.P.; Bassirirad, H. Compensatory roles of nitrogen uptake and photosynthetic N-use efficiency in determining plant growth response to elevated CO$_2$: Evaluation using a functional balance model. *Ann. Bot.* **2000**, *86*, 723–730. [CrossRef]

77. Chu, C.C.; Coleman, J.S.; Mooney, H.A.; Chu, C.C.; Coleman, J.S.; Mooney, H.A. Controls of biomass partitioning between roots and shoots: Atmospheric CO$_2$ enrichment and the acquisition and allocation of carbon and nitrogen in wild radish. *Oecologia* **1992**, *89*, 580–587. [CrossRef] [PubMed]

78. Gloser, V.; Frehner, M.; Luscher, A.; Nosberger, J.; Hartwig, U.A. Does the response of perennial ryegrass to elevated CO$_2$ concentration depend on the from of the supplied nitrogen? *Biol. Plant.* **2002**, *45*, 51–58. [CrossRef]

79. Locke, A.M.; Ort, D.R. Diurnal depression in leaf hydraulic conductance at ambient and elevated [CO$_2$] reveals anisohydric water management in fi eld-grown soybean and possible involvement of aquaporins. *Environ. Exp. Bot.* **2015**, *116*, 39–46. [CrossRef]

80. Mcgrath, J.M.; Lobell, D.B. Reduction of transpiration and altered nutrient allocation contribute to nutrient decline of crops grown in elevated CO$_2$ concentrations. *Plant Cell Environ.* **2013**, *36*, 697–705. [CrossRef]

81. Chavan, S.G.; Duursma, R.A.; Tausz, M.; Ghannoum, O. Elevated CO$_2$ alleviates the negative impact of heat stress on wheat physiology but not on grain yield. *J. Exp. Bot.* **2019**, *70*, 6447–6459. [CrossRef]

82. Yang, J.; Sears, R.G.; Gill, B.S.; Paulsen, G.M. Genotypic differences in utilization of assimilate sources during maturation of wheat under chronic and heat shock stresses: Utilization of assimilate sources by wheat under heat stresses. *Euphytica* **2002**, *125*, 179–188. [CrossRef]

83. Wrigley, C.; Blumenthal, C.; Gras, P.; Barlow, E. Temperature Variation during Grain Filling and Changes in Wheat-Grain Quality. *Funct. Plant Biol.* **1994**, *21*, 875. [CrossRef]

84. Castro, M.; Peterson, C.J.; Rizza, M.D.; Dellavalle, P.D.í.; Vázquez, D.; IbáÑez, V.; Ross, A. Influence of Heat Stress on Wheat Grain Characteristics and Protein Molecular Weight Distribution. *Mol. Breed. Forage Turf* **2007**, *12*, 365–371.

85. Spiertz, J.H.J.; Hamer, R.J.; Xu, H.; Primo-Martin, C.; Don, C.; van der Putten, P.E.L. Heat stress in wheat (*Triticum aestivum* L.): Effects on grain growth and quality traits. *Eur. J. Agron.* **2006**, *25*, 89–95. [CrossRef]

86. Sofield, I.; Evans, L.; Cook, M.; Wardlaw, I. Factors Influencing the Rate and Duration of Grain Filling in Wheat. *Funct. Plant Biol.* **1977**, *4*, 785. [CrossRef]

87. Corbellini, M.; Canevar, M.G.; Mazza, L.; Ciaffi, M.; Lafiandra, D.; Borghi, B. Effect of the duration and intensity of heat shock during grain filling on dry matter and protein accumulation, technological quality and protein composition in bread wheat and durum wheat. *Funct. Plant Biol.* **1997**, *24*, 245–260. [CrossRef]

88. Bhullar, S.; Jenner, C. Differential Responses to High Temperatures of Starch and Nitrogen Accumulation in the Grain of Four Cultivars of Wheat. *Funct. Plant Biol.* **1985**, *12*, 363. [CrossRef]

89. Randall, P.J.; Moss, H.J. Some effects of temperature regime during grain filling on wheat quality. *Aust. J. Agric. Res.* **1990**, *41*, 603–617. [CrossRef]

90. Ferris, R.; Ellis, R.H.; Wheeler, T.R.; Hadley, P. Effect of High Temperature Stress at Anthesis on Grain Yield and Biomass of Field-grown Crops of Wheat. *Ann. Bot.* **1998**, *82*, 631–639. [CrossRef]

91. Stone, P.J.; Nicolas, M.E. The effect of duration of heat stress during grain filling on two wheat varieties differing in heat tolerance: Grain growth and fractional protein accumulation. *Aust. J. Plant Physiol.* **1998**, *25*, 13–20. [CrossRef]

92. Tahir, I.S.A.; Nakata, N. Remobilization of nitrogen and carbohydrate from stems of bread wheat in response to heat stress during grain filling. *J. Agron. Crop Sci.* **2005**, *191*, 106–115. [CrossRef]

93. Viswanathan, C.; Khanna-Chopra, R. Effect of heat stress on grain growth, starch synthesis and protein synthesis in grains of wheat (*Triticum aestivum* L.) varieties differing in grain weight stability. *J. Agron. Crop Sci.* **2001**, *186*, 1–7. [CrossRef]

94. Wardlaw, F.; Sofieldb, I.; Cartwrightc, P.M. Factors Limiting the Rate of Dry Matter Accumulation in the Grain of Wheat Grown at High Temperature. *Funct. Plant Biol.* **1980**, *7*, 387–400. [CrossRef]

95. Bechtel, D.B.; Zayas, I.; Kaleikau, L.; Pomeranz, Y. Size-Distribution of Wheat Starch Granules during Endosperm Development.pdf. *Cereal Chem.* **1990**, *67*, 59–63.

96. Hurkman, W.J.; McCue, K.F.; Altenbach, S.B.; Korn, A.; Tanaka, C.K.; Kothari, K.M.; Johnson, E.L.; Bechtel, D.B.; Wilson, J.D.; Anderson, O.D.; et al. Effect of temperature on expression of genes encoding enzymes for starch biosynthesis in developing wheat endosperm. *Plant Sci.* **2003**, *164*, 873–881. [CrossRef]

97. Ball, S.G.; Van De Wal, M.H.; Visser, R.G.F. Progress in understanding the biosynthesis of amylose. *Trends Plant Sci.* **1998**, *3*, 462–467. [CrossRef]
98. Myers, A.M.; Morell, M.K.; James, M.G.; Ball, S.G. Update on Biochemistry Recent Progress toward Understanding Biosynthesis of the Amylopectin Crystal 1. *Plant Physiol.* **2000**, *122*, 989–997. [CrossRef]
99. Nuttall, J.G.; O'Leary, G.J.; Panozzo, J.F.; Walker, C.K.; Barlow, K.M.; Fitzgerald, G.J. Models of grain quality in wheat—A review. *Field Crop. Res.* **2017**, *202*, 136–145. [CrossRef]
100. Keeling, P.L.; Bacon, P.J.; Holt, D.C. Elevated temperature reduces starch deposition in wheat endosperm by reducing the activity of soluble starch synthase. *Planta* **1993**, *191*, 342–348. [CrossRef]
101. Rijven, A.H.G.C. Heat Inactivation of Starch Synthase in Wheat Endosperm Tissue. *Plant Physiol.* **1986**, *81*, 448–453. [CrossRef] [PubMed]
102. Blumenthal, C.; Stone, P.J.; Gras, P.W.; Bekes, F.; Clarke, B.; Barlow, E.W.R.; Appels, R.; Wrigley, C.W. Heat-shock protein 70 and dough-quality changes resulting from heat stress during grain filling in wheat. *Cereal Chem.* **1998**, *75*, 43–50. [CrossRef]
103. Majoul, T.; Bancel, E.; Triboi, E.; Ben Hamida, J.; Branlard, G. Proteomic analysis of the effect of heat stress on hexaploid wheat grain: Characterization of heat-responsive proteins from total endosperm. *Proteomics* **2003**, *3*, 175–183. [CrossRef]
104. Blumenthal, C.S.; Batey, I.L.; Bekes, F.; Wrigley, C.W.; Barlow, E.W.R. Gliadin genes contain heat-shock elements: Possible relevance to heat-induced changes in grain quality. *J. Cereal Sci.* **1990**, *11*, 185–188. [CrossRef]
105. Blumenthal, C.; Batey, I.; Wrigley, C.; Barlow, E. Involvement of a Novel Peptide in the Heat Shock Response of Australian Wheats. *Funct. Plant Biol.* **1990**, *17*, 441. [CrossRef]
106. Ciaffi, M.; Colapricoz, B.M.G.; De Stefanis, E. Effect of high temperature during grain filling on the amount of insoluble proteins in durum wheat. *J. Genet. Breed.* **1995**, *49*, 285–296.
107. Kumar, S.; Kumar, R.; Kumar, P.; Singh, S.K. Comparative study of Fe and Mn micronutrient accumulation in flag leaf and spike of wheat (*Triticum aestivum* L.) grown under heat stress. *J. Pharmacogn. Phytochem.* **2018**, *7*, 2979–2982.
108. Dias, A.S.; Lidon, F.C.; Ramalho, J.C. IV. Heat stress in triticum: Kinetics of Fe and Mn accumulation. *Braz. J. Plant Physiol.* **2009**, *21*, 153–164. [CrossRef]
109. Narendra, M.C.; Roy, C.; Kumar, S.; Virk, P.; De, N. Effect of terminal heat stress on physiological traits, grain zinc and iron content in wheat (*Triticum aestivum* L.). *Czech J. Genet. Plant Breed.* **2021**, *57*, 43–50.
110. Peck, A.W.; Mcdonald, G.K.Ā.; Graham, R.D. Zinc nutrition influences the protein composition of flour in bread wheat (*Triticum aestivum* L.). *J. Cereal Sci.* **2008**, *47*, 266–274. [CrossRef]
111. Kadam, N.N.; Xiao, G.; Melgar, R.J.; Bahuguna, R.N.; Quinones, C.; Tamilselvan, A.; Prasad, P.V.V.; Jagadish, K.S.V. Agronomic and physiological responses to high temperature, drought, and elevated CO_2 interactions in cereals. *Adv. Agron.* **2014**, *127*, 111–156.
112. LONG, S.P. Modification of the response of photosynthetic productivity to rising temperature by atmospheric CO_2 concentrations: Has its importance been underestimated? *Plant. Cell Environ.* **1991**, *14*, 729–739. [CrossRef]
113. Abdelhakim, L.O.A.; Palma, C.F.F.; Zhou, R.; Wollenweber, B.; Ottosen, C.O.; Rosenqvist, E. The effect of individual and combined drought and heat stress under elevated CO_2 on physiological responses in spring wheat genotypes. *Plant Physiol. Biochem.* **2021**, *162*, 301–314. [CrossRef] [PubMed]
114. Macabuhay, A.; Houshmandfar, A.; Nuttall, J.; Fitzgerald, G.J.; Tausz, M.; Tausz-Posch, S. Can elevated CO_2 buffer the effects of heat waves on wheat in a dryland cropping system? *Environ. Exp. Bot.* **2018**, *155*, 578–588. [CrossRef]
115. Kannojiya, S.; Singh, S.D.; Prasad, S.; Kumar, S.; Malav, L.C.; Kumar, V. Effect of elevated temperature and carbon dioxide on wheat (Triticum aestivum) productivity with and without weed interaction. *Indian J. Agric. Sci.* **2019**, *89*, 751–756.
116. Wieser, H.; Manderscheid, R.; Erbs, M.; Weigel, H.-J. Effects of Elevated Atmospheric CO_2 Concentrations on the Quantitative Protein Composition of Wheat Grain. *Agric. Food Chem.* **2008**, *56*, 6531–6535. [CrossRef] [PubMed]
117. Chaturvedi, A.K.; Bahuguna, R.N.; Pal, M.; Shah, D.; Maurya, S.; Jagadish, K.S.V. Elevated CO_2 and heat stress interactions affect grain yield, quality and mineral nutrient composition in rice under field conditions. *Field Crop. Res.* **2017**, *206*, 149–157. [CrossRef]

Article

Role of Tuber Developmental Processes in Response of Potato to High Temperature and Elevated CO$_2$

Chien-Teh Chen [1] and Tim L. Setter [2,*]

[1] Department of Agronomy, National Chung Hsing University, Taichung 402, Taiwan; ctchen41@dragon.nchu.edu.tw

[2] Section of Soil and Crop Sciences, School of Integrative Plant Science, Cornell University, Ithaca, NY 14853, USA

* Correspondence: TLS1@cornell.edu

Abstract: Potato is adapted to cool environments, and there is concern that its performance may be diminished considerably due to global warming and more frequent episodes of heat stress. Our objectives were to determine the response of potato plants to elevated CO$_2$ (700 µmol/mol) and high temperature (35/25 °C) at tuber initiation and tuber bulking, and to elucidate effects on sink developmental processes. Potato plants were grown in controlled environments with treatments at: Tuber initiation (TI), during the first two weeks after initiating short-day photoperiods, and Tuber bulking (TB). At TI, and 25 °C, elevated CO$_2$ increased tuber growth rate, while leaves and stems were not affected. Whole-plant dry matter accumulation rate, was inhibited by high temperature about twice as much at TI than at TB. Elevated CO$_2$ partially ameliorated high temperature inhibition of sink organs. At TI, with 25 °C, elevated CO$_2$ primarily affected tuber cell proliferation. In contrast, tuber cell volume and endoreduplication were unaffected. These findings indicate that the TI stage and cell division is particularly responsive to elevated CO$_2$ and high temperature stress, supporting the view that attention should be paid to the timing of high-temperature stress episodes with respect to this stage.

Keywords: *Solanum tuberosum*; tuber; sink organ; ambient temperature; cell proliferation

check for **updates**

Citation: Chen, C.-T.; Setter, T.L. Role of Tuber Developmental Processes in Response of Potato to High Temperature and Elevated CO$_2$. *Plants* **2021**, *10*, 871. https://doi.org/10.3390/plants10050871

Academic Editor: James Bunce

Received: 11 April 2021
Accepted: 24 April 2021
Published: 26 April 2021

Publisher's Note: MDPI stays neutral with regard to jurisdictional claims in published maps and institutional affiliations.

1. Introduction

Atmospheric CO$_2$ levels and average global temperature have risen in the last decades and earth air temperature is predicted to continue to increase as a result of the rise in the levels of CO$_2$ and other greenhouse gases [1,2]. While it is recognized that these trends will impact crop production and food security [3], crop growth models of potato (*Solanum tuberosum*) response to environmental variables, suggest that our understanding of the effects of elevated atmospheric CO$_2$ and temperature remains far from complete and in need of improvement [4]. Findings from eight global climate models that were used to simulate potato agronomic and climate responses indicate that model variation was modest for predicted CO$_2$ but was particularly uncertain for temperature [5]. Furthermore, climate predictions indicate that accompanying an increase in global average temperatures of 1.5 °C, will be a sharp rise in the likelihood of extreme heat stress events [6].

Atmospheric CO$_2$ enrichment can affect potato growth and productivity by directly increasing leaf photosynthesis, in accordance with leaf photosynthetic CO$_2$ response, and also by eliciting partial stomatal closing and increasing water use efficiency [7–9]. In addition, altered rates of photosynthate availability can modify partitioning of photosynthate among plant parts, which in turn can have feedback effects on photosynthesis [10–12]. These effects may lessen the potential benefit of elevated CO$_2$ on potato photosynthesis, as has been found in season-long free air CO$_2$ enrichment (FACE) [11,13,14]. In potato, elevated CO$_2$ can lead to greater stimulation of tuber development than above-ground plant parts such that harvest index is increased [15].

Consistent with its Andean origin, potato is a heat-sensitive crop, which performs best at relatively cool temperatures [3]. For example, tuber dry matter partitioning in the temperate-adapted cultivar Katahdin was decreased 65% when day/night temperature was increased from 24/15 °C (max/min) to 30/21 °C [16]. Increases in temperature have the potential of affecting growth and development in several ways. High temperature inhibits photosynthesis in potato, as is generally the case in cool-adapted C3 plants [17]. In support of this explanation, some studies have reported that above-ground temperature has a greater effect than either root or stolon temperature on tuber growth [18]. Moreover, studies to determine the cause of poor yield of heat-susceptible potato genotypes grown at high temperature suggest it is related to inhibited photosynthesis and insufficient availability of the transportable sugar, sucrose [19]. In addition, warm temperature can have direct effects that inhibit tuber initiation and growth [20,21], thereby diverting photosynthate from tubers to shoots [15,22,23]. High temperature in the tuber zone inhibits tuber development [24], and studies have indicated that below-ground high temperature has greater effects than above-ground temperature due to inhibition of tuber development and sink strength, which shifts partitioning to above-ground shoots [25]. Studies have indicated that the negative effects of high temperature on potato growth and yield is dependent on the growth stage and is especially great at the early stages of tuber growth [26].

Potato is adapted to relatively cool growing environments, and it has been predicted that its yield will diminish due to global warming and more frequent episodes of heat stress [3]. It is possible that elevated atmospheric CO_2 concentration, may partially ameliorate negative effects of heat stress by enhancing photosynthesis and lessening negative effects on potato tuber growth. Our objectives were to determine the response of potato plants to elevated CO_2 and high temperature treatment at tuber initiation and tuber bulking, and to elucidate effects on sink developmental processes, including tuber cell division and expansion, levels of sugars, and certain carbohydrate-metabolizing enzymes. Our findings indicate that cell proliferation is particularly responsive to elevated CO_2 and high temperature treatment, and that elevated CO_2 partially ameliorates the effects of high temperature.

2. Results

2.1. The Effects of High Temperature and Elevated CO_2 on Potato Plant Growth

2.1.1. Tuber Initiation Stage

Dry matter accumulation rate (DMAR) during each treatment period was estimated from the difference in dry matter of plants harvested at the onset of a treatment, and those harvested 2 weeks later at the conclusion of the treatment. At the tuber initiation stage, high temperature (35 °C) substantially decreased tuber growth compared to normal temperature (25 °C) in both low (Low_{CO_2}) and elevated ($Elev_{CO_2}$) CO_2 environments (Figure 1). High temperature under Low_{CO_2} decreased tuber DMAR by 82%, and had similar effects (about 60% decrease) on DMAR of leaves and stems. Under $Elev_{CO_2}$, high temperature decreased DMAR somewhat less: by 63% in tubers and by about 25% in leaves and stems. On a whole plant basis (sum of the three plant parts), high temperature decreased DMAR by 56 and 35% in plants at Low_{CO_2} and $Elev_{CO_2}$, respectively (Table 1). In contrast to temperature, elevated CO_2 provided distinctly greater benefit to tubers compared to other plant parts. Elevated CO_2 at 25 °C increased DMAR 60% in tubers, but did not affect DMAR in leaves and stems. At 35 °C, elevated CO_2 appeared to increase DMAR, and thereby partially ameliorate high temperature inhibition, in all three plant parts, but did not achieve statistical significance ($p \leq 0.05$).

2.1.2. Tuber Bulking Stage

At the tuber bulking stage, tuber growth predominated, as there was negligible leaf and stem growth (Figure 2). Elevated CO_2 increased tuber DMAR about 30 to 40% at both temperatures (Figure 2), and this was reflected in whole plant DMAR (the sum of tuber, leaf and stem) (Table 1). Comparing across growth stages, high temperature had greater effects on whole plant DMAR (Table 1) at the tuber initiation stage than at the tuber bulking stage (overall about -45 and -24%, respectively). In contrast, the effects of elevated CO_2 on whole-plant DMAR were fairly similar at the two stages. Elevated CO_2 at 25 °C increased whole plant DMAR by 29% at tuber initiation and by 34% at tuber bulking. Given that whole plant DMAR is a measure of net whole-plant photosynthesis in excess of respiration, this indicates that elevated CO_2 provided consistent benefit throughout these phases of development. The main difference for CO_2 effects across development stages was the case with high temperature during tuber initiation, where elevated CO_2 benefitted leaf and stem growth in addition to tuber growth, whereas only tubers benefitted at the tuber bulking stage. This suggests that growing plant parts have a development window within which they are capable of responding to improved photosynthate supply.

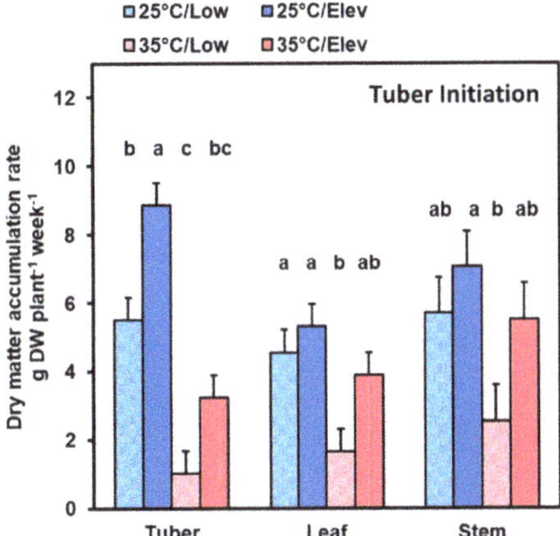

Figure 1. Tuber, leaf, and stem dry matter accumulation rate at the tuber initiation stage in plants that were given temperature (25 or 35 °C) and CO_2 (Low_{CO_2} or $Elev_{CO_2}$) treatments during the 2-week tuber initiation period. To initiate tuberization, plants that were grown in long days were transferred to short-day growth chambers and exposed to treatments for 2 weeks, then harvested. Bars represent averages $\pm$ SEM of 6 replicates; values labeled with different letters are significantly ($p \leq 0.05$).

Table 1. Total plant dry-matter accumulation rate during temperature and CO_2 treatments at the tuber initiation or tuber bulking stage. CO_2 treatments were Low and Elevated (Elev). The pooled standard error of the design (SE) are shown ($n = 6$); values labeled with different letters are significantly ($p \leq 0.05$) different.

Treatment		Stage of Development When Treatment Was Imposed			
temperature	atm CO_2	Tuber Initiation		Tuber Bulking	
				dry-matter accumulation rate	
°C	[CO_2]			g DW plant^{-1} wk^{-1}	
25	Elev	24.3	c	37.7	c
25	Low	18.9	cb	28.0	b
35	Elev	15.8	ba	31.9	cb
35	Low	8.3	a	19.0	a
	SE	1.6		1.8	

2.2. Tuber Numbers and Size

High temperature at the tuber initiation stage substantially decreased the number of tubers per plant (-37% at Low_{CO_2}, and -53% at $Elev_{CO_2}$) (Figure 3). Plants receiving treatments during tuber bulking were exposed to 25 °C and Low CO_2 during their tuber initiation stage, and they did not further increase their tuber numbers during tuber bulking. Hence, there were no treatment effects on tuber numbers at tuber bulking (Figure 3). Elevated CO_2 did not affect tuber number at either stage. Histograms of the distribution of biomass among individual tubers at tuber initiation and tuber bulking, appeared to show that temperature $\times$ CO_2 treatment interactions affected the proportions of tubers in various size classes (Supplementary Figure S1).

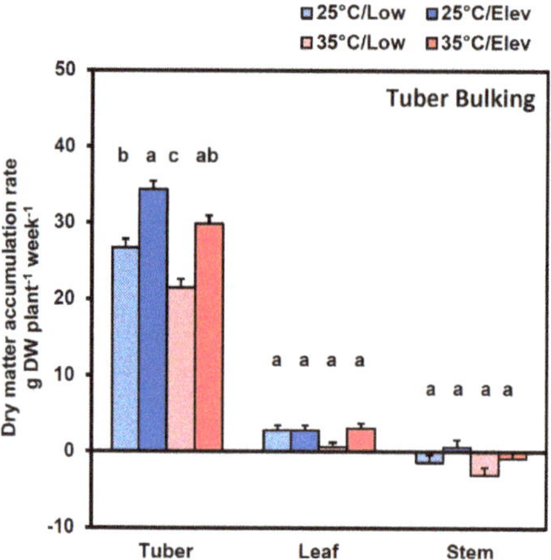

Figure 2. Tuber, leaf, and stem biomass accumulation rate at the tuber bulking stage following temperature (25 or 35 °C) and CO_2 (Low$_{CO_2}$ or Elev$_{CO_2}$) treatments during the 2 week tuber bulking stage. Plants were given two weeks of shortday conditions at 25 °C with low CO_2 to induce tuberization, then maintained at short days and exposed to treatments for 2 weeks, and then harvested. Bars represent averages $\pm$ SEM of 6 replicates; values labeled with different letters are significantly ($p \leq 0.05$) different.

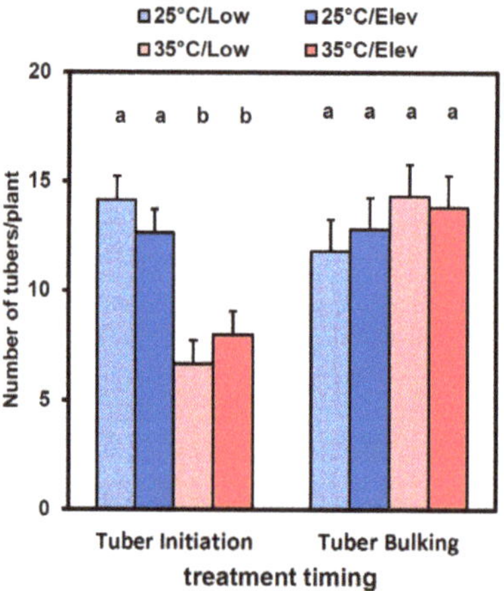

Figure 3. Tuber number in response to temperature (25 or 35 °C) and CO_2 (Low$_{CO_2}$ or Elev$_{CO_2}$) treatments at stages of tuber initiation and tuber bulking. Plants were harvested at 2 weeks after starting short-day exposure (tuber initiation period), or 4 weeks after starting short-day exposure and receiving temperature and CO_2 treatments for the latter 2 weeks of the period (tuber bulking period). Tubers exceeding 1 cm diameter were counted. Bars represent averages $\pm$ SEM of 6 replicates; values labeled with different letters are significantly ($p \leq 0.05$) different.

2.3. The Effects of High Temperature and Elevated CO$_2$ on Potato Tuber Cell Properties

Given that plant growth responses to CO$_2$ and temperature treatments primarily involved changes in tuber growth, we determined the extent to which each aspect of tuber growth was altered. Three aspects were evaluated: (1) cell division, which we evaluated by counting cell nuclei with flow cytometry; (2) cell expansion growth, which we evaluated by calculating average cell volume; and (3) post-mitotic DNA replication (endoreduplication), which we evaluated from the nuclear DNA analysis as the count of nuclei with DNA content $\geq$ 8C (where C is the haploid DNA content). The data in Figures 4 and 5 are shown as a proportion of the Low$_{CO_2}$/25 °C control so that treatment effects on various tuber properties can be readily compared. At tuber initiation (Figure 4), high temperature substantially decreased tuber fresh weight, and correspondingly, total tuber cell number per plant. Elevated CO$_2$ significantly ($p \leq 0.05$) increased cell proliferation at 25 °C, while at 35 °C the number of cells produced was very low and effects of elevated CO$_2$ could not be discerned (Figure 4). In contrast, cell volume was not affected by CO$_2$ and temperature treatments, except at high temperature and elevated CO$_2$, where average cell volume was increased. The extent of endoreduplication, which reflects nuclear DNA reduplication in the absence of intervening mitosis, was not affected by treatments. Similarly, at tuber bulking (Figure 5), elevated CO$_2$ and high temperature affected cell proliferation in some treatment combinations, and at elevated CO$_2$ and temperature cell volume increased; however, the magnitude of treatment effects was less than during tuber initiation.

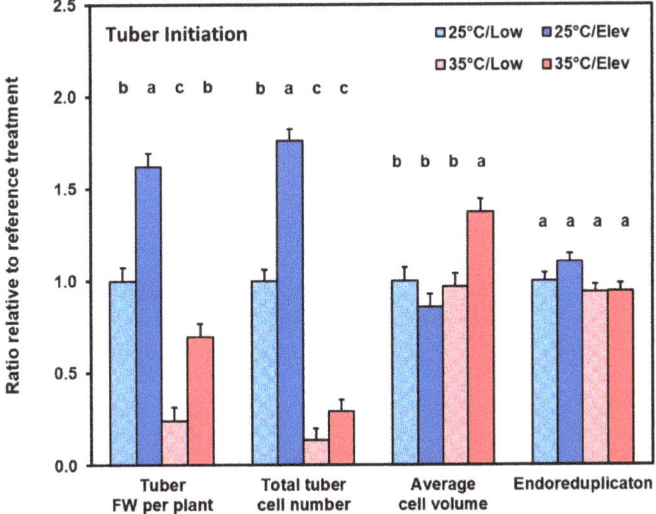

Figure 4. Relative increase in tuber properties with respect to 25 °C/Low$_{CO_2}$ control in response to temperature (25 or 35 °C) or CO$_2$ treatments (Low$_{CO_2}$ or Elev$_{CO_2}$) at the tuber initiation stage. Tuber weight and tuber cell number were summed for each plant and represent the increments of tuber fresh-weight growth and cell proliferation during the treatment periods. Data are expressed relative to the 25 °C/Low$_{CO_2}$ reference which was: 80.5 g tuber fresh weight plant^{-1}, 1.25 × 10^8 tuber cells plant^{-1}, 7.9 × 10^5 μm^3 average tuber cell volume, and 31.3% average proportion of tuber cells in endoreduplication categories ($\geq$ 8C nuclear DNA content). See Materials and Methods for details. Bars represent averages ± SEM of 6 replicates; values labeled with different letters are significantly ($p \leq 0.05$) different.

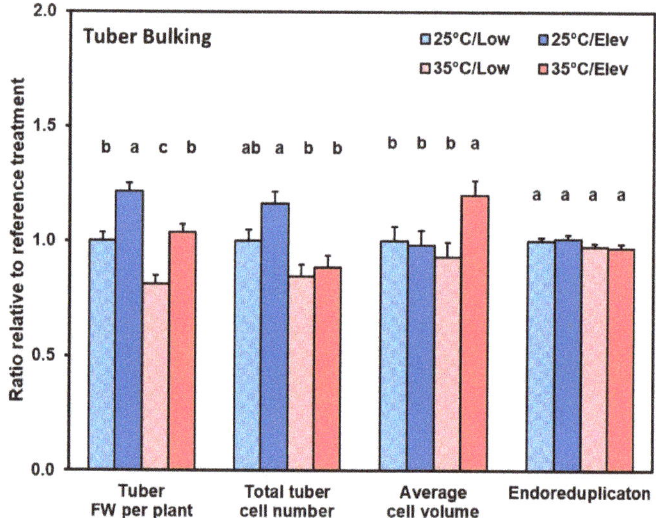

Figure 5. Relative increase in tuber properties with respect to 25 °C/Low$_{CO_2}$ control in response to temperature (25 or 35 °C) or CO_2 treatments (Low$_{CO_2}$ or Elev$_{CO_2}$) at the tuber bulking stage. Tuber weight, and tuber cell number were summed for each plant and represent the increments of tuber fresh-weight growth and cell proliferation during the treatment periods. Data are expressed relative to the 25 °C/Low$_{CO_2}$ reference which was: 529 g tuber fresh weight plant^{-1}, 6.12×10^8 tuber cells plant^{-1}, 1.17×10^6 µm^3 average tuber cell volume, and 48.2% average proportion of tuber cells in endoreduplication categories ($\geq$8C nuclear DNA content). See Materials and Methods for details. Bars represent averages ± SEM of 6 replicates; values labeled with different letters are significantly ($p \leq 0.05$) different.

2.4. The Effects of High Temperature and Elevated CO_2 on Sugar Levels and Invertases

2.4.1. Sugars

Given the evidence that cell proliferation in tubers responded substantially to treatments, we considered the hypothesis that CO_2 and temperature treatments could affect photosynthate supply or consumption in the phloem-rich zone of tubers. Phloem-containing perimedullar-zone tissues were sampled to provide an indication of carbohydrate status in the region of abundant phloem and rapid cell proliferation. At the tuber initiation stage with elevated CO_2, high temperature treatment decreased hexose (glucose + fructose) concentrations (Figure 6a). A decrease in sugar concentration would be expected if high temperature increases the rate of metabolism and use of sugar. However, high temperature increased sucrose concentration at tuber initiation (Figure 6b), suggesting that high temperature did not have a net effect on sugar import/utilization, but might affect conversion of one form to the other. There were no treatment effects on sugars at the tuber bulking stage. Trends in sugar concentrations were toward higher values in elevated CO_2, though the differences were not significant ($p \leq 0.05$). This finding suggests that the conversion of sucrose to hexoses might be affected by the treatments.

2.4.2. Invertases

To evaluate whether treatments affected enzymes for converting imported sucrose into hexoses, we measured cell-wall-bound (CWB) and soluble invertase activities in tubers. Activities of CWB and soluble invertases were analyzed in perimedullary zones where phloem is abundant and is delivering photosynthate to growing tissues. In the elevated CO_2 treatment at tuber initiation, and at both CO_2 treatments at tuber bulking, CWB invertase activity was slightly decreased in response to high temperature (Figure 7a). Though slight, the direction of this inhibition is consistent with it playing a role in the observed increase in sucrose and decrease in hexose in response to high temperature (Figure 6a,b). In contrast, at tuber initiation with low temperature, the elevated CO_2 treatment increased both CWB invertase and soluble invertase activity (Figure 7a,b). The direction of this stimulation of invertase activity is consistent with it playing a role at low temperature for maintaining low sucrose and a tendency for higher hexose in response to elevated CO_2 (Figure 6).

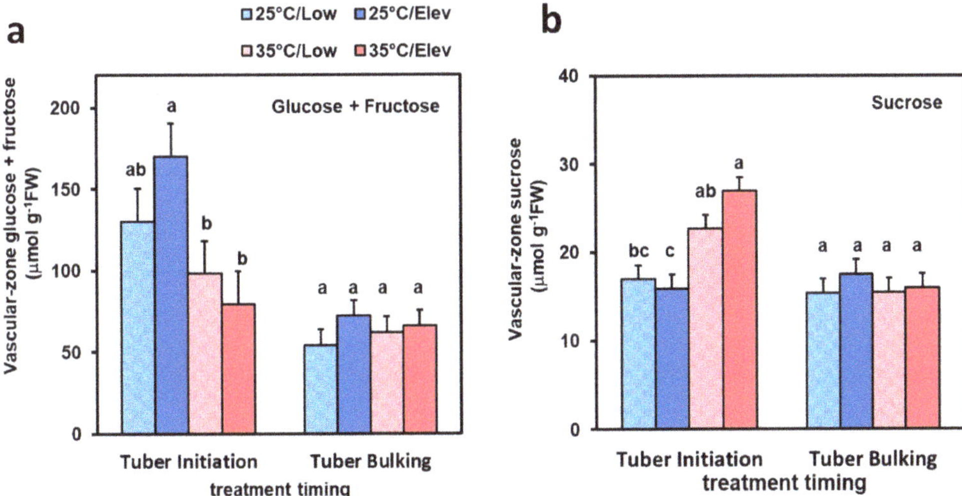

Figure 6. Tuber phloem-rich perimedullary-zone hexose (glucose + fructose) (**a**) and sucrose (**b**) in plants exposed to temperature (25 or 35 °C) or CO_2 (Low$_{CO_2}$ or Elev$_{CO_2}$) treatments at the tuber initiation stage (left of each figure) or tuber bulking stage (right). Plants were harvested after exposure to treatments for 2 weeks. Bars represent averages ± SEM of 6 replicates; values labeled with different letters are significantly ($p \leq 0.05$) different.

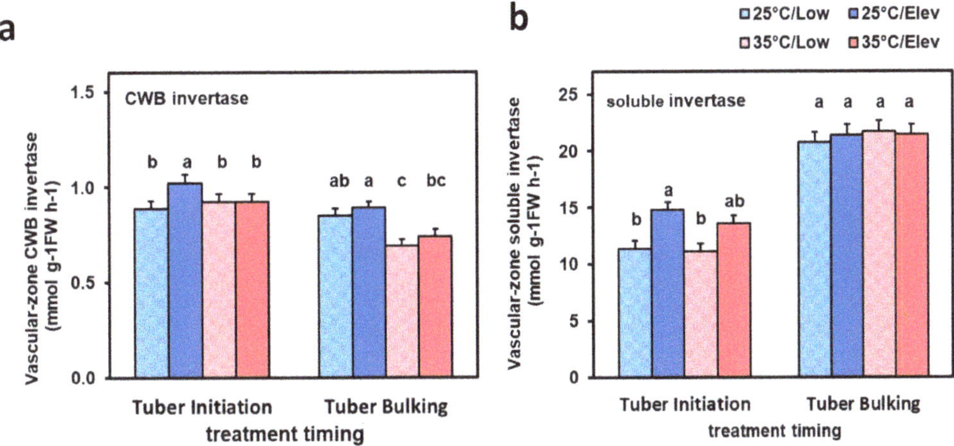

Figure 7. Tuber phloem-rich perimedullary-zone (**a**): cell wall bound (CWB) acid invertase activity and (**b**): soluble acid invertase activity in plants exposed to temperature (25 or 35 °C) or CO_2 (Low$_{CO_2}$ or Elev$_{CO_2}$) treatments at the tuber initiation stage (left of each figure) or tuber bulking stage (right). Plants were harvested after exposure to treatments for 2 weeks. Bars represent averages ± SEM of 6 replicates; values labeled with different letters are significantly ($p \leq 0.05$) different.

3. Discussion

3.1. Elevated Atmospheric CO_2 Effect on Whole Plant Biomass Accumulation

The current findings indicated that two-week treatments involving elevated atmospheric CO_2 concentration at 25 °C increased dry matter accumulation rate by about 30%. This is consistent with previous findings involving short-term elevated CO_2 treatments which have been conducted in growth chambers and controlled environments [27,28]. However, studies involving long-term exposure to elevated CO_2 in open-top and FACE experiments have indicated that such increases in whole-plant photosynthesis are not generally sustained, such that benefit from CO_2 enrichment is typically only about half as much in the long term [11,13,14]. Several factors are involved in this phenomenon, among them partial stomatal closing, limitations in nitrogen supply, and acclimation as plants adjust their development of sink organs in response to an incremental increase in photosynthate availability, and feedback inhibition of photosynthesis [7,9,29]. One of the objectives of the current study was to determine the extent to which potato plants respond to short-term elevated CO_2 by stimulating growth of sink organs: tubers, stems, and expanding leaves. In previous full-season studies of potato exposed to CO_2 enrichment with open-top chambers, elevated CO_2 increased tuber growth and yield, and tubers were preferentially favored, as the partitioning ratio of below-ground to above-ground biomass was increased [15]. Similar findings have been reported for experiments in growth chambers where precise control of the timing of treatments can be achieved [30]. Consistent with this, the current study indicated that elevated CO_2 at 25 °C during tuber initiation increased tuber dry matter accumulation rate (DMAR) by more than 60%, while leaf and stem DMAR was not significantly ($p \leq 0.05$) affected, even though these organs collectively represented almost two thirds of the sink-organ DMAR (Figure 1). At the tuber bulking phase, when most of plant photosynthate is used for starch storage in tubers, leaves had very low DMAR and stems had a net loss in dry matter, so almost all the benefit from elevated CO_2 was obtained by tubers.

3.2. High Temperature Stress Effect on Whole Plant Biomass Accumulation

Potato grows optimally in cool climates (15 to 22 °C) whereas its growth is inhibited in warm or hot environments [10,21,31]. Heat stress can exert detrimental effects on many growth and development processes, so the current studies, involving 2-week treatments at two discrete developmental stages, provide information on the relative susceptibilities of various developmental processes in tubers. Whole-plant dry matter accumulation rate, a measure of whole plant photosynthesis in excess of respiration, was inhibited by high temperature about twice as much at the stage of tuber initiation than at tuber bulking (Table 1; Figures 1 and 2). Similar findings were obtained in studies of elevated temperature under field conditions with temporary transparent chambers in which temperature was increased by about 4 °C for discrete periods of time [32,33]. Moreover, studies that involved treatments with 35 °C stress, similar to the present investigation, showed that tuber growth was strongly decreased by exposure to 35 °C at early phase of tuberization but had a diminishing effect at later stages [26]. In contrast, in a greenhouse study to test the effect of high temperature (30 °C) before tuber initiation versus after tubers had begun bulking, both treatments reduced tuber growth to a similar extent [34]. However, this experiment involved plants that were exposed starting at the date of planting, prior to exposure to tuber-inducing photoperiods, whereas in the current study, plants were first grown with long days, which is a non-inductive photoperiod, then tuber induction was begun at a discrete experimentally determined time. To separate high temperature effects on above-ground and below-ground plant parts, investigators have used growth chambers with root-zone temperature control [25]. After tuber induction, high temperature whole-plant or below-ground treatments both resulted in a reduction in tuber growth and decreased photosynthetic rates [25]. This suggests that communication of root-zone environment or sink-strength status is conveyed to leaves and this may determine the extent to which plant biomass is diminished. Hence, potato response to high temperature appears to depend on differences in sensitivity of various plant parts and of the developmental processes at different stages.

Amelioration of high temperature effects by Elevated CO_2. In the current study, the percentage inhibition of dry matter accumulation rate (DMAR) by high temperature was somewhat greater in Low_{CO_2} than in $Elev_{CO_2}$ treatments (56 and 35%, respectively, at tuber initiation; and 32 and 15%, respectively, at tuber bulking), consistent with earlier findings [33]. In several studies, elevated CO_2 has been found to partly ameliorate high temperature stress [35,36]. Although the 35 °C environment decreased both tuber and whole plant DMAR at both tuber initiation and tuber bulking stages (Figures 1 and 2), the elevated CO_2 treatment (35 °C/$Elev_{CO_2}$) at tuber initiation eased some of the impact of heat stress so that the DMAR of tubers and whole plants recovered to rates comparable to the normal treatment (25 °C/Low_{CO_2}). At tuber bulking, elevated CO_2 eased most of the heat stress, such that tuber DMAR at 35 °C /$Elev_{CO_2}$ was mid-way between 25 °C/Low_{CO_2} and 25 °C/$Elev_{CO_2}$.

3.3. Tuber Growth and Development

3.3.1. High Temperature Effects on Components of Tuber Development

Both tuber cell proliferation and cell expansion (and associated organelle and enzyme development) are needed to develop starch storage capacity for tuber bulking in potato tubers [37]. In previous studies, high temperature suppressed tuber initiation so that tuber numbers per plant were decreased [33,38,39]. Consistent with this, we found that tuber numbers were decreased by high temperature when applied at tuber initiation (Figure 3). However, the lower number of tubers may have been due, in part, to the poorer cell division and expansion growth of tubers at high temperature (Figure 4) such that more of the initiated tubers were smaller than the 1 cm diameter cutoff used for tuber counts. Outcomes similar to this were observed in studies with other treatments that were unfavorable to tuber development and growth, such as shade and long-day photoperiods during tuber bulking [30,40]. Moreover, in the present study photoperiod was switched from 14 to 10 h day length, which provides a very strong tuber inductive signal that may have been sufficient to overcome inhibitory influences due to high temperature [41–44].

3.3.2. Elevated CO_2 Effects on Components of Tuber Development

We obtained data on three components of early tuber development: cell proliferation, cell expansion, and post-mitotic DNA endoreduplication (Figures 3 and 4). At the tuber initiation stage, with 25 °C, elevated CO_2 had a substantial stimulative effect on cell proliferation such that tuber cell numbers in the elevated CO_2 treatment were 62% greater than with the corresponding low CO_2 control (Figure 4). In contrast, cell volume and the proportion of cells that had undergone endoreduplication were unaffected. Cell proliferation continued during the tuber bulking stage and cell numbers responded to $Elev_{CO_2}$ in proportion to tuber weight, while average cell volumes and endoreduplication were not affected (Figure 5). These data are in agreement with previous studies which have shown that cell proliferation is highly responsive to elevated CO_2 while cell expansion growth is less responsive at the early phase of tuber development [30]. Transcriptome studies on potato tubers show that genes associated with cell division are highly upregulated during tuber initiation [45].

3.3.3. Sugar Signaling in Tuber Development

It is plausible that sugar status in tubers serves as a component of signaling pathways for stimulating enhanced growth in the tubers. At both tuber initiation and tuber bulking, high temperature decreased tuber dry matter (Figures 1 and 2) and fresh weight (Figures 4 and 5) growth. The percentage inhibition at each stage was approximately the same as that for respective whole-plant DMAR, so it is plausible that the inhibitory response in tubers involved photosynthate status as a cue. This interpretation is also supported by the approximately similar pattern of CO_2 benefit in tubers compared with that in whole plants in response to the $Elev_{CO_2}$ treatments, although the magnitude of response was greater in tubers than leaves and stems, as discussed above. Studies have identified sugar signaling in potato tubers that regulate patatin storage-protein synthesis [46] and expression of starch pathway enzymes [25,47,48], which are thought to involve sucrose sensing. Consistent with this, when the levels of Trehalose-6-phosphaste (T6P), a sugar-signaling metabolite, was elevated with tuber-specific overexpression of the T6P synthase, or lowered with expression of T6P phosphatase, expression of genes involved in tuber cell proliferation were affected [47]. Studies have shown linkages between sugar signaling pathways and enhanced cell proliferation [49]. In the current study, tuber hexose concentrations tended to increase in the 25 °C/$Elev_{CO_2}$ treatment while sucrose was unaffected (Figure 6). Invertase activities were increased in response to $Elev_{CO_2}$ (Figure 7), consistent with potential involvement in increasing hexose levels and sugar signaling through a hexose signaling pathway [50]. In tomato (*Solanum lycopersicum*) fruit, evidence from gene silencing indicates that cell wall invertase is essential for sugar signaling and development [51]. While sugar signaling has been studied with respect to potato tuber carbon flux and starch synthesis, it merits further attention with respect to regulation of cell proliferation.

Contrary to the general pattern with $Elev_{CO_2}$, in the high temperature treatments, at both tuber initiation and tuber bulking, partial amelioration provided by elevated CO_2 was substantially due to enhanced cell expansion in the high temperature treatments, whereas cell expansion was not involved in the normal temperature treatments (Figures 4 and 5). In studies of the effects of shade on tuber development, the extent of endoreduplication usually increased where average cell volume was higher [40]. However, in the current study, even though high temperature enhanced cell expansion, it did not affect endoreduplication (Figures 4 and 5). Endoreduplication is common in plants, and in the present study a high proportion of cells were endoreduplicated to nuclear DNA contents $\geq$ 8C; they were 35 and 49% at tuber initiation and tuber bulking, respectively. The roles of endoreduplication are still a matter of speculation, though in other plant systems it has been suggested that endoreduplication has a role in stress response [52]. In the present system there was no evidence for an endoreduplication response to $Elev_{CO_2}$ or high temperature. So, the mechanism by which environment affects tuber growth and development may be different from that involved in other systems.

3.4. Global Climate Change

In relation to global climate change, the current study provides insight into potato's sensitivity to various combinations of CO_2 and temperature at two discrete developmental phases. In the field, although elevated CO_2 is a constant condition throughout a crop season, high temperature stress often occurs for relatively short periods of time, such as one or several days. Given the short time over which heat stress episodes commonly occur in the field, it is of interest to determine whether there are particular stages of potato development that are more sensitive to heat stress. Evaluations of simulation models for potato agronomic responses and global climate change indicate that model variation is modest for predicted CO_2 but is particularly uncertain for temperature [5]. This suggests that more knowledge about the nature of heat susceptibility in potato would be valuable in guiding future efforts to improve the crop genetically and to optimize management [12,21,53]. In regard to whole plant DMAR, the current studies indicate that the tuber initiation stage may be somewhat more sensitive to stress than tuber bulking. This may be a consequence of the considerably greater sensitivity of cell proliferation and organelle developmental processes at tuber initiation, whereas starch accumulating cells at tuber bulking may be more robust and less responsive to stress. Moreover, plants have numerous feedbacks in their regulatory systems such that changes in tuber sink capacity can provide feedback regulation of photosynthesis [25].

4. Materials and Methods

4.1. Plant Material

Potato plants (cv. Katahdin) were grown from tuber cuttings in a greenhouse with 14 h supplemental lighting and were watered with nutrient solution (Peter's 15-16-17 fertilizer, W.R. Grace and Co., Fogelsville, PA, USA) which supplied the following elemental nutrients: 33 mg L^{-1} of NO$_3$-N, 86 mg L^{-1} of NH$_4$-N, 42 mg L^{-1} of urea-N, 74 mg L^{-1} of P, 151 mg L^{-1} of K. Pots (6 L) contained peat, vermiculite, and perlite (1:1:1 *v/v/v*) with 3 g of pulverized limestone, 17 g of CaSO$_4$, 21 g of powdered FeSO$_4$, 0.5 g of fritted trace elements (Peters FTE 555, Scotts Co., Marysville, OH, USA), and 1.5 g of wetting agent (AquaGro G, Aquatrols, Pennsauken, NJ, USA). After two weeks, young plants were trimmed to include one shoot and re-sown in new 12 L pots with one plant/pot. After transplanting, plants were grown in the greenhouse for an additional 4 weeks, then the plants were transferred to four matched growth chambers where short days (10 h photoperiod) were imposed to initiate tuber formation, as described below. Nutrient solution and watering regime were the same as in the greenhouse.

On the day that plants were transferred to growth chambers for imposition of temperature × CO$_2$ treatments, plants from a set of uniform material were randomly assigned to the treatments, as described below; concurrently, a subset was designated for immediate harvest at time = 0 for estimating the dry matter of plants parts at the onset of treatments.

4.2. Temperature and CO$_2$ Control

Controlled-environment chambers (Model CEL-63-10, Sherer Inc, Marschall, MI, USA) had interior dimension of 112 × 74 cm (width × depth) and 600 μmol photons (photosynthetically active radiation, 400–700 nm) m^{-2} s^{-1} at the top of the canopy, supplied by fluorescent lamps. Temperature regime was 25/18 °C (day/night) in the normal temperature treatment and 35/25 °C (day/night) in the high temperature treatment. CO$_2$ treatments were either 700 μmol CO$_2$ mol^{-1} air (elevated CO$_2$) or between 350 and 400 μmol CO$_2$ mol^{-1} air (low CO$_2$). Chamber CO$_2$ concentration was monitored with a calibrated infrared gas analyzer [30]. The four growth chambers were used to impose a 2 × 2 matrix of temperature × CO$_2$ treatments at two stages of development: (1) during the first 2 weeks after switch from 14 to 10 h days (weeks after short d; WASD), which stimulates tuber production [hereinafter described as the tuber initiation stage (TI)], or (2) during the next two weeks after switch to short days, when starch accumulation predominates [hereinafter described as the tuber bulking (TB) stage]. Plants for the tuber bulking treatments were first given a short-day (10 h photoperiod) treatment for two weeks in chambers at the normal conditions of 25/18 °C (day/night) and low CO$_2$.

4.3. Tuber Sampling Method

To representatively sample tubers for analyses, all tubers within each plant were characterized into three groups by size, each group containing about one third of the tuber fresh biomass in the entire plant. The grouping started by ranking tubers within a plant by size and, starting with the largest tuber, assigning it to the largest size-category, then continuing to the next smaller tuber until the sum of tuber fresh biomass in the category accounted for about one third of the tuber biomass of the entire plant. In this way, at least two tubers were chosen for this size category. Then tubers were assigned to the next size category until the summed biomass in each of the three categories accounted for one third of the total. In each category, three tubers (or two, depending on availability) were randomly chosen for sugar levels or enzymes activities determination.

4.4. Assay Methods

From representative tubers, two cores, longitudinal and transverse, were immediately withdrawn with a 0.5 cm diameter cylindrical borer and fixed in ethanol/acetic acid (3:1) solution for flow cytometry. Then the apical-bud end (rose end) of the tuber was cut off 1 cm proximal from the apex, and a 0.5 cm thick slice was obtained from the apical end, frozen in liquid nitrogen, and stored at $-20\ ^\circ$C for further determination of sugars and related enzyme activities. Leaves, stems, and remaining portions of tubers were dried at $50\ ^\circ$C to constant weight and weighed. Tuber dry weights were corrected for the fresh weight removed for cell cytometry and enzyme activity.

For flow cytometry analysis, core slices were digested with 0.5% (w/v) pectinase (EC 3.2.1.15, MP Biomedicals, Solon, OH, USA) for 18 h without shaking at $37\ ^\circ$C, then cooled in a $4\ ^\circ$C refrigerator overnight before mechanically disrupting cell walls and releasing nuclei into the medium. After about 18 h of shaking (150 rpm) the tubers were stored in a $4\ ^\circ$C refrigerator then filtered through nylon-mesh fabric with 100 μm openings (Nitex, Teco Inc, Briarcliff Manor, NY, USA). The DNA-binding fluorochrome propidium iodide was added to the filtrate to a final concentration of 90 μM in a 10 mM Tris-HCl buffer (pH 7.4). A 100 μL aliquot of filtrate was counted by flow cytometry using either a FACScan analyzer (Becton Dickinson, Mountain View, CA, USA), or an Epic Profile (Coulter Electronics, Hialeah, FL, USA) operating with an argon-neon laser (488 nm).

For determination of sugars and related enzymes, 0.5 cm diameter disks were cut out from the zone rich in vascular bundles in the perimedullary zone near the perimedulla/cortex interface of slices with a cylindrical borer and then homogenized with 50 mM Hepes-KOH (pH 7.4) buffer (tissue:buffer 1:2(w/v)) containing 5 mM $MgCl_2$, 1 mM EGTA, 1 mM EDTA, 40% (v/v) glycerol, 0.1% bovine serum albumin, 0.5 mM dithiotheritol and 2% insoluble polyvinylpyrrolidone and stored at $-20\ ^\circ$C. Ten microliter aliquots of supernatant were mixed with 90 μL ethanol and stored for sugar assay. The concentrations of glucose, fructose and sucrose were determined using an enzyme-coupled assay based on hexokinase (EC 2.7.1.1, 0.14 unit) and glucose-6-phosphate dehydrogenase (EC 1.1.1.49, 0.07 unit), and using phosphoglucoisomerase (EC 5.3.1.9, 0.2 unit) and invertase (EC 3.2.1.26, 80 units) for fructose and sucrose interconversions [54].

Acid invertase activities were determined in the sucrose hydrolysis direction. For soluble acid invertase, 0.5 mL of buffer-extract was centrifuged at $16,000\times g$ and desalted on a Sephadex G-25M to obtain enzyme extract. For cell-wall-bound acid invertase, 0.5 mL of buffer-extract was centrifuged at $16,000\times g$ and, then, the pellet was resuspended in 1 mL of 1 M NaCl and incubated at $4\ ^\circ$C at least 12 h for salt extraction. After centrifugation at $16,000\times g$, the salt extract was desalted on a Sephadex G-25M to obtained enzyme extract. Both soluble and cell-wall-bond acid invertase activities were determined by mixing enzyme extract (150 μL) with 50 μL of 1 M sucrose to start the enzyme reaction. The reaction was incubated for 75 min at $24\ ^\circ$C and the glucose + fructose produced was assayed as before [54]. All enzyme assays were linear with time and the amount of enzyme extract added.

4.5. Data Analysis

The experiments were conducted in three batches with two plants in each (6 biological replicates). A plant was considered an experimental unit. Each batch contained a complete set of treatments, which were randomly assigned to four matched chambers set at the two CO_2 concentrations and temperatures (a 2×2 set of treatments), as described above. Data were modeled using a simple linear model with batch, temperature treatment (T), CO_2 treatment, and T $\times$ CO_2 interaction as sources of variation. Analysis of variance was conducted in R version 3.6.0 [55] using the *lm* function. Tukey HSD multiple-range significant-difference tests were used for multiple comparisons.

5. Conclusions

The current study elucidated the effects of elevated CO_2 and high temperature on potato tuber development at two stages: tuber initiation and tuber bulking. While whole-plant dry matter accumulation rate was increased by elevated CO_2 about 30% at both stages, the responses of plant parts differed considerably. At the tuber initiation stage, and at normal day-time temperature of $25\ ^\circ$C, elevated CO_2 increased tuber dry matter accumulation rate by more than 60%, while leaf and stem DMAR was not affected, even though these organs collectively represented almost two thirds of the sink-organ DMAR. At the tuber bulking phase, when most of plant photosynthate is used for starch storage in tubers, leaves had very low DMAR and stems had a net loss in dry matter, so almost all the benefit from elevated CO_2 was obtained by tubers. The extent to which high temperature stress affected plant growth and sink development differed considerably between stages. Whole-plant dry matter accumulation rate, was inhibited by high temperature treatment about twice as much at the stage of tuber initiation than at tuber bulking. In contrast to CO_2 effects at 25 °C, where tubers primarily benefited, at high temperature stress, elevated CO_2 benefitted leaf and stem growth in addition to tuber growth, and hence partially ameliorated high temperature inhibition. This suggests that growing plant parts have a development window within which they are capable of responding to improved photosynthate supply. At the tuber initiation stage, with 25 °C, elevated CO_2 had its primary effect on cell proliferation such that tuber cell numbers in the elevated CO_2 treatment were 62% greater than with the corresponding low CO_2 control. In contrast, cell volume and the proportion of cells that had undergone endoreduplication were unaffected. These findings indicate that the tuber initiation stage and cell division process are particularly responsive to elevated CO_2 and high temperature stress. To fully understand impacts of climate change and to develop improved crops, attention should be paid to the timing of high-temperature stress episodes with respect to this stage.

Supplementary Materials: The following are available online at https://www.mdpi.com/article/10.3390/plants10050871/s1, Suppl. Figure S1. Distribution of whole-plant tuber production among individual tuber size-classes in plants exposed to temperature (25 or 35 °C) or CO_2 treatments (Low$_{CO_2}$ or Elev$_{CO_2}$). (a) Tuber initiation stage; (b) tuber bulking stage. Plants were harvested after exposure to treatment for 2 weeks. Tubers exceeding 1 cm diameter were counted. Tuber production was the sum of tuber fresh weight within the defined size-range category. Data was from 6 replicate plants per stage of development.

Author Contributions: C.-T.C. and T.L.S. contributed to all elements of the work. All authors have read and agreed to the published version of the manuscript.

Funding: This research received no external funding.

Data Availability Statement: The data presented in this study are available on request from the corresponding author.

Acknowledgments: We thank the Section of Crop and Soil Sciences, Cornell University, for growth chamber and lab facilities, and the flow cytometry facilities of the Cornell Institute of Biotechnology and the College of Veterinary Medicine.

Conflicts of Interest: The authors declare no conflict of interest.

References

1. IPCC. Global Warming of 1.5 °C. An IPCC Special Report on the Impacts of Global Warming of 1.5 °C above Pre-Industrial Levels and Related Global Greenhouse Gas Emission Pathways, in the Context of Strengthening the Global Response to the Threat of Climate Change, Sustainable Development, and Efforts to Eradicate Poverty. Masson-Delmotte, V., Zhai, P., Pörtner, H.O., Roberts, D., Skea, J., Shukla, P.R., Pirani, A., Moufouma-Okia, W., Péan, C., Pidcock, R., et al., Eds.; 2018. Available online: https://www.ipcc.ch/site/assets/uploads/sites/2/2019/06/SR15_Full_Report_Low_Res.pdf (accessed on 4 April 2021).
2. King, A.W.; Emanuel, W.R.; Post, W.M. Projecting future concentrations of atmospheric carbon dioxide with global carbon cycle models the importance of simulating historical changes. *Environ. Manag.* **1992**, *16*, 91–108. [CrossRef]
3. Jarvis, A.; Ramirez-Villegas, J.; Herrera Campo, B.; Navarro-Racines, C. Is cassava the answer to African climate change adaptation? *Trop. Plant Biol.* **2012**, *5*, 9–29. [CrossRef]
4. Raymundo, R.; Asseng, S.; Prassad, R.; Kleinwechter, U.; Concha, J.; Condori, B.; Bowen, W.; Wolf, J.; Olesen, J.E.; Dong, Q.; et al. Performance of the SUBSTOR-potato model across contrasting growing conditions. *Field Crop. Res.* **2017**, *202*, 57–76. [CrossRef]
5. Fleisher, D.H.; Condori, B.; Quiroz, R.; Alva, A.; Asseng, S.; Barreda, C.; Bindi, M.; Boote, K.J.; Ferrise, R.; Franke, A.C.; et al. A potato model intercomparison across varying climates and productivity levels. *Glob. Chang. Biol.* **2017**, *23*, 1258–1281. [CrossRef]
6. Sun, Q.; Miao, C.; Hanel, M.; Borthwick, A.G.L.; Duan, Q.; Ji, D.; Li, H. Global heat stress on health, wildfires, and agricultural crops under different levels of climate warming. *Environ. Int.* **2019**, *128*, 125–136. [CrossRef]

7. Engineer, C.B.; Hashimoto-Sugimoto, M.; Negi, J.; Israelsson-Nordström, M.; Azoulay-Shemer, T.; Rappel, W.-J.; Iba, K.; Schroeder, J.I. CO_2 Sensing and CO_2 Regulation of Stomatal Conductance: Advances and Open Questions. *Trends Plant Sci.* **2016**, *21*, 16–30. [CrossRef]

8. Drake, B.G.; Gonzalez-Meler, M.A.; Long, S.P. More efficient plants: A consequence of rising atmospheric CO_2? *Annu. Rev. Plant Physiol. Plant Mol. Biol.* **1997**, *48*, 609–639. [CrossRef]

9. Ainsworth, E.A.; Rogers, A. The response of photosynthesis and stomatal conductance to rising [CO_2]: Mechanisms and environmental interactions. *Plant Cell Environ.* **2007**, *30*, 258–270. [CrossRef]

10. Timlin, D.; Rahman, S.M.L.; Baker, J.; Reddy, V.R.; Fleisher, D.; Quebedeaux, B. Whole plant photosynthesis, development, and carbon partitioning in potato as a function of temperature. *Agron. J.* **2006**, *98*, 1195–1203. [CrossRef]

11. Leakey, A.D.B.; Ainsworth, E.A.; Bernacchi, C.J.; Rogers, A.; Long, S.P.; Ort, D.R. Elevated CO_2 effects on plant carbon, nitrogen, and water relations: Six important lessons from FACE. *J. Exp. Bot.* **2009**, *60*, 2859–2876. [CrossRef]

12. Tausz, M.; Tausz-Posch, S.; Norton, R.M.; Fitzgerald, G.J.; Nicolas, M.E.; Seneweera, S. Understanding crop physiology to select breeding targets and improve crop management under increasing atmospheric CO_2 concentrations. *Environ. Exp. Bot.* **2013**, *88*, 71–80. [CrossRef]

13. Long, S.P.; Ainsworth, E.A.; Leakey, A.D.B.; Nosberger, J.; Ort, D.R. Food for Thought: Lower-Than-Expected Crop Yield Stimulation with Rising CO_2 Concentrations. *Science* **2006**, *312*, 1918–1921. [CrossRef]

14. Miglietta, F.; Magliulo, V.; Bindi, M.; Cerio, L.; Vaccari, F.P.; Loduca, V.; Peressotti, A. Free Air CO_2 Enrichment of potato (*Solanum tuberosum* L.): Development, growth and yield. *Glob. Chang. Biol.* **1998**, *4*, 163–172. [CrossRef]

15. Högy, P.; Fangmeier, A. Atmospheric CO_2 enrichment affects potatoes: 1. Aboveground biomass production and tuber yield. *Eur. J. Agron.* **2009**, *30*, 78–84. [CrossRef]

16. Manrique, L.A.; Bartholomew, D.P. Growth and Yield Performance of Potato Grown at Three Elevations in Hawaii: II. Dry Matter Production and Efficiency of Partitioning. *Crop Sci.* **1991**, *31*, 367–372. [CrossRef]

17. Sage, R.F.; Way, D.A.; Kubien, D.S. Rubisco, Rubisco activase, and global climate change. *J. Exp. Bot.* **2008**, *59*, 1581–1595. [CrossRef]

18. Struik, P.C.; Geertsema, J.; Custers, C.-H.M. Effects of shoot; root and stolon temperature on the development of the potato (*Solanum tuberosum* L.) plant. III. Development of tubers. *Potato Res.* **1989**, *32*, 151–158. [CrossRef]

19. Basu, P.S.; Minhas, J.S. Heat tolerance and assimilate transport in different potato genotypes. *J. Exp. Bot.* **1991**, *42*, 861–866. [CrossRef]

20. Kaminski, K.P.; Korup, K.; Nielsen, K.L.; Liu, F.; Topbjerg, H.B.; Kirk, H.G.; Andersen, M.N. Gas-exchange, water use efficiency and yield responses of elite potato (*Solanum tuberosum* L.) cultivars to changes in atmospheric carbon dioxide concentration, temperature and relative humidity. *Agric. For. Meteorol.* **2014**, *187*, 36–45. [CrossRef]

21. Singh, B.; Kukreja, S.; Goutam, U. Impact of heat stress on potato (*Solanum tuberosum* L.): Present scenario and future opportunities. *J. Hortic. Sci* **2020**, *95*, 407–424. [CrossRef]

22. Ewing, E.E. Heat stress and the tuberization stimulus. *Am. Potato J.* **1981**, *58*, 31–49. [CrossRef]

23. Mohabir, G.; John, P. Effect of temperature on starch synthesis in potato tuber tissue and in amyloplasts. *Plant Physiol.* **1989**, *88*, 1222–1228. [CrossRef]

24. Reynolds, M.P.; Ewing, E.E. Effects of high air and soil temperature stress on growth and tuberization in Solanum tuberosum. *Ann. Bot.* **1989**, *64*, 241–248. [CrossRef]

25. Hastilestari, B.R.; Lorenz, J.; Reid, S.; Hofmann, J.; Pscheidt, D.; Sonnewald, U.; Sonnewald, S. Deciphering source and sink responses of potato plants (*Solanum tuberosum* L.) to elevated temperatures. *Plant Cell Environ.* **2018**, *41*, 2600–2616. [CrossRef]

26. Rykaczewska, K. The Effect of High Temperature Occurring in Subsequent Stages of Plant Development on Potato Yield and Tuber Physiological Defects. *Am. J. Potato Res.* **2015**, *92*, 339–349. [CrossRef]

27. Poorter, H. Interspecific variation in the growth response of plants to an elevated ambient CO_2 concentration. *Vegetatio* **1993**, *104*, 77–97. [CrossRef]

28. Cure, J.D.; Acock, B. Crop responses to carbon dioxide doubling: A literature survey. *Agric. For. Meteorol.* **1986**, *38*, 127–145. [CrossRef]

29. Bryant, J.; Taylor, G.; Frehner, M. Photosynthetic acclimation to elevated CO_2 is modified by source: Sink balance in three component species of chalk grassland swards grown in a free air carbon dioxide enrichment (FACE) experiment. *Plant Cell Environ.* **1998**, *21*, 159–168. [CrossRef]

30. Chen, C.-T.; Setter, T.L. Response of potato dry matter assimilation and partitioning to elevated CO_2 at various stages of tuber initiation and growth. *Environ. Exp. Bot.* **2012**, *80*, 27–34. [CrossRef]

31. Midmore, D.J.; Prange, R.K. Growth responses of two Solanum species to contrasting temperatures and irradiance levels relations to photosynthesis dark respiration and chlorophyll fluorescence. *Ann. Bot.* **1992**, *69*, 13–20. [CrossRef]

32. Kim, Y.-U.; Lee, B.-W. Differential Mechanisms of Potato Yield Loss Induced by High Day and Night Temperatures During Tuber Initiation and Bulking: Photosynthesis and Tuber Growth. *Front. Plant Sci.* **2019**, *10*, 300. [CrossRef] [PubMed]

33. Lee, Y.-H.; Sang, W.-G.; Baek, J.-K.; Kim, J.-H.; Shin, P.; Seo, M.-C.; Cho, J.-I. The effect of concurrent elevation in CO_2 and temperature on the growth, photosynthesis, and yield of potato crops. *PLoS ONE* **2020**, *15*, e0241081. [CrossRef] [PubMed]

34. Obiero, C.O.; Milroy, S.P.; Bell, R.W. Importance of whole plant dry matter dynamics for potato (*Solanum tuberosum* L.) tuber yield response to an episode of high temperature. *Environ. Exp. Bot.* **2019**, *162*, 560–571. [CrossRef]

35. Ahmed, F.E.; Hall, A.E.; Madore, M.A. Interactive effects of high temperature and elevated carbon dioxide concentration on cowpea (*Vigna unguiculata* L. Walp.). *Plant Cell Environ.* **1993**, *16*, 835–842. [CrossRef]

36. Wayne, P.M.; Reekie, E.G.; Bazzaz, F.A. Elevated CO_2 ameliorates birch response to high temperature and frost stress: Implications for modeling climate-induced geographic range shifts. *Oecologia* **1998**, *114*, 335–342. [CrossRef]

37. Xu, X.; Vreugdenhil, D.; Lammeren, A.A.M.V. Cell division and cell enlargement during potato tuber formation. *J. Exp. Bot.* **1998**, *49*, 573–582. [CrossRef]

38. Manrique, L.A.; Bartholomew, D.P.; Ewing, E.E. Growth and yield performance of several potato clones grown at three elevations hawaii usa i. Plant morphology. *Crop Sci.* **1989**, *29*, 363–370. [CrossRef]

39. Nooruddin, A.; Mehta, A.N.; Patel, H.R. Tuber production in relation to weather parameters and agrometeorological indices prevailing during different phenological stages of potato crop. *J. Indian Potato Assoc.* **1995**, *22*, 22–117.

40. Chen, C.-T.; Setter, T.L. Response of potato tuber cell division and growth to shade and elevated CO_2. *Ann. Bot.* **2003**, *91*, 373–381. [CrossRef]

41. Dutt, S.; Manjul, A.S.; Raigond, P.; Singh, B.; Siddappa, S.; Bhardwaj, V.; Kawar, P.G.; Patil, V.U.; Kardile, H.B. Key players associated with tuberization in potato: Potential candidates for genetic engineering. *Crit. Rev. Biotechnol.* **2017**, *37*, 942–957. [CrossRef]

42. Jackson, S.D. Multiple signaling pathways control tuber induction in potato. *Plant Physiol.* **1999**, *119*, 1–8. [CrossRef]

43. Van Dam, J.; Kooman, P.L.; Struik, P.C. Effects of temperature and photoperiod on early growth and final number of tubers in potato (*Solanum tuberosum* L.). *Potato Res.* **1996**, *39*, 51–62. [CrossRef]

44. Wallace, D.; Yan, W. A model of photoperiod x temperature interaction effects on plant development. In *Plant Breeding and Whole-System Crop Physiology: Improving Adaptation, Maturity and Yield*; CAB International: New York, NY, USA, 1998; pp. 78–98.

45. Kloosterman, B.; De Koeyer, D.; Griffiths, R.; Flinn, B.; Steuernagel, B.; Scholz, U.; Sonnewald, S.; Sonnewald, U.; Bryan, G.J.; Prat, S.; et al. Genes driving potato tuber initiation and growth: Identification based on transcriptional changes using the POCI array. *Funct. Integr. Genom.* **2008**, *8*, 329–340. [CrossRef]

46. Yoon, J.; Cho, L.-H.; Tun, W.; Jeon, J.-S.; An, G. Sucrose signaling in higher plants. *Plant Sci.* **2021**, *302*, 110703. [CrossRef]

47. Debast, S.; Nunes-Nesi, A.; Hajirezaei, M.R.; Hofmann, J.; Sonnewald, U.; Fernie, A.R.; Börnke, F. Altering Trehalose-6-Phosphate Content in Transgenic Potato Tubers Affects Tuber Growth and Alters Responsiveness to Hormones during Sprouting. *Plant Physiol.* **2011**, *156*, 1754–1771. [CrossRef]

48. Ferreira, S.J.; Sonnewald, U. The mode of sucrose degradation in potato tubers determines the fate of assimilate utilisation. *Front. Plant Sci.* **2012**, *3*, 23. [CrossRef]

49. Wang, L.; Ruan, Y.-L. Regulation of cell division and expansion by sugar and auxin signaling. *Front. Plant Sci.* **2013**, *4*, 163. [CrossRef]

50. Li, L.; Sheen, J. Dynamic and diverse sugar signaling. *Curr. Opin. Plant Biol.* **2016**, *33*, 116–125. [CrossRef]

51. Liao, S.; Wang, L.; Li, J.; Ruan, Y.-L. Cell Wall Invertase Is Essential for Ovule Development through Sugar Signaling Rather Than Provision of Carbon Nutrients. *Plant Physiol.* **2020**, *183*, 1126–1144. [CrossRef]

52. Scholes, D.R.; Paige, K.N. Plasticity in ploidy: A generalized response to stress. *Trends Plant Sci.* **2015**, *20*, 165–175. [CrossRef]

53. Bhargava, S.; Mitra, S. Elevated atmospheric CO_2 and the future of crop plants. *Plant Breed.* **2021**, *140*, 1–11. [CrossRef]

54. Cairns, A.J. Colorimetric microtiter plate assay of glucose and fructose by enzyme-linked formazan production: Applicability to the measurement of fructosyl transferase activity in higher plants. *Anal. Biochem.* **1987**, *167*, 270–278. [CrossRef]

55. R_Core_Team. R: A Language and Environment for Statistical Computing. 2017. Available online: https://www.R-project.org/ (accessed on 22 March 2021).

Article

Elevated Carbon Dioxide and Chronic Warming Together Decrease Nitrogen Uptake Rate, Net Translocation, and Assimilation in Tomato

Dileepa M. Jayawardena [1], Scott A. Heckathorn [1,*], Krishani K. Rajanayake [2], Jennifer K. Boldt [3] and Dragan Isailovic [2]

[1] Department of Environmental Sciences, University of Toledo, Toledo, OH 43606, USA; dileepa.jayawardena@utoledo.edu

[2] Department of Chemistry and Biochemistry, University of Toledo, Toledo, OH 43606, USA; krishani.rajanayake@utoledo.edu (K.K.R.); dragan.isailovic@utoledo.edu (D.I.)

[3] U.S. Department of Agriculture, Agricultural Research Service, Toledo, OH 43606, USA; jennifer.boldt@usda.gov

* Correspondence: scott.heckathorn@utoledo.edu

Abstract: The response of plant N relations to the combination of elevated CO_2 (eCO_2) and warming are poorly understood. To study this, tomato (*Solanum lycopersicum*) plants were grown at 400 or 700 ppm CO_2 and 33/28 or 38/33 °C (day/night), and their soil was labeled with $^{15}NO_3^-$ or $^{15}NH_4^+$. Plant dry mass, root N-uptake rate, root-to-shoot net N translocation, whole-plant N assimilation, and root resource availability (%C, %N, total nonstructural carbohydrates) were measured. Relative to eCO_2 or warming alone, eCO_2 + warming decreased growth, NO_3^- and NH_4^+-uptake rates, root-to-shoot net N translocation, and whole-plant N assimilation. Decreased N assimilation with eCO_2 + warming was driven mostly by inhibition of NO_3^- assimilation, and was not associated with root resource limitations or damage to N-assimilatory proteins. Previously, we showed in tomato that eCO_2 + warming decreases the concentration of N-uptake and -assimilatory proteins in roots, and dramatically increases leaf angle, which decreases whole-plant light capture and, hence, photosynthesis and growth. Thus, decreases in N uptake and assimilation with eCO_2 + warming in tomato are likely due to reduced plant N demand.

Keywords: climate change; elevated CO_2; heat stress; nitrogen assimilation; nitrogen metabolism; nitrogen uptake; *Solanum*; tomato; warming

Citation: Jayawardena, D.M.; Heckathorn, S.A.; Rajanayake, K.K.; Boldt, J.K.; Isailovic, D. Elevated Carbon Dioxide and Chronic Warming Together Decrease Nitrogen Uptake Rate, Net Translocation, and Assimilation in Tomato. *Plants* **2021**, *10*, 722. https://doi.org/10.3390/plants10040722

Academic Editor: James Bunce

Received: 17 March 2021
Accepted: 6 April 2021
Published: 8 April 2021

Publisher's Note: MDPI stays neutral with regard to jurisdictional claims in published maps and institutional affiliations.

1. Introduction

Anthropogenic CO_2 and other greenhouse gas emissions have increased significantly since industrialization, warming the planet. Global climate models predict atmospheric CO_2 concentration will be about 420 to 935 ppm, and global mean surface temperature will increase by about 1.4 to 5.8 °C, by the end of this century [1,2]. This rise in temperature may cause both acute and chronic heat stress in plants, affecting both root and shoot functions [3]. Although CO_2 enrichment alone benefits plants (e.g., increased photosynthesis and water-use efficiency), these beneficial effects may disappear when eCO_2 is compounded with other climate-change variables, such as supra-optimal temperatures [4–7]. For example, the combination of eCO_2 and warming decreased growth and root N-uptake rate in tomato, relative to either factor alone [5]. In addition, eCO_2 alone is not always beneficial to plants, as it can result in a dilution of tissue N concentration (%N) due to increased photosynthetic assimilation of C, resulting in plant tissue of lower nutritional quality for food [8,9].

Since plants procure and assimilate nutrients through nutrient-specific-uptake and -assimilatory proteins, crop improvement under future climate conditions could be achieved by modification of these proteins using transgenic or genetic engineering, or traditional plant breeding approaches, the latter by identifying species or genotypes with better-adapted nutrient-uptake and -assimilation mechanisms [8]. However, in order to implement such efforts, we should identify which biochemical pathway (e.g., uptake vs. translocation vs. assimilation) and which biochemical component (e.g., uptake proteins vs. assimilatory proteins) to target. Nitrogen is often the most limiting nutrient for plants [10]. Plants can procure N either in inorganic or organic forms, but most plants procure the majority of their N primarily as inorganic N (NO_3^- and NH_4^+) [11]. A number of studies have investigated the effects of eCO_2, or warming alone, on plant N uptake, N translocation, and N assimilation, but studies examining the combined effects of eCO_2 and warming on these responses are scarce.

Though eCO_2 tends to decrease root N-uptake rate (N uptake per unit mass or length), past studies show that root N-uptake rate in response to eCO_2 can be highly variable and can depend on the form of N supplied [9,12,13]. Optimum temperature for plant growth and function is species specific, and the effect of warming on root N uptake depends on whether the temperature increase is from suboptimal to optimal or from optimal to supra-optimal. Warming from suboptimal to optimal often increases root N-uptake rate [14–17], while warming from optimal to supra-optimal (acute or chronic) often decreases root N uptake [17–21]. The limited evidence available suggests that the interactive effects of eCO_2 and warming on N uptake can be equivocal. Coleman and Bazzaz [22], using *Abutilon theophrasti*, and Jayawardena et al. [5], using *Solanum lycopersicum*, showed that the interactive effect of eCO_2 plus warming can inhibit root N-uptake rate in C_3 plants when compared with other treatments (i.e., eCO_2 or warming alone). In addition, in the C_4 species, *Amaranthus retroflexus*, N-uptake rates varied with plant ages in response to eCO_2 plus warming [22]. Using ^{15}N labeling, Arndal et al. [23] found no effect of eCO_2 plus warming on NO_3^- nor NH_4^+-uptake rates of *Calluna vulgaris* (an evergreen dwarf shrub) and *Deschampsia flexuosa* (a perennial grass). Dijkstra et al. [24] also found no interactive effect of eCO_2 and warming on NO_3^- uptake of grasses in a semiarid grassland.

Studies that investigated root-to-shoot N translocation in response to eCO_2 showed or suggested a consistent decreasing trend with eCO_2 [25–29]. As Cohen et al. [26] explained, one potential reason for decreased N translocation in response to eCO_2 could be the reduced size of xylem volume when plants are grown at eCO_2. Nitrogen translocation from roots to shoots in response to temperature has been studied in some detail, but results were highly variable. In most studies, the highest temperature examined was 30 °C or less, and the temperature was altered only in the root-zone while maintaining the shoots at a control temperature [15,30–34]. These studies showed that increased root-zone temperature can increase [15,31,32], decrease [32,34], or have no effect [33] on, root-to-shoot N translocation. Moreover, a study conducted by Hungria and Kaschuk [35] showed that whole-plant heat stress (39 vs. 28 or 34 °C) can reduce xylem organic-N translocation in *Phaseolus vulgaris*, while Mainali et al. [21] suggested whole-plant acute heat stress (40 vs. 30 °C) did not affect N translocation in *Andropogon geradii*. As with N-uptake rate, data on root-to-shoot N translocation in response to eCO_2 plus warming are scarce. Rufty et al. [31] studied the interactive effect of root-zone temperature (18, 24, and 30 °C) and eCO_2 (1000 vs. 400 ppm) on N translocation of *Glycine max* supplied NO_3^- as the sole N source, and they noted an increase with eCO_2 plus root warming. In contrast, based on lower transpiration and leaf ^{15}N isotopic composition observed in *Triticum durum* at eCO_2 (700 vs. 400 ppm) plus warming (ambient vs. ambient + 4 °C), Jauregui et al. [4] concluded that eCO_2 plus warming can reduce root-to-shoot N translocation.

Plant N assimilation in response to eCO_2 has been extensively studied. A number of studies have shown that eCO_2 can inhibit shoot NO_3^-, but not NH_4^+, assimilation in C_3 plants [36–39]. However, challenging this view, Andrews et al. [40] showed that eCO_2 does not inhibit NO_3^- assimilation in C_3 plants, and the assimilation of both forms of N take place in a similar way in response to eCO_2. In response, Bloom et al. [41] stated that eCO_2 inhibits shoot NO_3^- assimilation, but enhances NO_3^- assimilation in roots of C_3 plants. This is consistent with results of Jauregui et al. [27], who suggested that eCO_2

favored N assimilation in roots over shoots in *T. durum*, based on the low shoot-to-root NR activity ratio observed at eCO_2. As with N uptake, one could expect N assimilation to increase as temperature rises from suboptimal-to-optimal, and decrease with optimal-to-supra-optimal temperatures, because N assimilation is carried out by enzymes which have temperature optima. The majority of studies that have investigated temperature effects on shoot N assimilation have looked at the effect of heat stress on NR activity only, and they reveal that NR activity diminished with heat stress [35,42–44]. Using two *Agrostis* species, Rachmilevitch et al. [45] investigated the effects of root-zone temperature (37 vs. 20 °C) on the rate of plant NO_3^- assimilation, and noted it decreased with increased temperature. The limited evidence from these studies suggests that optimal to supra-optimal temperature increases are most likely to reduce plant N or NO_3^- assimilation. Nitrogen assimilation is a key process that influences the nutritional quality of food and, recently, researchers have started investigating how it responds to eCO_2 plus warming. To date, three reports showed a decreasing trend for N assimilation in response to eCO_2 plus warming. Vicente et al. [46] investigated the effects of eCO_2 (700 vs. 370 ppm) and warming (ambient vs. ambient + 4 °C) at two levels of N supply on C and N metabolism of *T. durum*, using gene expression analysis. Based on decreased soluble protein, amino acids, and NR activity in flag leaves, they showed that N assimilation can be inhibited by eCO_2 and warming. Since most of the genes involved in N metabolism are post-transcriptionally or post-translationally regulated [10,47], gene expression measurements alone do not necessarily reflect phenotypic effects on N metabolism. A study conducted by Jauregui et al. [4] also reported that eCO_2 (700 vs. 400 ppm) plus warming (ambient + 4 °C) inhibited N assimilation in flag leaves of *T. durum*, based on decreased levels of amino acids, total soluble protein, and NR activity. Root N assimilation in response to eCO_2 (700 vs. 400 ppm) plus chronic warming (37 vs. 30 °C) was indirectly investigated by Jayawardena et al. [5] in *S. lycopersicum*, and they suggested eCO_2 plus warming can inhibit root N assimilation, which could have resulted from the observed decreases in levels of N-assimilatory proteins that were measured in roots. Notably, none of the previous studies investigated whole-plant N assimilation in response to eCO_2 plus warming.

The aforementioned review reveals a lack of studies have investigated the combined influence of eCO_2 and chronic warming on plant N metabolism. Therefore, the objective of this study was to determine the individual and interactive effects of eCO_2 and chronic warming on NO_3^- and NH_4^+ uptake rates, net N translocation, and whole-plant N and NO_3^- assimilation, using tomato (*S. lycopersicum*) as a model. The information resulting from this study will be helpful for crop scientists, plant breeders, and molecular biologists to understand how N metabolism of tomato and other plants will respond to future climates, and how to develop new tomato genotypes with improved N use under future climate conditions.

2. Results

Total plant dry mass was significantly decreased with chronic warming (Table S1, Figure 1). In contrast, the effect of eCO_2 on plant dry mass was dependent on the treatment temperature. Elevated CO_2 significantly and non-significantly increased the plant dry mass of $^{15}NH_4^+$-supplied and $^{15}NO_3^-$-supplied plants at 33 °C, respectively, while it significantly decreased the plant dry mass of both sets of plants at 38 °C. Plants grown at eCO_2 plus chronic warming had the lowest dry mass.

Both NO_3^- and NH_4^+-uptake rates were significantly affected by the interaction of $CO_2 \times$ temperature (Table S1). Though chronic warming significantly increased NO_3^--uptake rate at ambient CO_2 (aCO_2), it did not influence NH_4^+-uptake rate at aCO_2. Elevated CO_2 did not influence either NO_3^- or NH_4^+-uptake rates at 33 °C, but it tended to decrease NO_3^--uptake rate and significantly decreased NH_4^+-uptake rate at 38 °C (relative to 33 °C and aCO_2). Hence, as with total plant dry mass, plants grown at eCO_2 plus chronic warming had the lowest NO_3^- and NH_4^+-uptake rates (Figure 2). Notably, the NH_4^+-uptake rate was greater than the NO_3^--uptake rate at 33 °C, regardless of the CO_2

treatment (approximately $\times 1.4$). However, when the temperature increased from 33 °C to 38 °C, $NO_3{}^-$-uptake rate surpassed the $NH_4{}^+$-uptake rate, regardless of the CO_2 treatment (approximately $\times 1.1$–1.2).

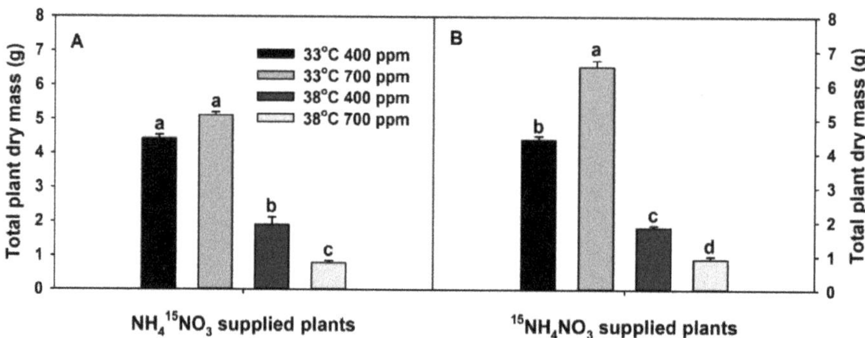

Figure 1. Effects of ambient (400 ppm) vs. elevated (700 ppm) CO_2 and near-optimal (33 °C) vs. chronic warming (38 °C) day-time temperatures on total plant dry mass (g) of *Solanum lycopersicum* labeled for 3 days with 1 mM (**A**) $NH_4{}^{15}NO_3$ or (**B**) $^{15}NH_4NO_3$ and grown for 21 or 23 days, respectively. Each bar represents mean ($n = 5$) + 1 standard error of mean (SEM). Within each panel, bars not sharing the same letters are significantly different ($p < 0.05$, Tukey's test).

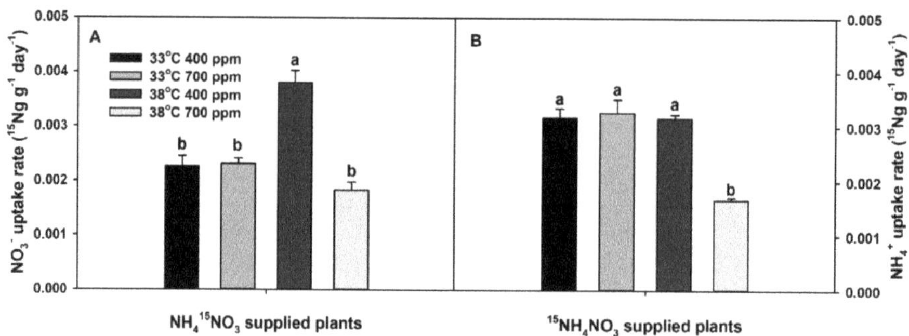

Figure 2. Effects of ambient (400 ppm) vs. elevated (700 ppm) CO_2 and near-optimal (33 °C) vs. chronic warming (38 °C) day-time temperatures on (**A**) $NO_3{}^-$ uptake rate ($^{15}Ng\ g^{-1}\ day^{-1}$) and (**B**) $NH_4{}^+$ uptake rate ($^{15}Ng\ g^{-1}\ day^{-1}$) of *Solanum lycopersicum* labeled for 3 days with 1 mM (**A**) $NH_4{}^{15}NO_3$ or (**B**) $^{15}NH_4NO_3$ and grown for 21 or 23 days, respectively. Each bar represents mean ($n = 5$) + 1 SEM. Within each panel, bars not sharing the same letters are significantly different ($p < 0.05$, Tukey's test).

The ratio of total ^{15}N content in the shoots vs. roots of $^{15}NO_3{}^-$-supplied plants was significantly affected only by the temperature, while that of $^{15}NH_4{}^+$-supplied plants was significantly affected by both temperature and the interaction of $CO_2 \times$ temperature (Table S1). Chronic warming tended to decrease the ratio of total ^{15}N content in the shoots vs. roots of $^{15}NO_3{}^-$-supplied plants at aCO_2. However, it did not influence the ratio of $^{15}NH_4{}^+$-supplied plants at aCO_2. Though eCO_2 tended to increase the ratio of total ^{15}N content in the shoots vs. roots of plants treated with isotopes of both N forms at 33 °C, it significantly or non-significantly decreased the ratio of total ^{15}N content in the shoots vs. roots in both $^{15}NO_3{}^-$ and $^{15}NH_4{}^+$-supplied plants at 38 °C. As with plant dry mass and N-uptake rate, plants grown at eCO_2 plus chronic warming had the lowest ratio of total ^{15}N content in the shoots vs. roots (Figure 3). The ratio of total $NO_3{}^-$ in shoots vs. roots was significantly affected by both individual and interactive effects of CO_2 and temperature, while the ratio of $NH_4{}^+$ in shoots vs. roots was affected only by CO_2 (Figure S1). Elevated CO_2 significantly increased shoot:root $NO_3{}^-$ ratio at 33 °C, but not at 38 °C (Figure S1A). Elevated CO_2 marginally decreased shoot:root NH_4^+ ratio at both temperatures (Figure S1B).

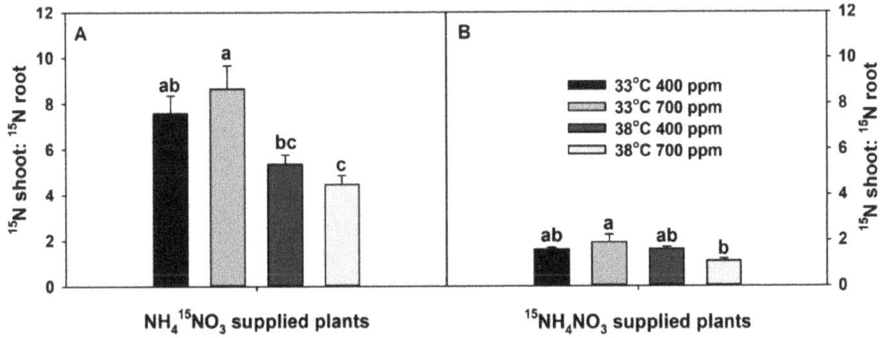

Figure 3. Effects of ambient (400 ppm) vs. elevated (700 ppm) CO_2 and near-optimal (33 °C) vs. chronic warming (38 °C) day-time temperatures on the ratio of total ^{15}N content in the shoots vs. roots of *Solanum lycopersicum* labeled for 3 days with 1 mM (**A**) $NH_4$$^{15}NO_3$ or (**B**) $^{15}NH_4NO_3$ and grown for 21 or 23 days, respectively. Each bar represents mean ($n = 5$) + 1 SEM. Within each panel, bars not sharing the same letters are significantly different ($p < 0.05$, Tukey's test).

The ratio between organic N to total N was significantly affected by both CO_2 and the interaction of $CO_2 \times$ temperature (Table S1). Elevated CO_2 tended to decrease the organic-N:total-N ratio at 33 °C (Figure 4A). In contrast, the effect of chronic warming on the ratio between organic N to total N was dependent on the CO_2 treatment; chronic warming tended to increase the ratio at aCO_2, while it significantly decreased the ratio at eCO_2. Though the ratio between total ^{15}N in plant proteins to total ^{15}N in the plant of $^{15}NO_3{}^-$-supplied plants was not significantly affected by CO_2 and/or temperature (Table S1), it responded in a similar way as the ratio between organic N to total N in response to the independent variables (Figure 4B). Furthermore, plants grown at eCO_2 plus chronic warming had the lowest ratio of ^{15}N in total plant protein to ^{15}N in the total plant. In addition, the ratio between total plant $NO_3{}^-$ to total plant N was greatest in the plants grown at eCO_2 plus chronic warming (Figure 4C). Neither eCO_2 at 33 °C nor chronic warming at aCO_2 affected this ratio significantly, relative to aCO_2 and 33 °C. However, eCO_2 significantly increased the ratio between total plant $NO_3{}^-$ to total plant N at 38 °C.

Root %C was significantly affected by both temperature and CO_2, while root total nonstructural carbohydrate (TNC) was significantly affected by temperature and the interaction of temperature $\times CO_2$ (Table S1). Notably, both root %C and TNC tended to increase with the combination of eCO_2 plus chronic warming, compared with the other treatments (Figure 5A,B). Root %N was significantly higher at 38 vs. 33 °C, while CO_2 had no significant effect on root %N (Figure 5C, Table S1).

In vitro activity (a measure of maximum potential activity) of NR in both leaves and roots was significantly increased at 38 °C vs. 33 °C. In addition, the in vitro activity of GS in roots also significantly increased at 38 °C. However, the in vitro activities of leaf GS and leaf and root GOGAT were unaffected by the higher temperature of 38 °C (Figure 6).

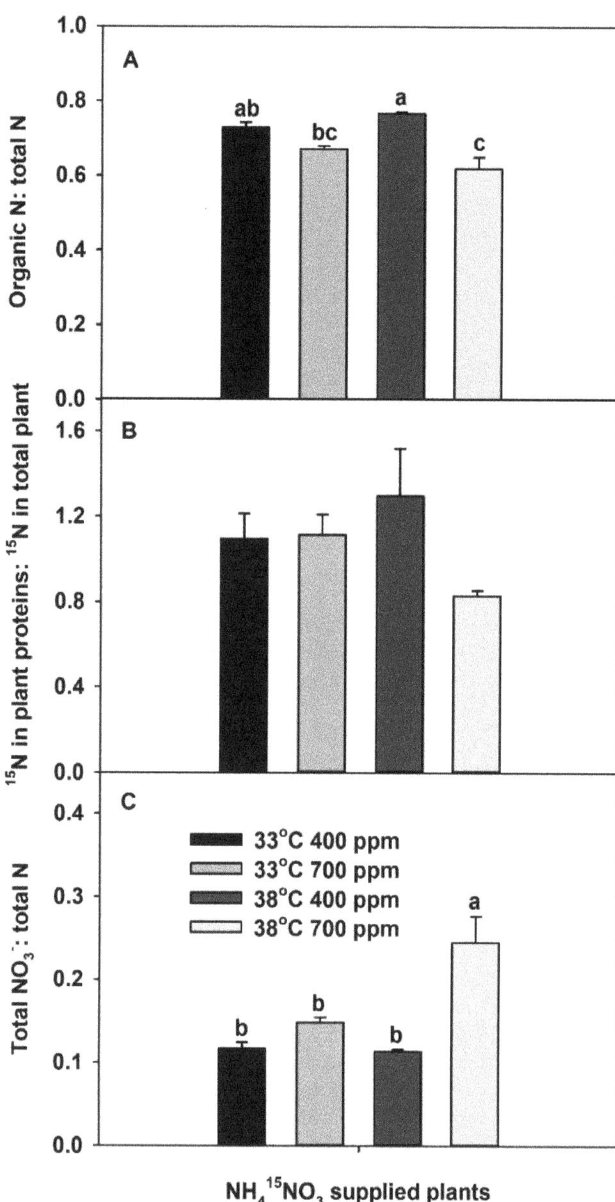

Figure 4. Effects of ambient (400 ppm) vs. elevated (700 ppm) CO_2 and near-optimal (33 °C) vs. chronic warming (38 °C) day-time temperatures on (**A**) organic N: total N, (**B**) ^{15}N in plant proteins: ^{15}N in total plant, and (**C**) total NO$_3^-$: total N ratios of *Solanum lycopersicum*, labeled for 3 days with 1 mM NH$_4$^{15}NO$_3$ and grown for 21 days. Each bar represents mean (*n* = 5) + 1 SEM. Within each panel, bars not sharing the same letters are significantly different (*p* < 0.05, Tukey's test); no letters are shown if there are no differences.

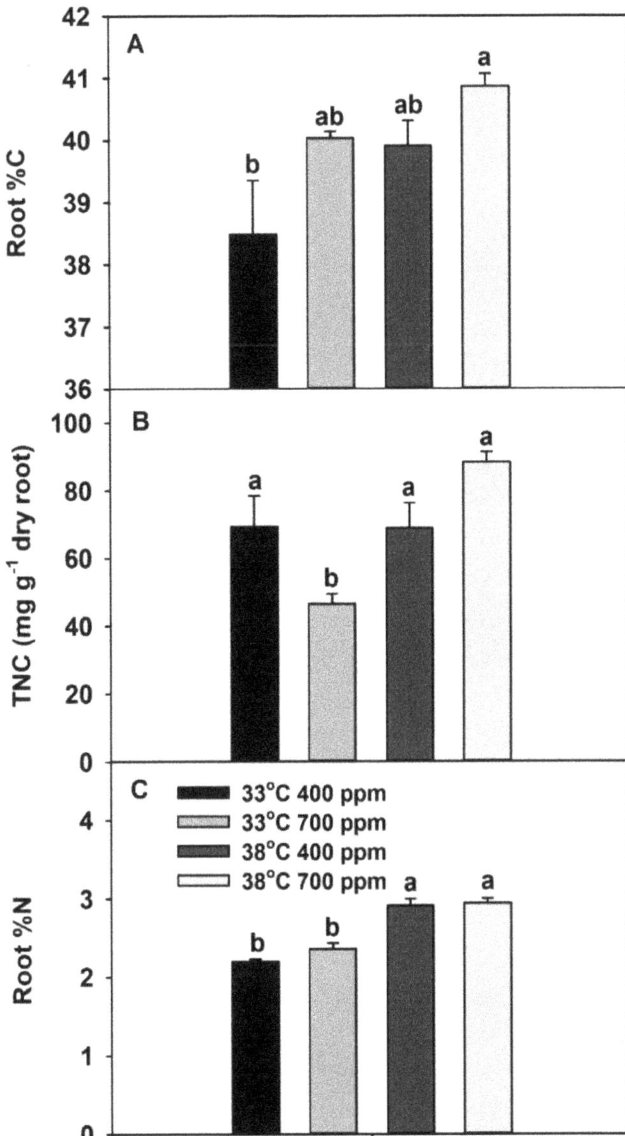

Figure 5. Effects of ambient (400 ppm) vs. elevated (700 ppm) CO_2 and near-optimal (33 °C) vs. chronic warming (38 °C) day-time temperatures on (**A**) root %C, (**B**) total nonstructural carbohydrates (mg g^{-1} dry root mass), and (**C**) root %N of *Solanum lycopersicum*, grown for 21 days. Each bar represents mean (n = 5) + 1 SEM. Within each panel, bars not sharing the same letters are significantly different ($p < 0.05$, Tukey's test).

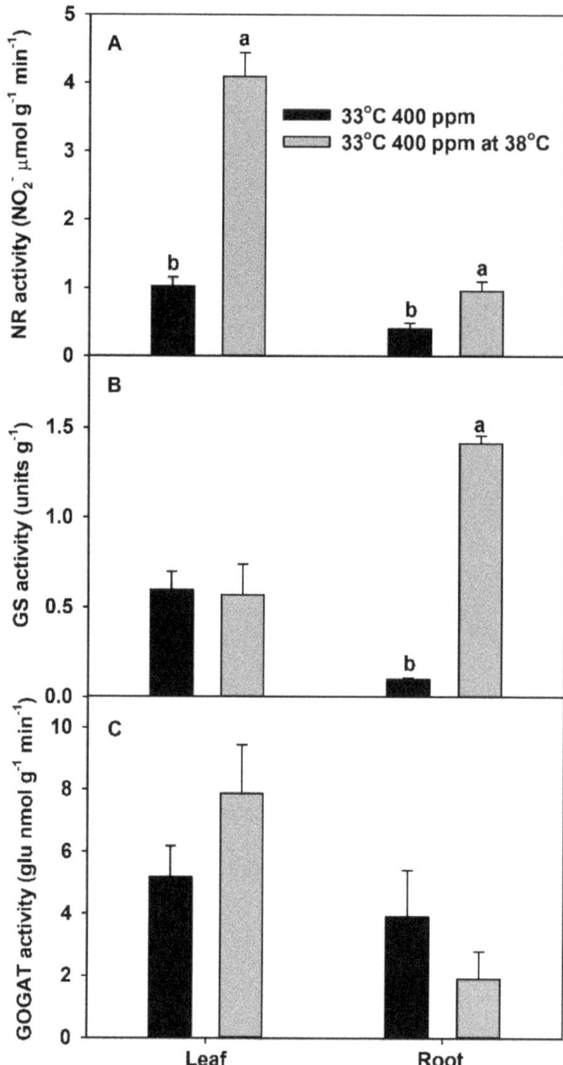

Figure 6. In vitro activities of (**A**) nitrate reductase (NR), (**B**) glutamine synthetase (GS), and (**C**) glutamine oxoglutarate aminotransferase (GOGAT), extracted from leaves or roots of *Solanum lycopersicum* plants grown at 33 °C and 400 ppm and measured at both 33 and 38 °C. Each bar represents mean (n = 4) + 1 SEM. Within each panel, bars not sharing the same letters are significantly different ($p < 0.05$, Tukey's test); no letters are shown if no differences.

3. Discussion

To date, most previous studies have focused on single-factor manipulation approaches when investigating the effects of global environmental changes (e.g., CO_2 enrichment, warming, drought, and N deposition) on plant N relations. However, as we expect these changes to occur concurrently in the future, multifactor manipulation approaches will be necessary to understand the impacts of climate change on plants. The combined effects of CO_2 enrichment and warming on plant N relations (uptake, translocation, and assimilation) were investigated in this study for this reason. As in our previous studies [5,48], the combination of eCO_2 plus chronic warming severely inhibited the growth of tomato, relative to either factor alone. Plants grown at eCO_2 plus chronic warming had the lowest N-uptake rate, net N translocation, and N assimilation when compared with plants grown at eCO_2 or chronic warming alone.

In general, it is expected that eCO_2 will increase plant biomass and supra-optimal temperatures (heat stress) will decrease plant biomass when water and nutrients are not limiting; this general trend was also observed in this study. However, the limited evidence from past studies shows that the interactive effects of eCO_2 plus warming on plant biomass can be highly variable. Some studies have shown eCO_2 plus warming to have a neutral or positive effect on plant growth [49], while others have shown eCO_2 plus warming to have a negative effect [5–7,48]. Our growth reduction can be partly explained by decreased light interception of leaves, and, thus, in situ photosynthesis, caused by a vertical growth orientation of leaves (hyponasty) that occurs when tomato is grown under eCO_2 plus warming [48]. The mechanism for eCO_2 + warming leaf hyponasty is not known, but, so far, appears to be restricted to compound-leaved species and is especially dramatic in tomato and potato. We have examined several tomato genotypes so far (hybrids and heirlooms, all indeterminant), and this hyponasty response occurs in all genotypes (with minor variation) [48].

The effect of eCO_2 on plant N uptake has not been consistent [12]. This is likely partly due to differences among studies in experimental protocols (e.g., length of CO_2 exposure, assaying intact vs. excised roots, differences in N source or amount) and partly due to naturally occurring variations among species [12]. At least two possible mechanisms can explain how eCO_2 can affect N uptake: (1) in the short term, eCO_2 increases plant growth and, hence, plant N demand. This, in turn, increases N-uptake capacity [12], and (2) eCO_2 induces stomatal closure and this decreases the transpiration-driven mass flow of N, which, in turn, decreases N uptake by roots [9,50]. However, in this study, eCO_2 did not affect either NO_3^- or NH_4^+-uptake rates at near-optimal growth temperature, but it did decrease uptake rates of both at 38 °C (and the effect eCO_2 + warming was non-additive). Further, plants grown at eCO_2 plus chronic warming had the lowest rates of NO_3^- and NH_4^+ uptake. Previously, we used sequential harvesting (vs. ^{15}N labeling in this study) to show that the combination of eCO_2 (700 vs. 400 ppm) plus chronic warming (37 vs. 30 °C) can reduce root N-uptake rate of plants treated with either NO_3^- or NH_4^+ singly (not NH_4NO_3 as in this study) [5]. This earlier study further showed that decreases in N-uptake rate may be due, at least in part, to decreased concentration or activity of N-uptake proteins (NRT1 and AMT1). Notably, chronic warming significantly increased NO_3^--uptake rate, but it did not influence NH_4^+-uptake rate at aCO_2. An enhanced rate of NO_3^- uptake with chronic warming could be due to the stimulation of uptake kinetics [12], and the neutral effect of chronic warming on NH_4^+-uptake rate could be to reduce excess accumulation of NH_4^+ and avoid NH_4^+ toxicity [16]. Moreover, Bassirirad [12] showed that the $NH_4^+:NO_3^-$-uptake ratio depends on soil or root temperature, and this ratio decreased as soil or root temperature increased from suboptimal to optimal levels (sufficient data were lacking to examine whether this statement holds true for optimal to supra-optimal temperature rises). Our results indicate this statement holds true even when temperature increases from optimal to supra-optimal levels. At 33 °C, NH_4^+-uptake rate was approximately 1.4 times greater than NO_3^--uptake rate, regardless of the CO_2 treatment. However, when temperature increased from 33 °C to 38 °C, NO_3^--uptake rate increased to approximately 1.1–1.2 times that of NH_4^+-uptake rate, regardless of the CO_2 treatment.

The ratio of total ^{15}N in the shoots vs. the roots was used as a proxy for net N translocation from roots to shoots. Here, we were careful to use the term "net N translocation" instead of "N translocation", as N can continuously circulate between roots and shoots. For example, foliar feeding of leaves with $^{15}NO_3^-$ has shown that leaf NO_3^- can be translocated to every part of the plant, including the root system [51]. Warm-season species, such as tomato, prefer shoot over root NO_3^- assimilation, so most of the soil-derived NO_3^- is translocated from roots to shoots and assimilated there [52]. This is consistent with the high shoot-to-root ^{15}N ratios (>4) for $^{15}NO_3^-$-supplied plants in all four treatments in this study. Since NH_4^+ assimilation generates H^+, and shoots have a limited capacity for proton disposal, nearly all soil-derived NH_4^+ is assimilated in roots [53]. Again, our results (low shoot-to-root ^{15}N ratio of <2) are consistent with limited translocation of NH_4^+ from roots to shoots in all four treatments. Although eCO_2 did not influence NO_3^- or NH_4^+-uptake rates, it tended to increase net N translocation of both $^{15}NO_3^-$ and $^{15}NH_4^+$ in plants at 33 °C. This caused a significant increase in the ratio of total NO_3^- in shoots-to-roots (Figure S1A). The ratio of NH_4^+ content in shoot-to-roots was significantly affected only by CO_2 ($p = 0.0285$), and there was a trend for slightly lower shoot:root NH_4^+ with eCO_2 regardless of the temperature (Figure S1B). Though we did not measure photorespiration in this study, the slight decrease in shoot:root NH_4^+ ratio with eCO_2 could be due to an inhibition of photorespiration by eCO_2. In C_3 plant leaves, NH_4^+ flux from photorespiration is 5 to 10-fold higher than that from NO_3^- reduction [47]. Since limited soil-derived NH_4^+ is typically translocated from roots to shoots, a higher percentage of shoot NH_4^+ is likely represented by NH_4^+ derived from photorespiration. Though chronic warming significantly increased NO_3^--uptake rate, it tended to decrease net N translocation in $^{15}NO_3^-$-supplied plants at aCO_2. This resulted in a low shoot:root NO_3^- ratio (Figure S1A). As with NO_3^- and NH_4^+ uptake rate, the interactive effect of eCO_2 plus chronic warming caused a decrease in net N translocation from roots-to-shoots in both $^{15}NO_3^-$ and $^{15}NH_4^+$-supplied plants. This also resulted in a low shoot:root NO_3^- ratio in plants grown at eCO_2 plus warming (Figure S1A).

The two ratios, total-plant organic N to total N, and total-plant ^{15}N in proteins to total-plant ^{15}N, were used as proxies for whole-plant N assimilation (decreases of these ratios denote inhibition), while the total NO_3^-:total N ratio was used as a proxy for whole-plant NO_3^- assimilation (increases of this ratio denote inhibition). Previous studies showed that eCO_2 can inhibit shoot NO_3^- assimilation and enhance root NO_3^- assimilation [38,41]. In this study, eCO_2 did not influence the ratio of ^{15}N in proteins to total plant, but it decreased the organic N to total N ratio at 33 °C, indicating some inhibition of plant N assimilation. Moreover, total NO_3^-:total N ratio was marginally increased with eCO_2 at 33 °C, which also indicates that eCO_2 may have a tendency to inhibit total-plant NO_3^- assimilation. In contrast, chronic warming may slightly increase both organic N:total N and protein ^{15}N:plant ^{15}N at aCO_2, indicating a possible marginal stimulation of total-plant N assimilation. Chronic warming did not have an effect on total NO_3:total N ratio at aCO_2. However, these plants had the lowest total NH_4^+:total N ratio (data not shown). Therefore, the slight stimulation of N assimilation by chronic warming at aCO_2 could have been due to the stimulation of NH_4^+ rather than NO_3^- assimilation. The combined effect of eCO_2 and chronic warming decreased both organic N:total N (significantly) and plant ^{15}N:total ^{15}N (marginally), indicating an inhibition of total-plant N assimilation. Moreover, eCO_2 plus chronic warming significantly increased total NO_3^-:total N ratio, indicating an inhibition of NO_3^- assimilation by eCO_2 plus chronic warming. Our results suggest the inhibition of whole-plant N assimilation by eCO_2 plus chronic warming was mainly due to inhibition of whole-plant NO_3^- assimilation. However, based on root %N and total-root protein data, Jayawardena et al. [5] suggested that eCO_2 plus chronic warming can inhibit root N assimilation in plants provided only NO_3^- or only NH_4^+. They further showed that, when plants were grown at eCO_2 plus warming, the roots had decreased levels of N-assimilatory proteins per gram root (i.e., nitrate reductase, NR, glutamine synthetase, GS, and glutamine oxoglutarate aminotransferase, GOGAT). Based on those results, we can assume that the inhibition of total-plant N assimilation by eCO_2 plus chronic warming could be due to decreased levels of these N assimilatory proteins. Further, we assessed the in vitro activities of assimilatory proteins extracted from plants grown at 33 °C, at both 33 and 38 °C. For all three proteins, the activities at 38 °C were greater or similar to 33 °C (Figure S2), which confirms that these assimilatory proteins are not damaged by chronic warming temperature.

Previously, using wheat as the model species, Jauregui et al. [4] reported that eCO$_2$ (700 vs. 400 ppm) plus warming (ambient + 4 °C) reduced leaf N assimilation by reducing energy availability. Since plants grown at eCO$_2$ plus chronic warming had the lowest N-uptake rates and N assimilation, we hypothesized that this could be due to the lower energy or resource availability in roots to perform root functions, which we tested by measuring root %C, TNC, and %N. Relative to the other treatments; plants grown at eCO$_2$ plus warming had the highest root %C and TNC concentrations, while they also had the co-highest root %N. Based on these results, we concluded that the low rates of N uptake and N assimilation in plants grown at eCO$_2$ plus warming were not due to limited energy or resource availability in roots.

In summary, tomato plants grown at eCO$_2$ plus chronic warming had the lowest plant dry mass, NO_3^- and NH_4^+-uptake rates, net N translocation, and whole-plant N (and NO_3^-) assimilation when compared with other treatments. Previously, we showed that N uptake can be inhibited by eCO$_2$ (700 vs. 400 ppm) plus warming (37 vs. 30 °C), using a sequential harvesting technique [5]; furthermore, in this study, we showed that both NO_3^- and NH_4^+-uptake rates can be inhibited by eCO$_2$ plus warming using ^{15}N labeling. Moreover, in this study, we showed that net N translocation can be inhibited by eCO$_2$ plus warming using two different methods: (1) the ^{15}N ratio between shoots and roots, and (2) the NO_3^- ratio between shoots and roots. Finally, we showed that the whole-plant N assimilation can be inhibited by eCO$_2$ plus warming using two methods: (1) the ratio between organic N and total N, and (2) the ratio between ^{15}N in proteins and ^{15}N in the plant). Inhibition of whole-plant N assimilation was mainly due to the inhibition of NO_3^- assimilation by eCO$_2$ plus chronic warming. In addition, the decreased rates of N uptake and N assimilation were not due to the resource limitation (N or C) for root functions, but, probably, due to the decreased levels of enzymes involved in N metabolism (NR, GS, GOGAT), as shown previously [5]. Overall, this study has shown that the interactive effects of CO$_2$ enrichment and global warming can negatively affect plant N metabolism in tomato, which will have serious consequences for the production and nutrient quality of tomato, one of the world's most-important non-grain food crops. Given global human population is expected increase by 1.4 to 3.9 billion by 2050, global crop production will need to increase by then by ca. 70% to meet global food demand [54]. This study provides valuable information regarding weak links in N metabolism in response to CO$_2$ enrichment and global warming that can be targeted for improvement, in order to improve yield and nutrient quality of tomato, and, perhaps, other crops in the future.

4. Materials and Methods

4.1. Plant Material, Growth Conditions, and Treatments

Tomato (*S. lycopersicum* L. cv. Big Boy), a heat-tolerant warm-season C$_3$ vegetable crop, was used as the model species. Seeds were germinated in trays filled with calcined clay in a greenhouse and watered daily. In the greenhouse, temperature fluctuated between 25–30 °C, and supplementary lighting (<15% of full sunlight) was provided by 250-W high-pressure sodium (GE Lighting Inc., Cleveland, OH, USA) and 400-W metal-halide (Osram Sylvania Products Inc., Manchester, NH, USA) lamps to provide a 15-h photoperiod. When seeds were germinating, quarter-strength complete nutrient solution was added to the trays only after cotyledons began turning yellow (nutrient concentrations of the full-strength solution: 2 mM MgSO$_4$, 1 mM KH$_2$PO$_4$, 1 mM K$_2$HPO$_4$, 2 mM CaCl$_2$, 71 μM Fe-DTPA, 10 μM MnCl$_2$, 50 μM H$_3$BO$_3$, 6 μM CuSO$_4$, 6 μM ZnSO$_4$, 1 μM Na$_2$MoO$_4$, and 1 mM NH$_4$NO$_3$; pH = 6.0). The nutrient solution used in this study provided N at an intermediate but limiting amount, and was designed to follow up on a previous study of ours examining effects of nitrate vs. ammonium [5], but the nutrient solution provided other nutrients at near-optimal levels e.g., [53,55,56]. When seedlings were at the first-adult-leaf stage, uniform seedlings were transplanted into 3.1-L cylindrical pots [10-cm diameter × 40-cm length PVC pipes; one plant per pot] filled with a mixture of calcined clay and perlite in a 3:1 (*v:v*) ratio (supported by mesh at the bottom to hold the substrate).

After transplant, all pots were transferred to four growth chambers (model E36HO, Percival Scientific Inc., Perry, IA, USA). Chambers were initially programmed to have the near-optimal growth temperature for this cultivar (33/28 °C day/night air temperatures), ambient CO_2 (400 ppm), ambient humidity (ca. 55–65%), 600 µmol m^{-2} s^{-1} PAR (supplied by 55-W Osram Dulux luminous lamps; Osram GmbH, Augsburg, Germany), and a 14-h (0600–2000 h) photoperiod. Each chamber received 14 pots; 10 were used for either $^{15}NO_3^-$ or $^{15}NH_4^+$ treatments (n = 5) while the other 4 were used as unlabeled controls. Pots were placed in individual shallow trays to retain any excess water and nutrient solution. Plants were kept inside the growth chambers for three days to acclimate to the new environment. During this period, quarter-strength complete nutrient solution (500 mL) was added to each pot once. When plants were free from transplant-stress and at their second adult-leaf stage, the temperature of the high-temperature-treatment chambers was increased gradually from near-optimal to chronic warming temperature (38/33 °C day/night) over three days to avoid potential heat shock. The temperatures used in this study were based on previous experiments that showed that optimal daytime temperature for this cultivar in our hands is 32–33 °C, as far as whole plant biomass of pre-reproductive plants, and that significant reductions in biomass only occur at temperatures of 36–38 °C [5,20,48]. Once the high-temperature-treatment chambers reached the chronic warming temperature, CO_2 treatment was started (day 0). A 2 × 2 factorial experimental design was used, with two levels of CO_2 [ambient (400 ppm) vs. elevated (700 ppm)] and two temperature treatments (33/28 °C vs. 38/33 °C day/night), with one growth chamber per treatment. Plants were fertilized with 500 mL of full-strength complete nutrient solution containing 2 mM N (1 mM NH_4NO_3) every other day. Plants were rotated within each chamber every 4–5 days to avoid potential position/edge effects and switched between chambers every 7–8 days to avoid potential chamber effects.

On day 18, a randomly selected subset of plants from each chamber (n = 5) was labeled with 1950 mL (this volume is equivalent to the estimated soil pore-volume in the 3.1-L pots) of full-strength complete nutrient solution containing 1 mM $NH_4{}^{15}NO_3$. On day 19, another randomly selected subset of plants (unlabeled controls, n = 4) from each chamber was treated with 1950 mL of nutrient solution containing 1 mM (unlabeled) NH_4NO_3. On day 20, the remaining plants (n = 5) were labeled with 1950 mL of nutrient solution containing 1 mM $^{15}NH_4NO_3$. The nutrient solution was isotopically labeled by adding either $NH_4{}^{15}NO_3$ or $^{15}NH_4NO_3$ with an isotopic purity of 98 atom % ^{15}N (Sigma-Aldrich Inc, St. Louis, MO). When making the isotopically labeled nutrient solutions, carbenicillin (an antibiotic) was added (at 20 mg L^{-1}) to avoid external nitrification, de-nitrification, and N immobilization by microbes, as in [57]. Addition of carbenicillin did not affect plant growth (based on preliminary experiments). Plants were harvested three days (based on preliminary experiments) after ^{15}N isotope labeling (day 21 for $^{15}NO_3^-$-labeled plants and day 23 for $^{15}NH_4^+$-labeled plants). Unlabeled plants were harvested on day 22, and these were used to determine background levels of ^{15}N in plant tissue. At the time of harvest, all plants were in the vegetative stage. This labeling experiment was repeated at a later time, with results from the two experiments being nearly identical, and results are presented for the first experiment only.

4.2. Plant N and Protein Analysis

Harvested plants were separated into shoots and roots, and then each shoot and root system was divided into two halves and weighed separately. One half of the shoots and roots were flash frozen in liquid N_2 and stored at −80 °C to be used for tissue NH_4^+ and protein analysis (fresh tissue). The other half of the shoots and roots were put in paper bags separately and placed in an oven at 70 °C for 72 h. Dry mass of shoots and roots were measured using these samples (the ratio between dry and fresh mass of these samples were used to calculate total shoot and root dry mass).

Dried shoot and root samples were pulverized separately using a mortar and pestle. A subset of each pulverized sample was used to measure shoot and root %C and %N using the combustion-MS technique [58]. Another subset of each pulverized sample was analyzed to measure ^{15}N (atom%) in solid samples using an elemental analyzer interfaced to a continuous-flow isotope-ratio mass spectrometer at the University of California, Davis, stable-isotope facility. Nitrate or NH_4^+-uptake rates were calculated as the total amount of ^{15}N taken up by plants treated with $^{15}NO_3^-$ or $^{15}NH_4^+$ per g of dry root per day (using root mass at the final harvest). Background ^{15}N was subtracted using ^{15}N (atom%) of unlabeled plants. Net N translocation was calculated as the ratio of total ^{15}N content in the shoots vs. roots of $^{15}NO_3^-$ or $^{15}NH_4^+$-supplied plants.

Since NO_3^- is the main source of N for most crop species [59], we studied the responses of $^{15}NO_3^-$ supplied plants in more detail. Therefore, shoot and root NO_3^- concentrations were measured only from $NH_4^{15}NO_3$-supplied plants, using the colorimetry-based salicylic-acid method, as explained in [60]. Briefly, 50 mg of pulverized tissue was suspended in 5 mL of deionized water, incubated at 45 °C for 1 h, and centrifuged at 10,000× g for 5 min. The supernatant was recovered and 0.2 mL of the extracted NO_3^- sample was reacted with 0.8 mL of 5% (w/v) salicylic acid–sulfuric solution and 19 mL of 2 N NaOH. Absorbance was measured at 410 nm using KNO_3^- as the standard. The NO_3^- (total pool size) ratio between shoots and roots was used as a metric of net NO_3^- translocation. Shoot and root NH_4^+ from $NH_4^{15}NO_3$ supplied plants were quantified using an NH_4^+ ion-selective electrode (Hach Co., Loveland, CO). Briefly, fresh tissue was ground to a fine powder using liquid N_2, mortar, and pestle. Then, 500 mg of tissue was suspended in a tube containing 30 mL of 0.001 M $CaSO_4$ solution (pH = 3) [61]. The tube was shaken for 30 min in a shaker and then centrifuged at 4000× g for 10 min [62]. The recovered supernatant was used for NH_4^+ quantification. The NH_4^+ (total pool size) ratio between shoots and roots was used as a metric of net-NH_4^+ translocation. Total-plant inorganic N content was estimated as the sum of total-NO_3^- and -NH_4^+ contents. Total-plant organic N content was estimated as the difference between total-plant N and total-plant inorganic N. The ratio between organic N and total N was used as a metric of total-plant N assimilation. The ratio between total NO_3^- and total N was used as a metric of total-plant NO_3^- assimilation.

Shoot and root proteins were extracted from fresh tissues of $NH_4^{15}NO_3$ supplied plants, as described in [5]. Briefly, a known amount of fresh shoot or root tissue was ground in liquid N_2 to a fine powder, then, with a known volume of extraction buffer (EB). The resulting mixture was transferred to a tube, an equal volume of buffer-saturated phenol was added, and the tube was centrifuged at 10,000× g for 10 min. Protein in the phenol phase was further purified by centrifugation after adding an equal volume of EB to the recovered phenol phase. Extracted protein was precipitated by adding five volumes of chilled 0.1 M ammonium acetate to the recovered phenol phase. Protein was pelleted down by centrifugation. The resulting pellet was washed several times with ammonium acetate and acetone. Following this method, proteins were extracted on two sets of samples per biological replicate. In the first extraction, protein pellets were freeze dried in a lyophilizer (Genesis 25 SQ Super ES; VirTis-SP Scientific, Gardiner, NY, USA) and analyzed to measure ^{15}N (atom%) in protein samples (as above). In the second extraction, protein pellets were dissolved in a resolubilizing buffer. Then, total shoot or root protein concentration per g of fresh shoot or root was measured using a colorimetric assay (DC protein assay; Bio-Rad Laboratories Inc., Hercules, CA, USA), using bovine serum albumin (BSA) as the protein standard. Using data on ^{15}N (atom%) in protein pellets, protein concentration (mg g^{-1} dry mass), ^{15}N (atom%) in oven-dried samples, and dry mass data, the ratio between ^{15}N in plant proteins and ^{15}N in total plant (pool size) was calculated as another metric of plant N assimilation.

4.3. Total Nonstructural Carbohydrate Assay

Total nonstructural carbohydrate concentration in roots was measured according to the method described by Chow and Landhäusser [63], with some modifications. Briefly, 1 mL of 0.1 N NaOH was added to 25 mg of pulverized dried tissue in a tube, incubated at 50 °C for 30 min, and then 1 mL of 0.1 N HCl was added. Then, 1 mL of an enzyme solution containing 3000 units/mL of α-amylase, plus 15 units/mL amyloglucosidase in 0.05 M sodium acetate, were added to the tube, and then tubes were incubated at 50 °C for 24 h. The digest was filtered through 0.22 μm polyvinylidene difluoride (PVDF) filters to remove digestion enzymes. The filtrate (0.5 mL) was reacted with 1 mL of 2% phenol and 2.5 mL of H_2SO_4, incubated at room temperature for 30 min, and absorbance was measured at 490 nm using glucose to generate a standard curve.

4.4. In Vitro N-Assimilatory-Protein Activity Assays

In an independent experiment, tomato (*S. lycopersicum* L. cv. Big Boy) seeds were germinated in a greenhouse, transplanted to pots, transferred to a growth chamber, and treated with ambient CO_2 and near-optimal temperature, as above. Chambers contained 12 pots for NR, GS, and GOGAT shoot and root in vitro activity assays ($n = 4$). Plants were grown for 19–23 days, but without isotope labeling. The third fully expanded leaf below the apical meristem was harvested for leaf NR, GS, or GOGAT activity assays, while the first 5–10 cm from the root tip was harvested for root NR, GS, or GOGAT activity assays. For all three enzymes, assays were performed as soon as the tissues were harvested (within 30 min).

Nitrate reductase activity was measured according to the protocol described by Sigma Aldrich for EC 1.6.6.1. A known amount of fresh tissue (0.5 g) was ground with 2 mL of reagent A (25 mM KH_2PO_4, 10 mM KNO_3, 0.05 ethylenediaminetetraacetic acid (EDTA), 0.5% polyvinylpolypyrrolidone (PVPP), 1 mM phenylmethylsulfonyl fluoride (PMSF), 1 mM leupeptin, and 3% (w/v) BSA). The mixture was transferred to a tube and centrifuged at 12,100× g for 5 min at 4 °C, and the supernatant (with enzyme) was recovered. In two other tubes, 1.8 mL of reagent A was mixed with 100 μL of 2 mM β-NADH and equilibrated at 33 °C or 38 °C temperature. The enzyme reaction was started by adding 100 μL of the enzyme mixture to each tube. Then, the reaction was stopped by adding 1 mL of 58 mM sulfanilamide in 3 M HCl at time = 0 min (tube 1) and after 5 min (tube 2). Finally, 1 mL of 0.77 mM N-(1-naphthyl) ethylenediamine dihydrochloride was added to each tube and absorbance was measured at 540 nm using $NaNO_2$ to generate a standard curve. In vitro NR activity was calculated as NO_2^- produced per g of tissue per minute.

Glutamine synthetase activity was measured according to the protocol described by Sigma Aldrich for EC 6.3.1.2. A reaction cocktail (pH = 7.1) was prepared by mixing 100 mM imidazole, 3 M sodium glutamate, 250 mM ATP, 900 mM $MgCl_2$, 1 M KCl, and 1.2 M NH_4Cl. A known amount of fresh tissue (0.5 g) was ground with 2 mL of an EB (50 mM Tris-HCl (pH = 7.1), 1 mM EDTA, 0.6% (w/v) PVPP, 1 mM PMSF, and 1 mM leupeptin). The mixture was transferred into a tube, centrifuged at 12,100× g for 10 min at 4 °C, and the supernatant (with enzyme) was recovered. In another tube, 2.7 mL of the reaction cocktail was mixed with 100 μL of 33 mM phosphoenol pyruvate (PEP), 60 μL of 12.8 mM β-NADH, and 40 μL of pyruvate kinase/L-lactic dehydrogenase enzyme mixture, and equilibrated at treatment temperature (33 °C or 38 °C). The first enzyme mixture (100 μL) was added to the second reaction mixture, and then the decrease in the absorbance at 340 nm (A_{340}) for 10 min was recorded. Using the maximum–slope linear range of the absorbance-time relationship, ΔA_{340}/min was calculated, after subtracting for background absorbance.

Glutamine oxoglutarate aminotransferase activity was measured according to the method described by Berteli et al. [64] and Lutts et al. [65]. A known amount of fresh tissue (0.5 g) was ground with 2 mL of an EB (100 mM KH_2PO_4, 0.05 mM EDTA, 0.5% (w/v) PVPP, 1 mM PMSF, 1 mM leupeptin, and 2 mM 2-oxoglutarate), the mixture was transferred into a tube and centrifuged at 12,100× g for 15 min at 4 °C, and then the supernatant (with enzyme) was recovered. Next, 100 μL of the enzyme mixture was mixed in a tube with 1.8 mL of a reaction mixture containing 125 mM phosphate buffer (pH = 7.5), 10 mM 2-oxoglutarate, 10 mM glutamine, 10 mM aminoxy-acetate, and 10 mM methyl viologen and equilibrated at treatment temperature (33 °C or 38 °C). The reaction was started by adding 100 μL of a mixture containing 4.7 μg of sodium dithionite and 5 μg of $NaHCO_3$. The tube was incubated at treatment temperature for 30 min. The reaction was stopped by boiling the tube in a water bath for 5 min. The tube was centrifuged at 5000 rpm for 5 min, and then the supernatant was recovered. In the supernatant, glutamate was separated from glutamine by anion-exchange chromatography. During anion-exchange chromatography, the column (Pasteur pipette) was packed with the anion-exchange resin, AG1X8 (acetate form), to a length of 4 cm and gradually washed with 5 mL of HPLC-grade water. Then, 2 mL of the recovered supernatant was passed through the column. The column was again washed with 12 mL of HPLC-grade water and glutamate was eluted with 8 mL of 1 M acetic acid [66]. To purify the glutamate, the eluent was completely dried using a vacuum evaporator at room temperature. The dried glutamate samples were redissolved, and the concentration was determined by electrospray ionization-mass spectrometry (ESI-MS, Orbitrap Fushion Tribid Mass Spectrometer, Thermo Fisher, Waltham, MA, USA) using sodium glutamate as the standard.

4.5. Statistical Analysis

The statistical software, RStudio version 3.6.1 [RCore Team (2013), Vienna, Austria], was used for all statistical analyses. Statistical assumptions of independence, normality, and equal variance were checked with residual vs. fitted, normal Q-Q, and S-L plots, respectively. If statistical assumptions were not met, data were transformed; log transformation proved to be sufficient. Data were analyzed using two-way (two levels of CO_2 × two levels of temperature) analysis of variance (ANOVA), with CO_2 and temperature as fixed factors. Results were considered significant if $p < 0.05$. If ANOVA results were significant, a Tukey's post hoc test was performed for multiple comparisons. Figures were generated using SigmaPlot version 14.0 (Systat Software, Inc., San Jose, CA, USA). Results presented in figures are untransformed means and standard errors of mean (SEM).

Supplementary Materials: The following are available online at https://www.mdpi.com/article/10.3390/plants10040722/s1. Table S1: Results from ANOVA statistical analysis. Figure S1: Effects of ambient (400 ppm) vs. elevated (700 ppm) CO_2 and near-optimal (33 °C) vs. chronic warming (38 °C) day-time temperatures on (A) shoot: root NO_3^- and (B) shoot to root NH_4^+ ratios of *Solanum lycopersicum*, grown for 21 days. Each bar represents mean ($n = 5$) + 1 SEM. Within each panel, bars not sharing the same letters are significantly different ($p < 0.05$, Tukey's test); no letters are shown if no differences.

Author Contributions: Conceptualization, D.M.J., S.A.H., and J.K.B.; methodology, D.M.J., S.A.H., J.K.B., K.K.R., and D.I.; investigation, D.M.J.; data curation, D.M.J.; writing—original draft preparation, D.M.J.; writing—review and editing, D.M.J., S.A.H., and J.K.B.; supervision, S.A.H.; project administration, S.A.H.; funding acquisition, S.A.H. and J.K.B. All authors have read and agreed to the published version of the manuscript.

Funding: This research was funded by the United States-Department of Agriculture, NACA 58-3607-4-026.

Institutional Review Board Statement: Not applicable.

Informed Consent Statement: Not applicable.

Data Availability Statement: All data are available on request.

Acknowledgments: We thank Mona-Lisa Banks and Douglas Sturtz for CN analyses and the UC-Davis Isotope Facility for ^{15}N isotope analyses.

Conflicts of Interest: The authors declare no conflict of interest. The funders had no role in the design of the study; in the collection, analyses, or interpretation of data; in the writing of the manuscript, or in the decision to publish the results.

References

1.   IPCC. *Climate Change 2007: The Scientific Basis. Contribution of Working Group I to the Fourth Assessment Report of the Intergovernmental Panel on Climate Change*; Solomon, S., Qin, D., Manning, M., Chen, Z., Marquis, M., Averyt, K., Tignor, M., Miller, H., Eds.; Cambridge University Press: Cambridge, UK, 2007.

2. IPCC. *Climate Change 2014: Impacts, Adaptation and Vulnerability. Contribution of Working Group II to the Fifth Assessment Report of the Intergovernmental Panel on Climate Change*; Field, C., Barros, V., Dokken, D., Mach, K., Mastrandrea, M., Bilir, T., Chatterjee, M., Ebi, K., Estrada, Y., Genova, R., et al., Eds.; Cambridge University Press: Cambridge, UK, 2014.

3. Heckathorn, S.A.; Giri, A.; Mishra, S.; Bista, D. Heat stress and roots. In *Climate Change and Plant Abiotic Stress Tolerance*; Tuteja, N., Gill, S., Eds.; Wiley-VCH Verlag GmbH & Co. KGaA: Weinheim, Germany, 2014; pp. 109–136.

4. Jauregui, I.; Aroca, R.; Garnica, M.; Zamarreño, Á.M.; García-Mina, J.M.; Serret, M.D.; Parry, M.; Irigoyen, J.J.; Aranjuelo, I. Nitrogen assimilation and transpiration: Key processes conditioning responsiveness of wheat to elevated [CO_2] and temperature. *Physiol. Plant.* **2015**, *155*, 338–354. [CrossRef] [PubMed]

5. Jayawardena, D.M.; Heckathorn, S.A.; Bista, D.R.; Mishra, S.; Boldt, J.K.; Krause, C.R. Elevated CO_2 plus chronic warming reduce nitrogen uptake and levels or activities of nitrogen-uptake and -assimilatory proteins in tomato roots. *Physiol. Plant.* **2017**, *159*, 354–365. [CrossRef] [PubMed]

6. Shaw, M.R.; Zavaleta, E.S.; Chiariello, N.R.; Cleland, E.E.; Mooney, H.A.; Field, C.B. Grassland responses to global environmental changes suppressed by elevated CO_2. *Science* **2002**, *298*, 1987–1990. [CrossRef]

7. Wang, D.; Heckathorn, S.A.; Wang, X.; Philpott, S.M. A meta-analysis of plant physiological and growth responses to temperature and elevated CO_2. *Oecologia* **2012**, *169*, 1–13. [CrossRef] [PubMed]

8. Schroeder, J.I.; Delhaize, E.; Frommer, W.B.; Guerinot, M.L.; Harrison, M.J.; Herrera-Estrella, L.; Horie, T.; Kochian, L.V.; Munns, R.; Nishizawa, N.K.; et al. Using membrane transporters to improve crops for sustainable food production. *Nature* **2013**, *497*, 60–66. [CrossRef]

9. Taub, D.R.; Wang, X. Why are nitrogen concentrations in plant tissues lower under elevated CO_2? A critical examination of the hypotheses. *J. Integr. Plant Biol.* **2008**, *50*, 1365–1374. [CrossRef] [PubMed]

10. Nacry, P.; Bouguyon, E.; Gojon, A. Nitrogen acquisition by roots: Physiological and developmental mechanisms ensuring plant adaptation to a fluctuating resource. *Plant Soil* **2013**, *370*, 1–29. [CrossRef]

11. Näsholm, T.; Kielland, K.; Ganeteg, U. Uptake of organic nitrogen by plants. *New Phytol.* **2009**, *182*, 31–48. [CrossRef]

12. Bassirirad, H. Kinetics of nutrient uptake by roots: Responses to global change. *New Phytol.* **2000**, *147*, 155–169. [CrossRef]

13. Rubio-Asensio, J.S.; Bloom, A.J. Inorganic nitrogen form: A major player in wheat and Arabidopsis responses to elevated CO_2. *J. Exp. Bot.* **2016**, *68*, 2611–2625. [CrossRef]

14. Atkin, O.; Cummins, W.R. The effect of root temperature on the induction of nitrate reductase activities and nitrogen uptake rates in arctic plant species. *Plant Soil* **1994**, *159*, 187–197. [CrossRef]

15. Clarkson, D.T.; Warner, A.J. Relationships between root temperature and the transport of ammonium and nitrate ions by Italian and perennial ryegrass (*Lolium multiflorum* and *Lolium perenne*). *Plant Physiol.* **1979**, *64*, 557–561. [CrossRef]

16. Cruz, C.; Lips, S.H.; Martins-Loucao, M.A. Effect of root temperature on carob growth: Nitrate versus ammonium nutrition. *J. Plant Nutr.* **1993**, *16*, 1517–1530. [CrossRef]

17. Tindall, J.A.; Mills, H.A.; Radcliffe, D.E. The effect of root zone temperature on nutrient uptake of tomato. *J. Plant Nutr.* **1990**, *13*, 939–956. [CrossRef]

18. Bassirirad, H.; Caldwell, M.M.; Bilbrough, C. Effects of soil temperature and nitrogen status on kinetics of $^{15}NO_3^-$ uptake by roots of field-grown *Agropyron desertorum* (Fisch. ex Link) Schult. *New Phytol.* **1993**, *123*, 485–489. [CrossRef]

19. Delucia, E.H.; Heckathorn, S.A.; Day, T.A. Effects of soil temperature on growth, biomass allocation and resource acquisition of *Andropogon gerardii* Vitman. *New Phytol.* **1992**, *120*, 543–549. [CrossRef]

20. Giri, A.; Heckathorn, S.; Mishra, S.; Krause, C. Heat stress decreases levels of nutrient-uptake and -assimilation proteins in tomato roots. *Plants* **2017**, *6*, 6. [CrossRef] [PubMed]

21. Mainali, K.P.; Heckathorn, S.A.; Wang, D.; Weintraub, M.N.; Frantz, J.M.; Hamilton, E.W. Impact of a short-term heat event on C and N relations in shoots vs. roots of the stress-tolerant C_4 grass, *Andropogon gerardii*. *J. Plant Physiol.* **2014**, *171*, 977–985. [CrossRef]

22. Coleman, J.S.; Bazzaz, F.A. Effects of CO_2 and temperature on growth and resource use of co-occurring C_3 and C_4 annuals. *Ecology* **1992**, *73*, 1244–1259. [CrossRef]

23. Arndal, M.F.; Schmidt, I.K.; Kongstad, J.; Beier, C.; Michelsen, A. Root growth and N dynamics in response to multi-year experimental warming, summer drought and elevated CO_2 in a mixed heathland-grass ecosystem. *Funct. Plant Biol.* **2014**, *41*, 1–10. [CrossRef]

24. Dijkstra, F.A.; Blumenthal, D.; Morgan, J.A.; Pendall, E.; Carrillo, Y.; Follett, R.F. Contrasting effects of elevated CO_2 and warming on nitrogen cycling in a semiarid grassland. *New Phytol.* **2010**, *187*, 426–437. [CrossRef]

25. Bassirirad, H.; Griffin, K.L.; Strain, B.R.; Reynolds, J.F. Effects of CO_2 enrichment on growth and root $^{15}NH_4^+$ uptake rate of loblolly pine and ponderosa pine seedlings. *Tree Physiol.* **1996**, *16*, 957–962. [CrossRef]

26. Cohen, I.; Halpern, M.; Yermiyahu, U.; Bar-Tal, A.; Gendler, T.; Rachmilevitch, S. CO_2 and nitrogen interaction alters root anatomy, morphology, nitrogen partitioning and photosynthetic acclimation of tomato plants. *Planta* **2019**, *250*, 1423–1432. [CrossRef] [PubMed]

27. Jauregui, I.; Aparicio-Tejo, P.M.; Avila, C.; Cañas, R.; Sakalauskiene, S.; Aranjuelo, I. Root-shoot interactions explain the reduction of leaf mineral content in Arabidopsis plants grown under elevated [CO_2] conditions. *Physiol. Plant.* **2016**, *158*, 65–79. [CrossRef] [PubMed]

28. Kruse, J.; Hetzger, I.; Hänsch, R.; Mendel, R.R.; Walch-Liu, P.; Engels, C.; Rennenberg, H. Elevated pCO$_2$ favours nitrate reduction in the roots of wild-type tobacco (*Nicotiana tabacum* cv. Gat.) and significantly alters N-metabolism in transformants lacking functional nitrate reductase in the roots. *J. Exp. Bot.* **2002**, *53*, 2351–2367. [CrossRef]

29. van Ginkel, J.H.; Gorissen, A.; van Veen, J.A. Carbon and nitrogen allocation in *Lolium perenne* in response to elevated atmospheric CO$_2$ with emphasis on soil carbon dynamics. *Plant Soil* **1997**, *188*, 299–308. [CrossRef]

30. Bigot, J.; Boucaud, J. Short-term responses of *Brassica rapa* plants to low root temperature: Effects on nitrate uptake and its translocation to the shoot. *Physiol. Plant.* **1996**, *96*, 646–654. [CrossRef]

31. Rufty, T.W.; Raper, C.D.; Jackson, W.A. Nitrogen assimilation, root growth and whole plant responses of soybean to root temperature, and to carbon dioxide and light in the aerial environment. *New Phytol.* **1981**, *88*, 607–619. [CrossRef]

32. Sattelmacher, B.; Marschner, H.; Kühne, R. Effects of root zone temperature on root activity of two potato (*Solanum tuberosum* L.) clones with different adaptation to high temperature. *J. Agron. Crop Sci.* **1990**, *165*, 131–137. [CrossRef]

33. Tachibana, S. The influence of root temperature on nitrate assimilation by cucumber and figleaf gourd. *J. Jpn. Soc. Hortic. Sci.* **1998**, *57*, 440–447. [CrossRef]

34. Ezeta, F.N.; Jackson, W.A. Nitrate translocation by detopped corn seedlings. *Plant Physiol.* **1975**, *56*, 148–156. [CrossRef]

35. Hungria, M.; Kaschuk, G. Regulation of N$_2$ fixation and NO$_3^-$/NH$_4^+$ assimilation in nodulated and N-fertilized *Phaseolus vulgaris* L. exposed to high temperature stress. *Environ. Exp. Bot.* **2014**, *98*, 32–39. [CrossRef]

36. Bloom, A.; Burger, M.A.; Kimball, B.J.; Pinter, J.P. Nitrate assimilation is inhibited by elevated CO$_2$ in field-grown wheat. *Nat. Clim. Chang.* **2014**, *4*, 477–480. [CrossRef]

37. Bloom, A.J.; Asensio, J.S.R.; Randall, L.; Rachmilevitch, S.; Cousins, A.B.; Carlisle, E.A. CO$_2$ enrichment inhibits shoot nitrate assimilation in C$_3$ but not C$_4$ plants and slows growth under nitrate in C$_3$ plants. *Ecology* **2012**, *93*, 355–367. [CrossRef] [PubMed]

38. Bloom, A.J.; Burger, M.; Asensio, J.S.R.; Cousins, A.B. Carbon dioxide enrichment inhibits nitrate assimilation in wheat and Arabidopsis. *Science* **2010**, *328*, 899–903. [CrossRef] [PubMed]

39. Bloom, A.J.; Smart, D.R.; Nguyen, D.T.; Searles, P.S. Nitrogen assimilation and growth of wheat under elevated carbon dioxide. *Proc. Natl. Acad. Sci. USA* **2002**, *99*, 1730–1735. [CrossRef] [PubMed]

40. Andrews, M.; Condron, L.M.; Kemp, P.D.; Topping, J.F.; Lindsey, K.; Hodge, S.; Raven, J.A. Elevated CO$_2$ effects on nitrogen assimilation and growth of C$_3$ vascular plants are similar regardless of N-form assimilated. *J. Exp. Bot.* **2019**, *70*, 683–690. [CrossRef]

41. Bloom, A.J.; Kasemsap, P.; Rubio-Asensio, J.S. Rising atmospheric CO$_2$ concentration inhibits nitrate assimilation in shoots but enhances it in roots of C$_3$ plants. *Physiol. Plant.* **2020**, *168*, 963–972. [CrossRef]

42. Amos, J.A.; Scholl, R.L. Effect of growth temperature on leaf nitrate reductase, glutamine synthetase and NADH-glutamate dehydrogenase of juvenile maize genotypes. *Crop Sci.* **1977**, *17*, 445–448. [CrossRef]

43. Chopra, R.K. Effects of temperature on the in vivo assay of nitrate reductase in some C$_3$ and C$_4$ species. *Ann. Bot.* **1983**, *51*, 617–620. [CrossRef]

44. Magalhães, A.C.; Peters, D.B.; Hageman, R.H. Influence of temperature on nitrate metabolism and leaf expansion in soybean (*Glycine max* L. Merr.) seedlings. *Plant Physiol.* **1976**, *58*, 12–16. [CrossRef]

45. Rachmilevitch, S.; Huang, B.; Lambers, H. Assimilation and allocation of carbon and nitrogen of thermal and nonthermal *Agrostis* species in response to high soil temperature. *New Phytol.* **2006**, *170*, 479–490. [CrossRef]

46. Vicente, R.; Pérez, P.; Martínez-Carrasco, R.; Usadel, B.; Kostadinova, S.; Morcuende, R. Quantitative RT–PCR platform to measure transcript levels of C and N metabolism-related genes in durum wheat: Transcript profiles in elevated [CO$_2$] and high temperature at different levels of N supply. *Plant Cell Physiol.* **2015**, *56*, 1556–1573. [CrossRef] [PubMed]

47. Masclaux-Daubresse, C.; Daniel-Vedele, F.; Dechorgnat, J.; Chardon, F.; Gaufichon, L.; Suzuki, A. Nitrogen uptake, assimilation and remobilization in plants: Challenges for sustainable and productive agriculture. *Ann. Bot.* **2010**, *105*, 1141–1157. [CrossRef]

48. Jayawardena, D.M.; Heckathorn, S.A.; Bista, D.R.; Boldt, J.K. Elevated carbon dioxide plus chronic warming causes dramatic increases in leaf angle in tomato, which correlates with reduced plant growth. *Plant Cell Environ.* **2019**, *42*, 1247–1256. [CrossRef] [PubMed]

49. Dieleman, W.I.J.; Vicca, S.; Dijkstra, F.A.; Hagedorn, F.; Hovenden, M.J.; Larsen, K.S.; Morgan, J.A.; Volder, A.; Beier, C.; Dukes, J.S.; et al. Simple additive effects are rare: A quantitative review of plant biomass and soil process responses to combined manipulations of CO$_2$ and temperature. *Glob. Chang. Biol.* **2012**, *18*, 2681–2693. [CrossRef] [PubMed]

50. McGrath, J.M.; Lobell, D.B. Reduction of transpiration and altered nutrient allocation contribute to nutrient decline of crops grown in elevated CO$_2$ concentrations. *Plant Cell Environ.* **2013**, *36*, 697–705. [CrossRef] [PubMed]

51. Pate, J.S. Uptake, assimilation and transport of nitrogen compounds by plants. *Soil Biol. Biochem.* **1973**, *5*, 109–119. [CrossRef]

52. Andrews, M. The partitioning of nitrate assimilation between root and shoot of higher plants. *Plant Cell Environ.* **1986**, *9*, 511–519. [CrossRef]

53. Marschner, H. *Mineral Nutrition of Higher Plants*; Academic Press: London, UK, 1995.

54. Alexandratos, N.; Bruinsma, J. *World Agriculture towards 2030/2050: The 2012 Revision*; ESA Working Paper, No. 12-03; Food and Agriculture Organization of the United Nations: Rome, Italy, 2012.

55. Sung, J.; Lee, S.; Lee, Y.; Ha, S.; Song, B.; Kim, T.; Waters, B.M.; Krishnan, H.B. Metabolomic profiling from leaves and roots of tomato (*Solanum lycopersicum* L.) plants grown under nitrogen, phosphorus or potassium-deficient condition. *Plant Sci.* **2015**, *241*, 55–64. [CrossRef]

56. Padilla, F.M.; Peña-Fleitas, M.T.; Gallardo, M.; Thompson, R.B. Threshold values of canopy reflectance indices and chlorophyll meter readings for optimal nitrogen nutrition of tomato. *Ann. Appl. Biol.* **2015**, *166*, 271–285. [CrossRef]
57. Smart, D.R.; Ferro, A.; Ritchie, K.; Bugbee, B.G. On the use of antibiotics to reduce rhizoplane microbial populations in root physiology and ecology investigations. *Physiol. Plant.* **1995**, *95*, 533–540. [CrossRef] [PubMed]
58. Mishra, S.; Heckathorn, S.; Frantz, J.; Yu, F.; Gray, J. Effects of boron deficiency on geranium grown under different nonphotoinhibitory light levels. *J. Am. Soc. Hortic. Sci.* **2009**, *134*, 183–193. [CrossRef]
59. Pilbeam, D. Nitrogen. In *Handbook of Plant Nutrition*; Barker, A., Pilbeam, D., Eds.; CRC Press: Boca Raton, FL, USA, 2015; pp. 17–64.
60. Cataldo, D.A.; Maroon, M.; Schrader, L.E.; Youngs, V.L. Rapid colorimetric determination of nitrate in plant tissue by nitration of salicylic acid1. *Commun. Soil Sci. Plant Anal.* **1975**, *6*, 71–80. [CrossRef]
61. Bloom, A.J.; Randall, L.; Taylor, A.R.; Silk, W.K. Deposition of ammonium and nitrate in the roots of maize seedlings supplied with different nitrogen salts. *J. Exp. Bot.* **2012**, *63*, 1997–2006. [CrossRef] [PubMed]
62. Kandeler, E.; Gerber, H. Short-term assay of soil urease activity using colorimetric determination of ammonium. *Biol. Fertil. Soils* **1988**, *6*, 68–72. [CrossRef]
63. Chow, P.S.; Landhäusser, S.M. A method for routine measurements of total sugar and starch content in woody plant tissues. *Tree Physiol.* **2004**, *24*, 1129–1136. [CrossRef]
64. Berteli, F.; Corrales, E.; Guerrero, C.; Ariza, M.J.; Pliego, F.; Valpuesta, V. Salt stress increases ferredoxin-dependent glutamate synthase activity and protein level in the leaves of tomato. *Physiol. Plant.* **1995**, *93*, 259–264. [CrossRef]
65. Lutts, S.; Majerus, V.; Kinet, J.M. NaCl effects on proline metabolism in rice (*Oryza sativa*) seedlings. *Physiol. Plant.* **1999**, *105*, 450–458. [CrossRef]
66. Darmaun, D.; Manary, M.J.; Matthews, D.E. A method for measuring both glutamine and glutamate levels and stable isotopic enrichments. *Anal. Biochem.* **1985**, *147*, 92–102. [CrossRef]

Article

Does Elevated [CO$_2$] Only Increase Root Growth in the Topsoil? A FACE Study with Lentil in a Semi-Arid Environment

Maryse Bourgault [1,2,3,*], Sabine Tausz-Posch [4], Mark Greenwood [5], Markus Löw [3], Samuel Henty [3,6], Roger D. Armstrong [7], Garry L. O'Leary [7], Glenn J. Fitzgerald [7] and Michael Tausz [4]

1 51 Campus Drive, College of Agriculture, University of Saskatchewan, Saskatoon, SK S7N 5A8, Canada
2 Northern Agricultural Research Center, Montana State University, 3710 Assinniboine Road, Havre, MT 59501, USA
3 Faculty of Veterinary and Agricultural Sciences, The University of Melbourne, 4 Water Street, Creswick, VIC 3363, Australia; llow@swin.edu.au (M.L.); sam.henty@agriculture.vic.gov.au (S.H.)
4 Department of Agriculture, Science and the Environment, CQ University Australia, 114-190 Yaamba Road, Norman Gardens, QLD 4701, Australia; s.tausz-posch@cqu.edu.au (S.T.-P.); m.tausz@cqu.edu.au (M.T.)
5 Department of Mathematical Sciences, Montana State University, Bozeman, MT 59717, USA; greenwood@montana.edu
6 Agriculture Victoria, The University of Melbourne Campus, Parkville, VIC 3053, Australia
7 Agriculture Victoria, Grains Innovation Park, 110 Natimuk Road, Horsham, VIC 3401, Australia; roger.armstrong@agriculture.vic.gov.au (R.D.A.); garry.oleary@agriculture.vic.gov.au (G.L.O.); glenn.fitzgerald@agriculture.vic.gov.au (G.J.F.)
* Correspondence: maryse.bourgault@usask.ca; Tel.: +1-306-966-4313

Citation: Bourgault, M.; Tausz-Posch, S.; Greenwood, M.; Löw, M.; Henty, S.; Armstrong, R.D.; O'Leary, G.L.; Fitzgerald, G.J.; Tausz, M. Does Elevated [CO$_2$] Only Increase Root Growth in the Topsoil? A FACE Study with Lentil in a Semi-Arid Environment. *Plants* **2021**, *10*, 612. https://doi.org/10.3390/plants10040612

Academic Editors: Rui Manuel Almeida Machado and James Bunce

Received: 6 February 2021
Accepted: 22 March 2021
Published: 24 March 2021

Publisher's Note: MDPI stays neutral with regard to jurisdictional claims in published maps and institutional affiliations.

Abstract: Atmospheric carbon dioxide concentrations [CO$_2$] are increasing steadily. Some reports have shown that root growth in grain crops is mostly stimulated in the topsoil rather than evenly throughout the soil profile by e[CO$_2$], which is not optimal for crops grown in semi-arid environments with strong reliance on stored water. An experiment was conducted during the 2014 and 2015 growing seasons with two lentil (*Lens culinaris*) genotypes grown under Free Air CO$_2$ Enrichment (FACE) in which root growth was observed non-destructively with mini-rhizotrons approximately every 2–3 weeks. Root growth was not always statistically increased by e[CO$_2$] and not consistently between depths and genotypes. In 2014, root growth in the top 15 cm of the soil profile (topsoil) was indeed increased by e[CO$_2$], but increases at lower depths (30–45 cm) later in the season were greater than in the topsoil. In 2015, e[CO$_2$] only increased root length in the topsoil for one genotype, potentially reflecting the lack of plant available soil water between 30–60 cm until recharged by irrigation during grain filling. Our limited data to compare responses to e[CO$_2$] showed that root length increases in the topsoil were correlated with a lower yield response to e[CO$_2$]. The increase in yield response was rather correlated with increases in root growth below 30 cm depth.

Keywords: *Lens culinaris*; climate change adaptation; root development; root depth distribution

1. Introduction

Atmospheric carbon dioxide concentrations [CO$_2$] are increasing steadily, rising from approximately 280 ppm before the industrial age to approximately 413 ppm today (December 2020; www.co2.earth (accessed on 25 January 2021)). In the absence of changes in temperatures and precipitation patterns, elevated [CO$_2$] (e[CO$_2$]) in C3 crops has been associated with higher photosynthetic rates and lower stomatal conductance, leading to greater biomass and grain yield, as well as greater transpiration efficiency [1,2]. While we have ample data on the effects of e[CO$_2$] on above-ground biomass albeit not always consistent, data on the effects of e[CO$_2$] on below-ground processes are more limited, especially in agricultural systems [3,4]. Considering the intimate relationship between resource acquisition and growth, understanding the above-ground response to e[CO$_2$]

and its variability might be better achieved through understanding the response of root systems [3].

Roots are the primary interface of water and nutrient acquisition from the soil, and yet are the most under-researched organ due to the difficulty in observing them in situ. A better understanding of root morphological and physiological traits is the new frontier in understanding adaptation to dryland cropping [5]. While the prevailing view has generally been to assume a large root system would prevent against the negative impacts of drought by providing better access to soil water, Passioura [6] suggested several decades ago that the extraction of too much water too early and excessive carbon allocation to roots under such scenario might be counterproductive. Palta et al. [7] and Lynch [8] recently expanded on this idea by suggesting the optimal size of a crop root system would depend on the typical pattern of water availability during the season [7] and input into the cropping system, with a parsimonious root system likely to be more efficient, especially in high-input systems [8].

A meta-analysis showed that e[CO_2] generally increases both shoot and root biomass [4]. However, some reports have shown that root growth of grain crops is preferentially increased in the topsoil rather than evenly through the soil profile [9–11]. For example, Qiao et al. [10] showed that although root length in the surface layers was increased 35 to 45% in wheat, root length at depth was reduced, leading to lower soil water extraction below 1 m with e[CO_2]. This change, if consistent, might not be beneficial for crops growing in semi-arid environments, especially if grown on stored water, and could lead to a lack of response to e[CO_2], or even potentially, lower grain yields.

Lentil is a cool-season, indeterminate crop whose global production has more than quadrupled since the 1960s. Expansion into new areas such as temperate regions of Australia, the Canadian Prairies and the Northern Great Plains of the United States has led to a global production of over 6.3 million metric tons in 2018 [12]. In Australia, it is grown as a winter crop and is subject to low temperatures during the vegetative stage followed by high temperature stress by flowering and pod filling, often combined with water stress. Bourgault et al. [13] showed that the grain yield response of lentil to e[CO_2] was considerable (average = 42%) but also highly variable (18–138%).

Therefore, the objective of this study was to investigate the effect of e[CO_2] on root growth and soil water extraction in lentil compared to growth at ambient [CO_2], with a special emphasis on the distribution of roots down the soil profile. Our hypothesis was that there might be disproportional root growth in the surface layers, leading to a lower response to e[CO_2] in semi-arid areas. This is the first report of season-long non-destructive root growth observations in lentils grown under Free Air CO_2 Enrichment (FACE).

2. Results

Root length was not increased by e[CO_2] equally across all depths. In 2014, e[CO_2] increased root length at depths 0–15 cm and 30–45 cm (Figure 1). Root length increases are particularly evident prior to flowering for depth 0–15 cm with an average increase of 65% between weeks 11 to 15 in the cultivar PBA Ace, and 72% for the cultivar HS3010 over the same period. At depth 30–45 cm, e[CO_2] increased root length by 266% and 170% for PBA Ace and HS3010, respectively, during grain filling (week 21) although the absolute value of the root increase was greater in HS3010 than PBA Ace. In 2015, e[CO_2] significantly increased root length at depth 0–15 cm for the genotype HS3010 but at 15–30 cm for the cultivar PBA Ace (Figure 1). At flowering (week 17), root length under e[CO_2] was increased 90% in the topsoil for HS3010. By contrast, root length in Ace was increased 15% at depth 15–30 cm at this same time point (Figure 1).

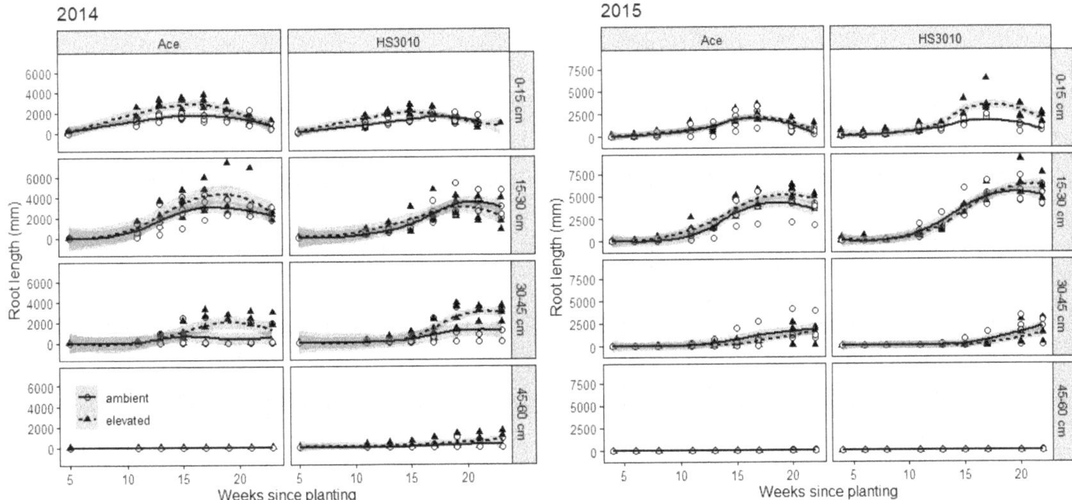

Figure 1. Root length captured by mini-rhizotrons for lentil by [CO_2], cultivar and depth in the Australian Grains Free Air CO_2 Enrichment in 2014/2015. A [CO_2] by time by depth interaction was significant in 2014 and a [CO_2] by cultivar by time by depth interaction was significant in 2015. Flowering occurred at week 17 in both years. Round white circles represent individual sub-plots in ambient main plot and black triangles individual sub-plots in elevated [CO_2] main plots. Grayed areas are the calculated confidence interval for treatment means as calculated by the ggplot2 R package. Symbols may be drawn on top of one another.

There were significant 4-way interactions in both 2014 and 2015 for the plant available water data, although this was possibly due to some tubes starting with higher values at the beginning of the season (Figure 2). Pre-flowering and post-flowering water use showed cultivar differences in 2014, with the cultivar HS3010 showing less water use pre-anthesis, but more post-anthesis (Table 1). These cultivar differences were not detected in 2015, and no [CO_2] treatment differences were obvious from these data in either year. Both seasons were quite dry, but the timing of soil water availability also differed with more soil water available pre-flowering in 2014, whereas in 2015 more soil water was available between flowering and grain filling compared to 2014, due mainly to emergency irrigation events (Table 1).

Table 1. Water use averaged over CO_2 treatments in lentil in 2014 and 2015 in the Australian Grains Free Air CO_2 Enrichment (AGFACE) facility.

Year	Cultivar	Pre-Flowering (mm)	Post-Flowering (mm)	Total (mm)
2014	PBA Ace	126 a [1]	27 b	153 a
	*HS3010	97 b	49 a	146 a
2015	PBA Ace	100 A	99 A	198 A
	*HS3010	97 A	87 A	184 A

[1] Data with the same letter in the same year are not statistically different from each other. *HS3010 is an abbreviation of line 05H010L-07HS3010.

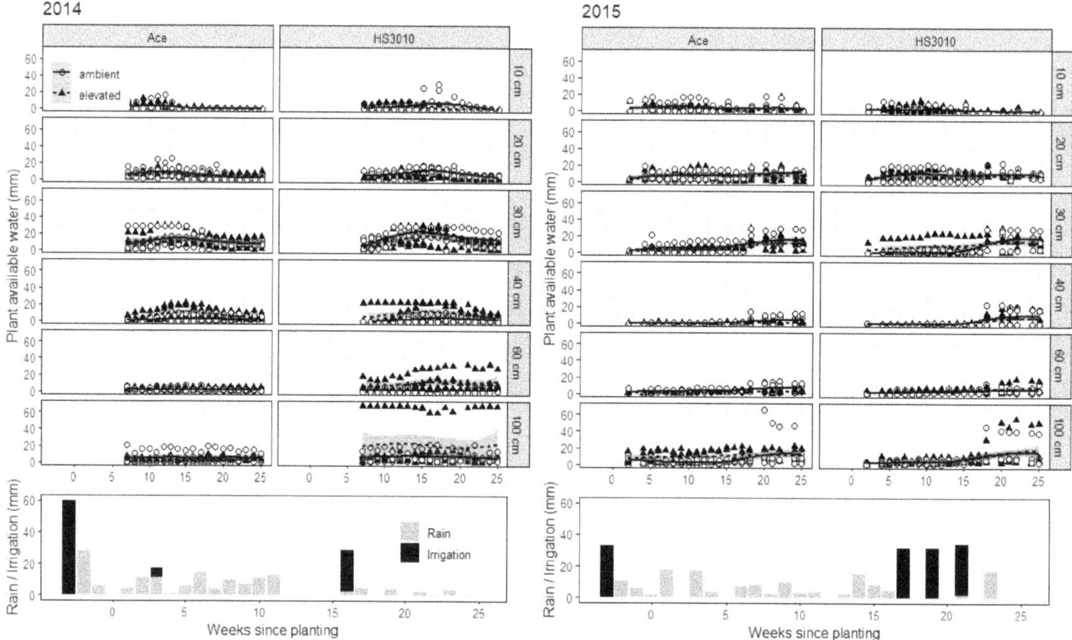

Figure 2. Plant available water for lentil by [CO$_2$] and depth in the Australian Grains Free Air CO$_2$ Enrichment in 2014–2015 with rainfall and irrigation events during the season. In both years, a [CO$_2$] by cultivar by time by depth interaction was significant in 2015. Planting occurred at week 0, flowering at week 17, and harvest at week 25 in both years. Round white circles represent individual sub-plots in ambient main plot, and black triangles individual sub-plots in elevated [CO$_2$] main plots. Grayed areas are the calculated confidence interval for treatment means as calculated by the ggplot2 R package. Symbols may be drawn on top of one another.

Since AGFACE rings were not paired, there is limited replication to directly compare responses. However, differences in root length between e[CO$_2$] and a[CO$_2$] were associated with different [CO$_2$] responses in grain yield, biomass at maturity, and harvest index (HI) depending on the depth where the increase in root growth occurred (Figure 3). There was marginally significant exponential decay relationships between increases in root length in the topsoil at the grain filling stage due to e[CO$_2$] and responses in grain yield ($p = 0.09$) and HI ($p = 0.08$), suggesting a trade-off in carbon (C) allocation between roots and above-ground biomass. By contrast, there were positive but non-significant linear relationships between increases in root length below 30 cm (again at the grain filling stage) and responses in these same parameters ($p = 0.14$, $p = 0.12$ and $p = 0.20$). Relationships between root length and grain [N] were less clear, but there was a marginally significant ($p = 0.08$) negative linear relationship between differences in total root length at grain filling and grain [N], so that increases in root length over the soil profile with e[CO$_2$] were associated with decreases in grain [N] with e[CO$_2$] (Figure 4).

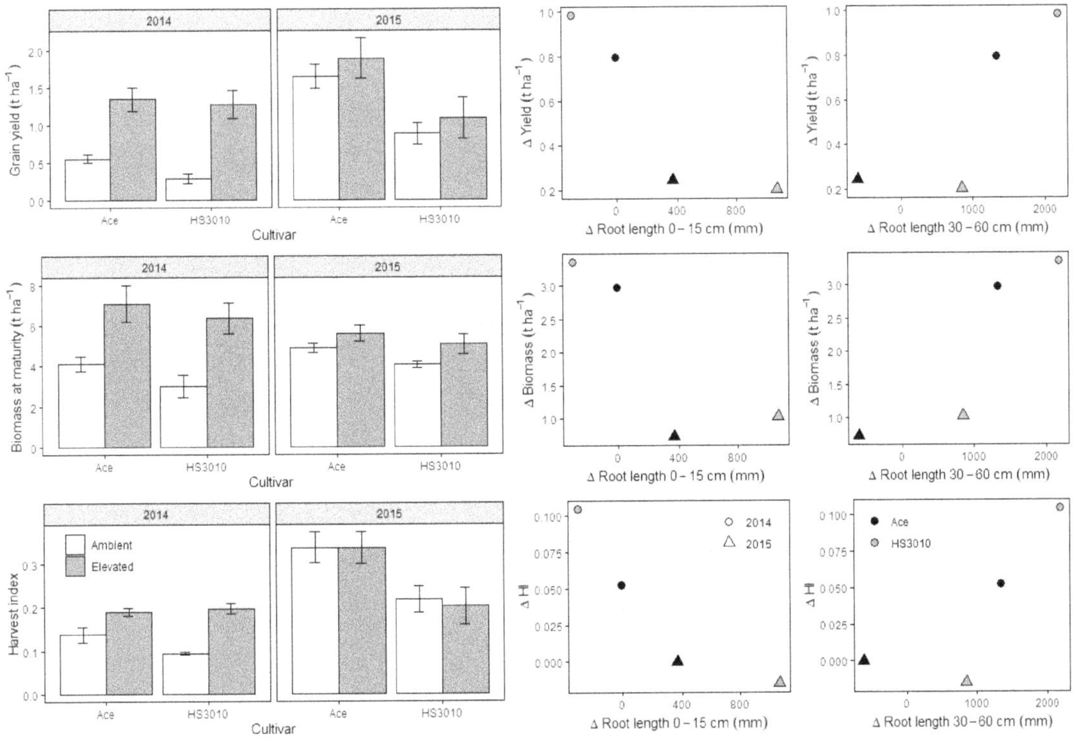

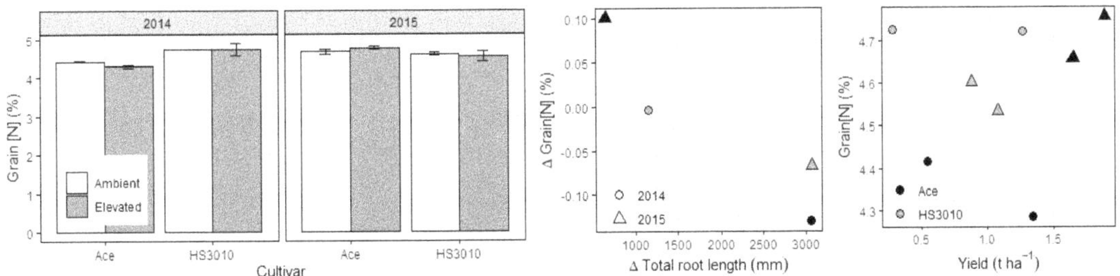

Figure 3. Grain yield, biomass at maturity and harvest index means over CO_2 treatments by years and cultivars in lentil and selected relationships in absolute differences between e[CO_2] and a[CO_2] in these parameters with increases in root length at 0–15 cm and 30–60 cm at grain fill. Barplots are a re-analysis of the data presented in Bourgault et al. [13].

Figure 4. Grain [N] means over CO_2 treatments by years and cultivars in lentil and selected relationships with increases in total root length at 0-15 cm and grain yield. Barplot is a re-analysis of the data presented in Bourgault et al. [13].

3. Discussion

The large increase in root length observed at the 30–45 cm depth in both genotypes in 2014 suggests that e[CO_2] did not, by itself, disproportionally increase root development in the top soil layers. Rather, the response in root growth due to e[CO_2] appeared to be strongly related to the availability of soil water in the soil profile. In fact, Nie et al. [4] showed that the proportion of roots in the topsoil in field experiments was generally reduced by e[CO_2] by approximately 8%. However, they also noted that out of the 17 studies that report this information, the 5 agricultural studies were clear outliers compared to grassland and forest studies and showed an increase of approximately 4%. It is worth noting that agricultural studies tend to be irrigated, often with frequent irrigation events, which may or may not fill the entire profile. For example, Saha et al. [11] applied "moderate irrigation (20 mm) [...] whenever required", and Wechsung et al. [9] used a subsurface (0.18–0.25 m) drip irrigation system with 50% replacement of potential evapotranspiration for their drought treatment. It is possible that soil water was more available at the surface, or near the drip line, and that root increases seen with e[CO_2] would be partly explained by soil water availability differences within the profile rather than depth per se.

Increases in root length in the top layer of the soil profile appeared to be related to a smaller response in grain yield, and characterized by an exponential decay (Figure 3). Although we have limited data points to observe this relationship from rings not being paired, this suggests that root length increases in the topsoil due to e[CO_2] may be a maladaptive response, at least in this environment. This is consistent with results from Benlloch-Gonzalez et al. [14] who demonstrated a lack of above-ground biomass response in a high vigor wheat line that showed greater root growth in the top layers and less root growth at depth under e[CO_2]. It suggests that allocation of the additional C to root growth in the top layer rather than to above-ground biomass is limiting the response to e[CO_2]; presumably greater leaf area at flowering is more important to continue to capture the growth benefits from e[CO_2] than additional access to water (if any) or nutrients in surface layers of the soil.

In contrast, root length increases at 30–60 cm due to e[CO_2] were related to an increasing yield and biomass response to e[CO_2], not just linearly, but exponentially. Again, more data are needed to confirm the shape of the relationship, if any, between above-ground parameter responses and increases in root growth at depth in agricultural crops. Other studies have reported comparisons that are consistent with our results: Uddin et al. [15], in the SoilFACE large soil core facility that is also part of AGFACE, showed that the canola genotype with the greater stimulation of roots at depth (41–80 cm) also showed the greater yield response to e[CO_2]. They also showed that in wheat, in the same environment as the current study, root growth at 45–60 cm depth was correlated with water use, biomass and yield, but root growth at 0–15 cm was not [16]. Roots deeper in the soil profile can extract water during grain filling and have been shown to be particularly useful for yield formation [17,18]. Manschadi et al. [17] for example showed that "each additional millimeter of water extracted [at depth] during grain filling generated an extra 55 kg ha^{-1} of [wheat] grain yield". However, the difference in soil water availability at the 30 and 40 cm depths (Figure 2) between the two years points to how the development of root growth at depth also depends on soil water availability in shallower depths. This raises concerns in semi-arid cropping systems that rely partially or fully on stored soil water, a dry soil layer due to continuous cropping and/or intermittent drought would likely reduce or eliminate crop responses to e[CO_2].

The greater yield response to e[CO_2] seen with less root development in the surface layers and more root development at depth echoes findings from retrospective studies looking at root growth, water use and water use efficiency in breeding populations of both maize [19] and wheat [20] for semi-arid environments. Bolanos et al. [19] showed that yield improvement in maize was associated with 33% less root mass at the 0–50 cm layer, but they did not find any differences in plant water status between old and new cultivars. Pask and Reynolds [20] similarly showed that over time the proportion of root biomass shifted towards deeper soil layers, which improved water use, but not water use efficiency. Selection for such a trait therefore appears useful in itself but might be particularly necessary to take advantage of the increasing atmospheric [CO_2].

We expected that increased root growth under e[CO_2] would have led to faster root exploration into deeper soil layers, with roots detected at depths 30–45 or 45–60 cm in e[CO_2] plots before ambient plots. For example, Chaudhuri et al. [21] investigating root growth with winter wheat in a rhizotron facility showed that roots reached the bottom of the pots faster under $e[CO_2]$ than under $a[CO_2]$. We could not however detect such differences in our experiment with scans being taken every second week.

Grain [N] is often decreased under future high CO_2, negatively affecting the nutritional and economic value of crops [22]. While results on cereals show universal decreases in grain [N], results on pulses are inconclusive. Under ideal growing conditions pulses may balance the increase in carbon source by increasing symbiotic N2 fixation, thus avoiding imbalances in the C:N ratio, but this benefit has proven elusive under more challenging growing conditions, especially under drought and heat [23,24]. This is consistent with the findings of the current study where grain [N] was generally maintained under $e[CO_2]$, but was significantly decreased in the drier 2014 season in cultivar Ace ($p = 0.0279$; Figure 4), indicating that the source of N (whether from N_2 fixation or from soil uptake) feeding into the grain can be affected by the amount and timing of water availability [25,26].

The relationship between the difference (in absolute values) in grain [N] due to $e[CO_2]$ and the difference in total root length was marginally significant ($p = 0.08$; Figure 4). The negative relationship indicates that increases in root length over the soil profile with $e[CO_2]$ were associated with decreases in grain [N]. This negative relationship coincides with the positive relationship between increases in root length and increases in yield (Figure 3), although the well-known negative relationship between grain yield and grain [N] (or grain protein more generally) is not significant in this case (Figure 4; $p = 0.70$). It is noteworthy that Bahrami et al. [27] did not find a relationship between grain protein and root traits in wheat and the authors suggested that root growth stimulation (or any changes in specific root uptake activity) under $e[CO_2]$ were insufficient to alleviate the negative effect of $e[CO_2]$ on wheat grain protein.

Root length and soil water data have inherently high variability, especially when measured under field conditions and this limits the ability to distinguish clear treatment differences. Before the beginning of this experiment, we were concerned that the images obtained with mini-rhizotrons might underestimate the root length in the surface layer and overestimate root length at depth, especially if roots took the path of least resistance and followed the outer wall of the tube down the profile. While we did not have the resources to take root biomass samples in addition to root images, it appears that the mini-rhizotrons did underestimate root length at the surface (based on published root mass distribution in various crops [21,28,29], but sufficient root growth was detected that the bias should be consistent between $[CO_2]$ and cultivar treatments. We also found that while roots would follow the tube down for some length, they did not generally go all the way down, but often turned back into the soil. Many images at depth had no roots visible at all, which might partly explain the lack of significance for some apparent differences between cultivars. For example, we might have expected a cultivar by depth interaction in 2014 due to the growth of roots at the 45–60 cm depth in HS3010 and lack of root growth in the cultivar Ace at this same depth, but the p-value for this interaction was 0.547. Some images were also lost in 2014 due to poor image quality arising from soil cracks (in this cracking clay soil) interfering with image capture. This problem was corrected in 2015 by covering the top of the tube with a carboard box. In a different experiment (at a different site), the coefficient of variation was around 40–48% for depths 0–15, 15–30 and 30–45 cm, and up to 87% for depth 45–60 cm for the check cultivar (Bourgault et al., unpublished). Ohashi et al. [30] suggested using at least 3 replications; we would rather suggest using 5 replications or more, and carefully evaluating the level of differences the researchers would like to be able to detect statistically.

Similarly, PAW showed some differences between tubes that remained consistent throughout the season, which suggests there might have been micro-environment differences between the tubes that were not captured in the depth-wide calibration (Figure 2). Freebairn and Ghahramani [31] showed that PAW is generally accurate in the ±50 mm range, and although errors are expected to go down proportionally as the absolute number gets lower, small $[CO_2]$ treatment differences in water use (in the range of 2–10 mm) might be too challenging to detect considering typical soil variability.

4. Materials and Methods

4.1. Site Location and Climate

The experiment was conducted in 2014 and 2015 at the Australian Grains Free Air CO_2 Enrichment (AGFACE) facility, near Horsham, VIC, Australia (36°45′07″ S, 142°06′52″ E, 127 m above sea level), on a Vertosol grey clay [32]. Long-term annual rainfall at this site is 435 mm (based on the 1981–2010 period), with approximately 320 mm falling during the winter growing season, i.e., from May to November inclusively [33]. However, 2014 and 2015 were considered very dry years, with only 221 and 214 mm of annual rainfall, respectively, and growing season rainfall (April to November), of 110 and 120 mm, respectively. Average maximum and minimum temperatures were 17.6 and 5.3 °C, respectively, during the growing season. In both years, warmer-than-average temperatures in October accelerated lentil development leading to early harvests for the region [34,35]. In addition, in 2015, a late frost on 1 October (-0.2 °C) was followed by a heat wave on October 4–6th (with maximum temperatures of 35–37 °C).

4.2. Site Management

Plots were planted on 12 May 2014 and 26 May 2015. Pre-emergence herbicides (simazine, trifluralin, isoxaflutole, and/or glyphosate) were applied to help manage weeds. Superphosphate was placed with the seed at sowing at a rate equivalent to 9 kg P ha^{-1} and 11 kg S per ha^{-1}. Seeds were inoculated with a granular pea and lentil inoculant in 2014 (Nodulator, BASF Corporation, Research Triangle Park, NC, USA) and a peat-based inoculant in 2015 (NoduleN, New Edge Microbials Pty Ltd., Albury, NSW, Australia). No nitrogen fertilizer was applied. In 2014, the insecticide dimetoate was applied to control aphid, and the fungicide chlorothalonil was applied to prevent against ascochyta blight. Weeds were controlled by hand during the season. Pre-planting irrigation events were applied (60 mm in 2014 and 33 mm in 2015), as well as emergency irrigation events during the season to alleviate the most severe water stress: in 2014, 6 mm irrigation was applied on 16 June to help emergence, and another 26 mm irrigation was applied on 3–5 September; in 2015, three 32 mm irrigations were applied on 21–22 September, 8–9 October, and 4–6 October (Figure 2).

4.3. FACE Technology

In AGFACE, elevated [CO_2] was achieved by injecting pure CO_2 into the air: stainless steel tubes were positioned about 50 cm above the canopy, and depending on wind speed and direction, CO_2 was released upwind so as to be carried across the ring. The target [CO_2] for the elevated treatment was 550 mmol mol^{-1} air. Concentrations were maintained within 90% of this target (i.e., 500–600 mmol mol^{-1} air) 93–98% of the time [36]. More details about the exposure equipment are given in Mollah et al. [36].

4.4. Experimental Design

The experiment was a randomized complete block split-plot design with 4 blocks, each containing one ambient and one elevated [CO_2] octagonal ring, as the main plots. Within the main plots, different genotypes were grown in subplots. In 2014, rings were 4 m in diameter with six subplots of 4 rows (0.244 m row spacing) by 2 m in length (total subplot size: 1.95 m^2) containing six lentil cultivars but only two equipped with mini-rhizotrons. Randomization within the main plot was restricted to have these cultivars end to end in a middle row. In 2015, lentil subplots were included in 12-m diameter rings containing primarily wheat, with subplots of 6 rows (0.25 m row spacing) by 4 m in length (total subplot size: 6 m^2). Randomization within the ring was also restricted to have lentil plots together end to end for ease of access to mini-rhizotron tubes. More details are available in Bourgault et al. [13,24].

4.5. Plant Material

The commercial cultivar PBA Ace and the breeding line 05H010L-07HS3010 (shortened to HS3010 hereafter) were selected for this study. Both lines are medium-sized red lentils with similar phenology. PBA Ace is a modern variety that performs well under a range of Australian conditions. By contrast, the breeding line HS3010 had previously showed large biomass accumulation in favorable growing conditions but showed a smaller harvest index than most commercial lines (M. Rodda, personal communication). Lentil is often thought of as source-limited, so these two contrasting lines were considered particularly interesting to examine in terms of the effect of e[CO_2] on both above-ground and below-ground processes.

4.6. Root Growth Observations

Clear acrylic tubes 105.5 cm long by 7.5 cm diameter (i.e., mini-rhizotrons) were installed at a 45° angle within 7 days of planting. These were installed between center rows of experimental plots, but towards the front of the plot (within 50–70 cm) to limit traffic inside the plots. Every 2–3 weeks, root scans were taken with a cylindrical scanner (CI-600 In Situ Root Imager, CID Bioscience, Camas, WA, USA). Four images were taken per tube with depths approximating 0–15, 15–30, 30–45, and 45–60. The length of the tube out of the ground was recorded, and the angle measured after installation. Overall, the angle varied by 5°, and the length out of the ground varied by at most 8 cm, so that the maximum difference in depth for images was 2–5 cm. Images were later processed with the *RootSnap!* software (CID Bioscience, Camas, WA, USA) and the root length for each image recorded.

4.7. Soil Water Monitoring

Soil water was monitored with a PR2/6 profile probe (Delta-T Devices, Cambridge, UK), which outputs soil water at depths 10, 20, 30, 40, 60, and 100 cm. Access tubes were installed between center rows of the experimental plots but not in the same row as the mini-rhizotrons. They were also installed further inside the plots (1–2.5 m from the front, depending on the size of the plot). Measurements were taken on a weekly basis for the duration of the experiment, except early in the season in 2014 when there were delays in receiving and installing the profile probe and access tubes. A manual calibration was performed for each depth using the soil samples taken at the time of the tube installation and assessed gravimetrically in both years. Plant available water (PAW) was calculated for each depth from the volumetric water content using values for the wilting point and for field capacity (by O'Leary et al. [37]). These data are presented in Figure 2, along with rainfall and irrigation events. Pre-flowering and growing season water use were calculated by taking the difference in PAW across depths between the first and last PR2 measurements of the period (week 16 for pre-flowering, therefore not including flowering), or season and adding rainfall and irrigation events. Post-flowering water use was calculated from the difference between season-long and pre-flowering water use. This water use is an approximation of evapotranspiration as it includes water lost through evaporation but assumes no runoff or deep drainage. In 2014, there was a delay in getting the equipment, and these data are therefore potentially under-estimated.

4.8. Other Measurements

Biomass, grain yield and yield components were previously published in Bourgault et al. [13,24]. Briefly, destructive samples were collected at flowering (full bloom) and maturity stages according to the description of phenological stages in lentil by Erskine et al. [38]. Samples were separated into leaf, stem and flower tissues at flowering, and grain and straw (dead leaves and stems) at maturity. These tissues were ground and analyzed separately for nitrogen concentration by LECO Tru Mac Elemental Analyser (LECO Corporation, St. Joseph, MI, USA).

4.9. Statistics

The root length data, by nature, contained many values of zero, especially at the beginning of the season and at depth. In addition, root length over the season followed a typical growth curve, but the curvature varied by depth. As such, the structure of the data included non-constant variance as well as time- and depth-related curvatures. Accordingly, the root length data were transformed by taking the square root, and the time factor was considered as a numerical cubic polynomial factor based on week from planting. Each year was analyzed separately because the timing of sampling differed slightly between years and because environmental conditions differed between the two years. Fixed factors included [CO_2], cultivar, time, depth, and all interactions between these four factors. A mixed model was used to account for the measurement structure to account for repeated measures at the same depth in the same year; the random effects included image location (depth), nested within plot, nested within ring. The model was run within R 3.6.3 [39] using the linear mixed model function (lme) within the nlme 3.1 package [40]. A similar approach was used for the PAW data, but the data were not transformed. The water use data over all depths (pre-flowering, post-flowering and season-long) were analyzed as a univariate model with [CO_2] and cultivars as fixed factors and plot nested within ring in the random term. Graphs were produced with the ggplot2 3.3.0 package [41]. Significance was determined at $\alpha = 0.05$, but $\alpha < 0.10$ was considered marginally significant and discussed in the text.

5. Conclusions

Our results suggested soil water availability might play an important role in the stimulation of root growth under e[CO_2]. In two dry seasons which had contrasting patterns of timing of water stress, root growth in deeper layers was indeed associated with increases in the yield response to e[CO_2], likely through improved access to soil water later in the season, while increases in root growth in shallow layers were not. Presumably, it is more beneficial for the crop, in lentil at least, to allocate the additional carbon towards above-ground biomass and greater leaf area before flowering. Our results also suggested that for semi-arid areas, breeding for improved water extraction at depth, while beneficial in itself, might be particularly important to take advantage of the benefits of increasing e[CO_2], provided there is sufficient plant available soil water at depth, and no intermediate dry soil layer to block root access to it.

Author Contributions: For research articles with several authors, a short paragraph specifying their individual contributions must be provided. Conceptualization: M.B., S.T.-P., R.D.A., G.L.O., G.J.F., and M.T.; methodology, M.B., S.T.-P., R.D.A., G.L.O., G.J.F., and M.T.; validation, M.B. and M.L.; formal analysis, M.B. and M.G.; investigation, M.B. and S.H.; data curation, M.B. and M.L.; writing—original draft preparation, M.B.; writing—review and editing, S.T.-P., M.G., R.D.A., G.L.O., G.J.F., and M.T.; visualization, M.B.; supervision, M.T., project administration, G.J.F. and M.T.; funding acquisition, S.T.-P., R.D.A., G.L.O., G.J.F., and M.T. All authors have read and agreed to the published version of the manuscript.

Funding: The AGFACE program was jointly run by The University of Melbourne and Agriculture Victoria (Victorian Department of Economic Development, Jobs, Transport and Resources), and received additional funding from the Australian Commonwealth Department of Agriculture and Water Resources (grant number FtRG 1193982-41) and the Grains Research and Development Corporation (grant numbers DAV00121 and DAV00137). Statistical Consulting and Research Services at Montana State University is supported by Institutional Development Awards (IDeA) from the National Institute of General Medical Sciences of the National Institutes of Health under Awards P20GM103474 and 2U54GM104944, and the content is solely the responsibility of the authors and does not necessarily represent the official views of the National Institutes of Health.

Institutional Review Board Statement: Not applicable.

Data Availability Statement: The data that support the findings of this study are available from the corresponding author (M.B.) upon reasonable request and approval from the former AGFACE program leaders (M.T. and G.J.F.).

Acknowledgments: The authors gratefully acknowledge Mahabubur Mollah for running the FACE technology, Russel Argall and Ash Purdue for help in installing mini-rhizotron tubes and for overall agronomic management of the AGFACE site, as well as Lee Warren and Shahnaj Parvin for technical help. Further, the contributions of Samuel Henty and Shelly Green in tracing roots is whole-heartedly acknowledged.

Conflicts of Interest: The authors declare no conflict of interest. The funders had no role in the design of the study; in the collection, analyses, or interpretation of data; in the writing of the manuscript; or in the decision to publish the results.

References

1. Kimball, B.A. Crop responses to elevated CO_2 and interactions with H_2O, N and temperature. *Curr. Opin. Plant Biol.* **2016**, *31*, 36–43. [CrossRef] [PubMed]
2. Leakey, A.D.B.; Ainsworth, E.A.; Bernacchi, C.J.; Rogers, A.; Long, S.P.; Ort, D.R. Elevated CO_2 effects on plant carbon, nitrogen, and water relations: Six important lessons from FACE. *J. Exp. Bot.* **2009**, *60*, 2859–2876. [CrossRef]
3. Madhu, M.; Hatfield, J.L. Dynamics of plant root growth under increased atmospheric carbon dioxide. *Agron. J.* **2013**, *105*, 657–669. [CrossRef]
4. Nie, M.; Lu, M.; Bell, J.; Raut, S.; Pendall, E. Altered root traits due to elevated CO_2: A meta-analysis. *Glob. Ecol. Biogeogr.* **2013**, *22*, 1095–1105. [CrossRef]
5. Hall, A.; Sadras, V. Whither Crop Physiology? In *Crop Physiology: Applications for Genetic Improvement and Agronomy*; Sadras, V.O., Calderini, D.F., Eds.; Academic Press: Burlington, MA, USA, 2009; pp. 545–570.
6. Passioura, J.B. Roots and Drought Resistance. *Agric. Water Manag.* **1983**, *7*, 265–280. [CrossRef]
7. Palta, J.A.; Chen, X.; Milroy, S.P.; Rebetzke, G.J.; Dreccer, M.F.; Watt, M. Large root systems: Are they useful in adapting wheat to dry environments? *Funct. Plant Biol.* **2011**, *38*, 347–354. [CrossRef]
8. Lynch, J.P. Rightsizing root phenotypes for drought resistance. *J. Exp. Bot.* **2018**, *69*, 3279–3292. [CrossRef] [PubMed]
9. Wechsung, G.; Wechsung, F.; Wall, G.W.; Adamsen, F.J.; Kimball, B.A.; Pinter, P.J., Jr.; Lamorete, R.L.; Garcia, R.L.; Kartschall, T.H. The effects of free-air CO_2 enrichment and soil water availability on spatial and seasonal patterns of wheat root growth. *Glob. Chang. Biol.* **1999**, *5*, 519–529. [CrossRef]
10. Qiao, Y.; Zhang, H.; Dong, B.; Shi, C.; Li, Y.; Zhai, H. Effects of elevated CO_2 concentration on growth and water use efficiency of winter wheat under two soil water regimes. *Agric. Water Manag.* **2010**, *97*, 1742–1748. [CrossRef]
11. Saha, S.; Chakraborty, D.; Lata; Pal, M.; Nagarajan, S. Impact of elevated CO_2 on utilization of soil moisture and associated soil biophysical parameters in pigeon pea (*Cajanus cajan* L.). *Agric. Ecosyst. Environ.* **2011**, *142*, 213–221. [CrossRef]
12. FAOSTAT, 2020. Crops. Available online: http://www.fao.org/faostat/en/#data/QC (accessed on 23 April 2020).
13. Bourgault, M.; Brand, J.; Tausz-Posch, S.; Armstrong, R.D.; O'Leary, G.L.; Fitzgerald, G.J.; Tausz, M. Yield, growth and grain nitrogen response to elevated CO_2 in six lentil (*Lens culinaris*) cultivars grown under Free Air CO_2 Enrichment (FACE) in a semi-arid environment. *Eur. J. Agron.* **2017**, *87*, 50–58. [CrossRef]
14. Benlloch-Gonzalez, M.; Berger, J.; Bramley, H.; Rebetzke, G.; Palta, J.A. The plasticity of growth and proliferation of wheat root system under elevated CO_2. *Plant Soil* **2014**, *374*, 963–976. [CrossRef]
15. Uddin, S.; Löw, M.; Parvin, S.; Fitzgerald, G.; Tausz-Posch, S.; Armstrong, R.; Tausz, M. Yield of canola (*Brassica napus* L.) benefits more from elevated CO_2 when access to deeper soil is improved. *Environ. Exp. Bot.* **2018**, *155*, 518–528. [CrossRef]
16. Uddin, S.; Löw, M.; Parvin, S.; Fitzgerald, G.; Bahrami, H.; Tausz-Posch, S.; Armstrong, R.; O'Leary, G.; Tausz, M. Water use and growth responses of dryland wheat grown under elevated [CO_2] are associated with root length in deeper, but not upper soil layer. *Field Crop. Res.* **2018**, *224*, 170–181. [CrossRef]
17. Manschadi, A.M.; Christopher, J.; Devoil, P.; Hammer, G.L. Dynamics of root architectural traits in adaptation of wheat to water-limited environments. *Funct. Plant Biol.* **2006**, *33*, 823–837. [CrossRef]
18. Kirkegaard, J.A.; Lilley, J.M.; Howe, G.N.; Graham, J.M. Impact of subsoil water use on wheat yield. *Aust. J. Agric. Res.* **2007**, *58*, 303–315. [CrossRef]
19. Bolanos, J.; Edmeades, G.O.; Martinez, L. Eight cycles of selection for drought tolerance in lowland tropical maize. III. Responses in drought adaptive physiological and morphological traits. *Field Crop. Res.* **1993**, *31*, 261–286. [CrossRef]
20. Pask, A.J.D.; Reynolds, M.P. Breeding for yield potential has increased deep soil water extraction capacity in irrigated wheat. *Crop Sci.* **2013**, *53*, 2090–2104. [CrossRef]
21. Chaudhuri, U.N.; Kirkham, M.B.; Kanemasu, E.T. Root growth of winter wheat under elevated carbon dioxide and drought. *Crop Sci.* **1990**, *30*, 853–857. [CrossRef]
22. Myers, S.S.; Zanobetti, A.; Kloog, I.; Huybers, P.; Leakey, A.D.B.; Bloom, A.J.; Carlisle, E.; Dietterich, L.H.; Fitzgerald, G.; Hasewaga, T.; et al. Increasing CO_2 threatens human nutrition. *Nature* **2014**, *510*, 139–142. [CrossRef] [PubMed]
23. Parvin, S.; Uddin, S.; Bourgault, M.; Roessner, U.; Tausz-Posch, S.; Armstrong, R.; O'Leary, G.; Tausz, M. Water availability moderates N2 fixation benefit from elevated [CO_2]: A 2-year free-air CO_2 enrichment study on lentil (Lens culinaris MEDIK.) in a water limited agroecosystem. *Plant Cell Environ.* **2018**, *41*, 2418–2434. [CrossRef] [PubMed]

24. Bourgault, M.; Löw, M.; Tausz-Posch, S.; Nuttall, J.G.; Delahunty, A.J.; Brand, J.; Panozzo, J.F.; McDonald, L.; O'Leary, G.J.; Armstrong, R.D.; et al. Effect of a heat wave on lentil grown under free-air CO_2 enrichment (FACE) in a semi-arid environment. *Crop Sci.* **2018**, *58*, 803–812. [CrossRef]
25. Parvin, S.; Uddin, S.; Tausz-Posch, S.; Fitzgerald, G.; Armstrong, R.; Tausz, M. Elevated CO_2 improves yield and N_2 fixation but not grain N concentration of faba bean (Vicia faba L.) subjected to terminal drought. *Environ. Exp. Bot.* **2019**, *165*, 161–173. [CrossRef]
26. Tausz-Posch, S.; Tausz, M.; Bourgault, M. Elevated [CO_2] effects on crops: Advances in understanding acclimation, nitrogen dynamics and interactions with drought and other organisms. *Plant Biol.* **2020**, *22* (Suppl. 1), 38–51. [CrossRef]
27. Bahrami, H.; De Kok, L.J.; Armstrong, R.; Fitzgerald, G.J.; Bourgault, M.; Henty, S.; Tausz, M.; Tausz-Posch, S. The proportion of nitrate in leaf nitrogen, but not changes in root growth, are associated with decreased grain protein in wheat under elevated [CO_2]. *J. Plant Physiol.* **2017**, *216*, 44–51. [CrossRef]
28. Gan, Y.T.; Campbell, C.A.; Janzen, H.H.; Lemke, R.; Liu, L.P.; Basnyat, P.; McDonald, C.L. Root mass for oilseed and pulse crops: Growth and distribution in the soil profile. *Can. J. Plant Sci.* **2009**, *89*, 883–893. [CrossRef]
29. Gorim, L.Y.; Vanderberg, A. Root traits, nodulation and root distribution in soil for five wild lentil species and *Lens culinaris* (Medik.) grown under well-watered conditions. *Front. Plant Sci.* **2017**, *8*, 1632. [CrossRef] [PubMed]
30. Ohashi, A.Y.P.; Pires, R.C.d.M.; Silva, A.L.B.d.O.; dos Sanots, L.N.S.; Matsura, E.E. Minirhizotron as an in-situ tool for assessing sugarcane root system growth and distribution. *Agric. Res. Technol.* **2019**, *22*, 556182. [CrossRef]
31. Freebairn, D.; Ghahramani, A. Challenges in estimating soil water. In Proceedings of the National Soils Conference, Canberra, ACT, Australia, 18–23 November 2018; Hulugalle, N., Biswas, T., Greene, R., Bacon, P., Eds.; pp. 429–430. Available online: https://www.soilscienceaustralia.org.au/wp-content/uploads/2019/10/Proceedings-Natl.-Soil-Sci-Conf-Canberra-18-23-Nov-2018-FINAL_reduced-size-1.pdf (accessed on 29 January 2021).
32. Isbell, R.F. *The Ausralian Soil Clasification*; CSIRO Publishing: Collingwood, VIC, Australia, 1996.
33. Bureau of Meteorology (Australian Government). Climate Data Online database (Horsham Polkemmet Rd VIC). 2016. Available online: http://www.bom.gov.au/climate/data/ (accessed on 5 July 2016).
34. Hollaway, K. *Victorian Winter Crop Summary 2015*; Department of Economic Development, Jobs, Transport and Resources: Victoria, Australia, 2015; 80p.
35. Couchman, J.; Hollaway, K. *Victorian Winter Crop Summary 2016*; Department of Economic Development, Jobs, Transport and Resources: Victoria, Australia, 2016; 80p.
36. Mollah, M.; Norton, R.; Huzzey, J. Australian grains free-air carbon dioxide enrichment (AGFACE) facility: Design and performance. *Crop Pasture Sci.* **2009**, *60*, 697–707. [CrossRef]
37. O'Leary, G.J.; Christy, B.; Nuttall, J.; Huth, N.; Cammarano, D.; Stockle, C.; Basso, B.; Shcherbak, I.; Fitzgerald, G.; Luo, Q.; et al. Response of wheat growth, grain yield and water use to elevated CO_2 under Free-Air CO_2 Enrichment (FACE) experiment and modelling in a semi-arid environment. *Glob. Chang. Biol.* **2015**, *21*, 2670–2686. [CrossRef]
38. Erskine, W.; Ellis, R.H.; Summerfield, R.J.; Roberts, E.H.; Hussain, A. Characterization of responses to temperature and photoperiod for time to flowering in a world lentil collection. *Theor. Appl. Genet.* **1990**, *80*, 193–199. [CrossRef] [PubMed]
39. R Core Team. *R: A Language and Environment for Statistical Computing*; R Foundation for Statistical Computing: Vienna, Austria, 2020; Available online: https://www.R-project.org/ (accessed on 27 April 2020).
40. Pinheiro, J.; Bates, D.; DebRoy, S.; Sarkar, D.; R Core Team. Nlme: Linear and Nonlinear Mixed Effects Models. R package version 3.1-144, 2020. Available online: https://CRAN.Rproject.org/package=nlme (accessed on 27 April 2020).
41. Wickham, H. *ggplot2: Elegant Graphics for Data Analysis*; Springer: New York, NY, USA, 2009; 213p.

Article

Cloning and Characterization of Three Sugar Metabolism Genes (*LBGAE*, *LBGALA*, and *LBMS*) Regulated in Response to Elevated CO$_2$ in Goji Berry (*Lycium barbarum* L.)

Yaping Ma [1,2,3], Mura Jyostna Devi [3,4,5,*], Vangimalla R. Reddy [3], Lihua Song [2,3], Handong Gao [1] and Bing Cao [2,*]

1   College of Forestry, Nanjing Forestry University, Nanjing 210037, China; YapingMa@njfu.edu.cn (Y.M.); gaohd@njfu.edu.cn (H.G.)
2   School of Agriculture, Ningxia University, Yinchuan 750021, China; slh382@126.com
3   USDA-ARS, Adaptive Cropping Systems Laboratory, 10300 Baltimore Ave, Beltsville, MD 20705, USA; Vangimalla.Reddy@ars.usda.gov
4   USDA-ARS, Vegetable Crops Research Unit, Madison, WI 53706, USA
5   Department of Horticulture, University of Wisconsin-Madison, Madison, WI 53705, USA
*   Correspondence: Jyostna.mura@usda.gov (M.J.D.); bingcao2006@126.com (B.C.)

Abstract: The composition and content of sugar play a pivotal role in goji berry (*Lycium barbarum* L.) fruits, determining fruit quality. Long-term exposure of goji berry to elevated CO$_2$ (eCO$_2$) was frequently demonstrated to reduce sugar content and secondary metabolites. In order to understand the regulatory mechanisms and improve the quality of fruit in the changing climate, it is essential to characterize sugar metabolism genes that respond to eCO$_2$. The objectives of this study were to clone full-length cDNA of three sugar metabolism genes—*LBGAE* (*Lycium barbarum* UDP-glucuronate 4-epimerase), *LBGALA* (*Lycium barbarum* alpha-galactosidase), and *LBMS* (*Lycium barbarum* malate synthase)—that were previously identified responding to eCO$_2$, and to analyze sequence characteristics and expression regulation patterns. Sugar metabolism enzymes regulated by these genes were also estimated along with various carbohydrates from goji berry fruits grown under ambient (400 μmol mol^{-1}) and elevated (700 μmol mol^{-1}) CO$_2$ for 90 and 120 days. Homology-based sequence analysis revealed that the protein-contained functional domains are similar to sugar transport regulation and had a high sequence homology with other Solanaceae species. The sucrose metabolism-related enzyme's activity varied significantly from ambient to eCO$_2$ in 90-day and 120-day samples along with sugars. This study provides fundamental information on sugar metabolism genes to eCO$_2$ in goji berry to enhance fruit quality to climate change.

Keywords: goji berry; sugar metabolism; elevated CO$_2$; functional domain; gene cloning and expression

check for **updates**

Citation: Ma, Y.; Devi, M.J.; R. Reddy, V.; Song, L.; Gao, H.; Cao, B. Cloning and Characterization of Three Sugar Metabolism Genes (*LBGAE*, *LBGALA*, and *LBMS*) Regulated in Response to Elevated CO$_2$ in Goji Berry (*Lycium barbarum* L.). *Plants* **2021**, *10*, 321. https://doi.org/10.3390/plants10020321

Academic Editor: James Bunce

Received: 19 January 2021
Accepted: 4 February 2021
Published: 7 February 2021

Publisher's Note: MDPI stays neutral with regard to jurisdictional claims in published maps and institutional affiliations.

1. Introduction

Goji berry (*Lycium barbarum*) is a perennial shrub and belongs to the genus *Lycium* of the family Solanaceae, also called wolfberry. China is the leading supplier of goji, with a yield of 95,000 tons and cultivating area of nearly 90,000 hectares. Currently, the goji berry is being used worldwide as popular nutraceutical food or dietary health supplement in various forms due to its rich chemical composition and healing properties [1]. With the increasing demand due to its medicinal properties, more attention has been paid to improve the fruit quality of the goji berry. The LBP (*Lycium barbarum* polysaccharide) is the most abundant group of active ingredients in goji berry fruit, containing several monosaccharides that are influenced by several environmental factors [2,3].

Climate change has profound global consequences with direct effects on agricultural production, livestock and environment quality, and indirect influence on food security and human survival [4]. Elevated atmospheric CO$_2$ concentration is beneficial for photosynthesis and can directly affect the growth and development of plants [5–7]. While plants exposed to increased levels of CO_2 are expected to benefit from carbon fertilization, the extent of advantage depends on the plant species. For example, in tomato, carbon exchange rates were significantly higher in CO_2-enriched plants for the first few weeks of treatment after that decreased due to acclimation to elevated CO_2, which attributed to an accumulation of sugar and/or starch [8]. Similarly, prolonged elevated atmospheric CO_2 levels were shown to negatively affect goji berry fruit quality with reduced flavonoid, carotenoid, total sugars and polysaccharides, which might be related to photosynthesis acclimation [9].

A long-term elevated CO_2 condition accelerates the vegetative growth and improves the morphology of the fruit, but it was also found that the contents of fructose, glucose, sucrose, polysaccharide, and total sugar were reduced. The sucrose-metabolizing enzymes changed significantly over the period across treatments [10]. Similar results were also observed in goji berry fruits grown under eCO_2, where the reduction in sugar and secondary metabolite levels was noticed, affecting fruit quality [9]. The sugar content in the fruit is determined by the sugar biosynthesis and accumulation, which is catalyzed by relating enzymes controlled by the expression of genes. The reduction in the sink and an increase in non-structural carbohydrate levels due to photosynthesis acclimation were related to a single gene mutation in soybean [11]. Hence, it is important to identify and characterize the genes involved in sugar metabolism to improve the quality of the goji berry fruit for future climatic conditions, especially with the projected increase in the atmospheric CO_2 concentration.

Sucrose, glucose, and fructose are the main components of plant fruits. Sucrose is a significant form of photosynthate component transportation from the source (leaves) and is distributed to sink organs (fruit). Sucrose supplies carbon and energy to plants by first relying on their cleavage to form hexoses. This process is catalyzed by at least two different classes of enzymes: one is sucrose synthase, which catalyzes the cleavage of sucrose to uridine diphosphate (UDP) and fructose; the other is invertase, catalyzing the cleavage of sucrose to form glucose and fructose [12–15]. The molecular regulation of sugar metabolism and genes involved in such regulation in goji berry fruit to different environmental conditions is still unknown, especially under elevated CO_2. Hence, the objective of the current study was to clone and analyze the sequence of three goji berry genes—*LBGAE* (*Lycium barbarum* UDP-glucuronate 4-epimerase), *LBGALA* (*Lycium barbarum* alpha-galactosidase), and *LBMS* (*Lycium barbarum* malate synthase)—that were identified in our previous study through transcriptome profiling [9]. The reason for selecting these three genes was that they were upregulated under eCO_2 and are involved in four primary sugar metabolic pathways. The three genes involved in sugar metabolism pathways and regulation, especially in sucrose metabolism, did not express under ambient conditions but increased their expression under elevated CO_2 treatment. The other objective was to study the activity of four sucrose metabolism-related enzymes (sucrose phosphate synthase, neutral invertase, acid invertase, sucrose synthase) regulated by the three genes. The end components of these three genes reducing sugars, starch, polysaccharide, and sucrose, were also estimated to observe the effects of elevated CO_2 on substance accumulation and metabolism in goji berry to improve the fruit quality.

2. Results

2.1. Identification and the Complete Sequence Cloning of Sugar Metabolism Genes

Three sugar metabolism-related genes *LBGAE* (*Lycium barbarum* UDP glucuronate 4-epimerase), *LBGALA* (*Lycium barbarum* alpha-galactosidase) and *LBMS* (*Lycium barbarum* malate synthase) upregulated to elevated CO_2 were selected from a previous transcriptome profiling study of goji berry [9]. These three genes were upregulated under eCO_2 and are involved in four primary sugar metabolic pathways. Subsequently, this study characterized the three genes since these genes are involved in four sugar pathways. The sequences obtained by the transcriptome profiling are usually incomplete; hence, cloning of cDNA of *LBGAE*, *LBGALA,* and *LBMS* was performed to analyze the complete sequence.

The full-length cDNA fragments of *LBGAE* (GenBank accession No. MH025911), *LBGALA* (GenBank accession No. MH025913), and *LBMS* (GenBank accession No. MH025912) obtained by RACE-PCR (rapid amplification of cDNA) were 1837 bp, 1233 bp, and 2449 bp, respectively. The *LBGAE* contains a 5′-untranslated region (UTR) of 466 bp and an open reading frame (ORF) of 1236 bp encoding 411 apiece, and a 3′-untranslated region (UTR) of 135 bp. Similarly, the *LBMS* contains a 5′- UTR of 400 bp and an ORF of 1716 bp encoding 571 amino acids apiece, and a 3′- UTR of 333 bp. The *LBGALA* only contained an ORF of 1233 bp encoding 410 amino acids. The nucleotide and deduced amino acid sequences of *LBGAE*, *LBGALA*, and *LBMS* are shown in Supplementary Materials Figure S1a–c.

2.2. Characterization and Sequences Analysis

Protein sequence analysis demonstrated that *LBGAE*, *LBGALA*, and *LBMS* contain a conservative structure domain of SDR (short-chain dehydrogenases/reductases; GenBank: CL25409, position: 73–401), AmyAc (Alpha-amylase catalytic; GenBank: cl07893, position: 38–409), and malate synthase (GenBank: PLN02626, position: 19–571), respectively (Table 1). The physicochemical properties of proteins, such as molecular weight, theoretical pI, instability index, aliphatic index, and grand average hydropathicity, were analyzed and listed in Table 1. In addition, the analysis indicated that the *LBGAE* protein sequence contains a transmembrane region (position: 7–29) and six protein functional domains (Figure 1a). The *LBGALA* protein sequence includes signal peptide cleavage sites in Ala²³ and two protein functional domains of glycoside hydrolase family (Figure 1b). The analysis also indicated that *LBGALA* is a secretory protein, while *LBGAE* and *LBMS* are soluble and non-secretory proteins synthesized in the cytoplasm without protein translocation ability.

Based on protein solvent accessibility composition (core and surface ration) analysis, the ration of residues exposed on the protein surface of *LBGAE*, *LBGALA*, and *LBMS* was 45.74%, 43.41%, and 44.13%, respectively. The residue ration in the protein core of *LBGAE*, *LBGALA*, and *LBMS* was 54.26%, 56.59% and 55.87%. This confirms that these three genes' hydrophilicity is weaker than that of hydrophobicity. The analysis showed different protein sites in the protein sequence (details are shown in Figure S1a–c). Similarly, the phosphorylation site prediction of the *LBGAE*, *LBGALA*, and *LBMS* protein sequences reveals phosphorylation sites of 44 (24 serine, 12 threonines, 8 tyrosine), 45 (25 serine, 14 threonines, 6 tyrosine), and 26 (12 serine, 7 threonines, 7 tyrosine), respectively, for these genes.

The subcellular localization prediction discloses that *LBGAE*, *LBGALA*, and *LBMS* have the highest possibility of located in mitochondria, chloroplast, and peroxisomes (Table 1). The prediction of protein secondary (Figure S2a–c) and tertiary structures (Figure 2a–c) indicated that *LBGAE*, *LBGALA*, and *LBMS* are mainly composed of alpha-helix, random coil and extended strand (Table 1). Additionally, no disulfide bond was formed among them.

2.3. Multiple Sequence Alignment and Phylogenetic Analysis

The multiple sequence alignment of *LBGAE*, *LBGALA*, and *LBMS* with amino acid sequences of several other species from NCBI was performed using ClustalX and MEGA 7.0 program. Results indicated that three genes are more than 80% identical to the *Capsicum annuum*, *Solanum tuberosum*, *Solanum lycopersicum* and *Nicotiana tabacum* (Figure S3a–c).

The phylogenetic tree was constructed with full-length proteins of the three genes showed that these three genes were clustered with other homologous genes of 20 species into three groups. The three genes were clustered in the same branch with members of *Nicotiana*, *Capsicum* and *Solanum*, but distantly related with members of *Momordica charantia*, *Erythranthe guttata*, and *Arabidopsis thaliana* (Figure 3).

Table 1. Analysis of protein characteristics of genes *LBGAE*, *LBGALA*, *LBMS*.

Gene	Sequence Length (bp)	Protein Length (aa)	Domains and Position	Physicochemical Properties					Secondary Structure			Subcellular Location
				Molecular Weight (kDa)	Theoretical pI	Instability Index	Aliphatic Index	Grand Average of Hydrophilicity	Alpha Helix (%)	Random Coil (%)	Extended Strand (%)	
LBGAE (MH025911)	1837	411	SDR family (73–401)	55.65	10.19	22.96	40.27	−1.019	45.99	39.17	14.84	mitochondria
LBGALA (MH025913)	1233	571	AmyAc family (38–409)	45.04	5.32	32.62	71.90	−0.369	29.27	53.66	17.07	chloroplast
LBMS (MH025912)	2449	410	MS (19–571)	65.00	8.45	40.23	81.65	−0.358	48.51	41.16	10.33	peroxisome

Note: SDR: short-chain dehydrogenases/reductases; AmyAc: Alpha-amylase catalytic; MS: malate synthase.

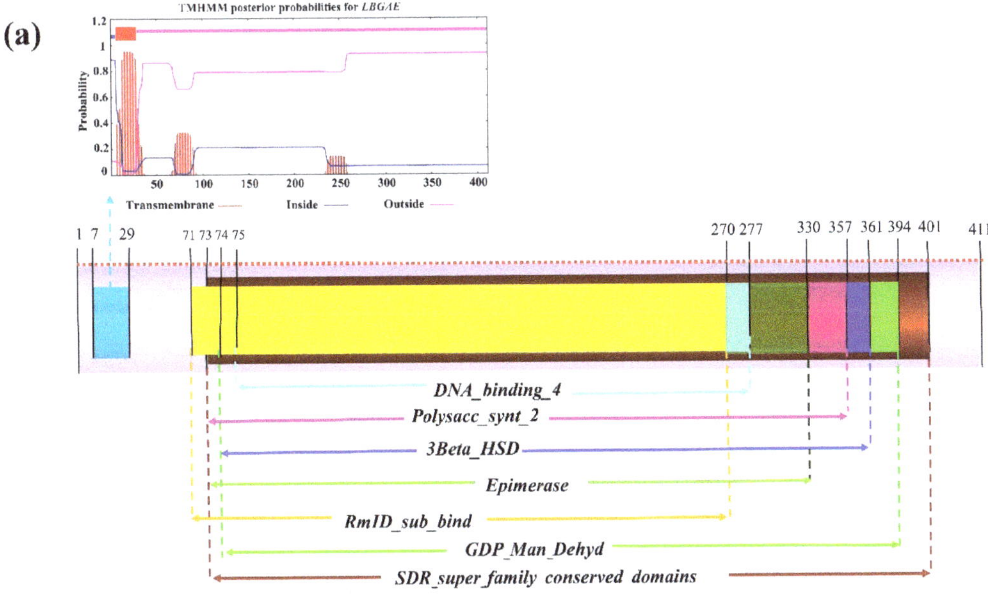

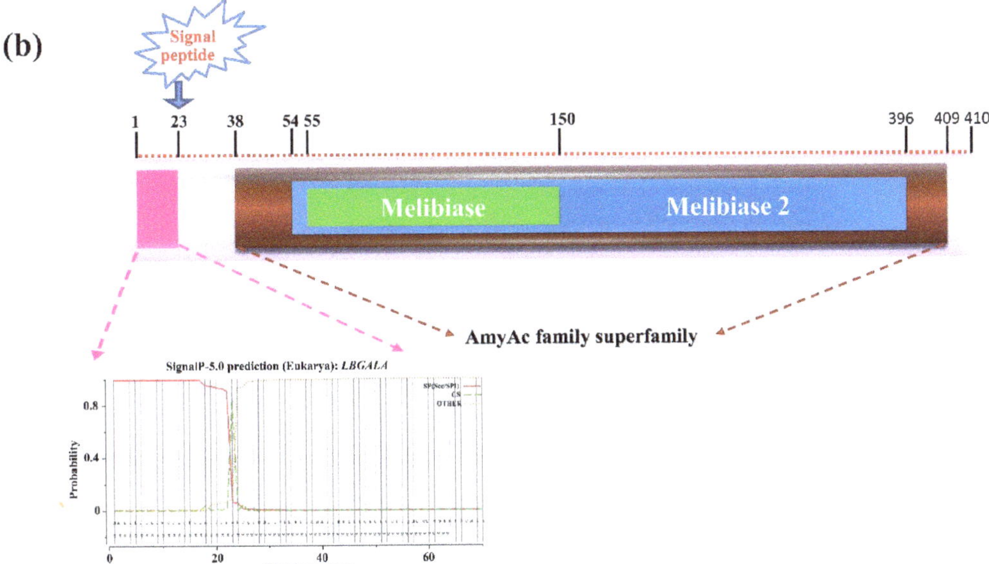

Figure 1. Protein functional domain analysis of *LBGAE* (**a**) and *LBGALA* (**b**). The *LBGAE*, including a *RmID_sub_bind* (RmID substrate binding domain, at position 71–270), an *Epimerase* (NAD-dependent epimerase/dehydratase family, 73–330), a *Polysacc_synt_2* (polysaccharide biosynthesis protein, 73–357), a *GDP_Man_Dehyd* (GDP-mannose4,6 dehydratase, 74–394), a *3Beta_HSD* (3-beta-hydroxysteroid dehydrogenase/isomerase family, 74–361), and a DNA_*binding_4* (male sterility protein, 75–277). The transmembrane region (position: 7–29) and the *LBGALA* including a *Melibiase_2* (54–396) and *Melibiase* (55–150). A signal peptide cleavage site in Ala[23].

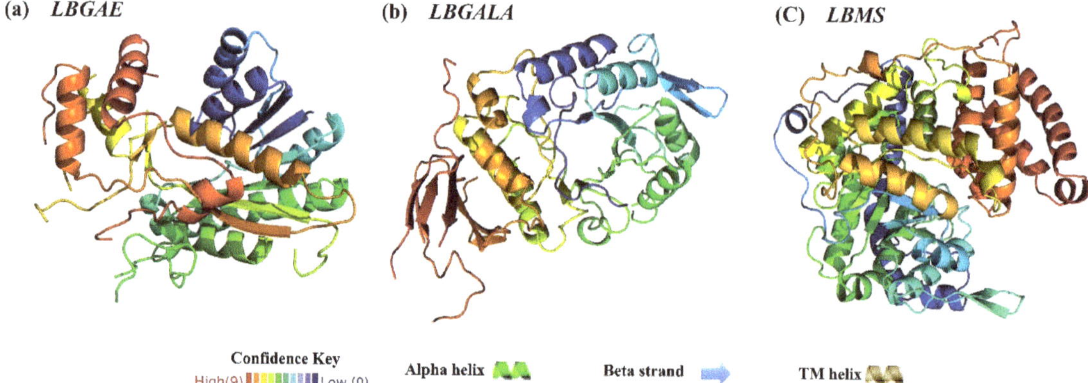

Figure 2. Predicted 3D structure model of protein *LBGAE* (**a**), *LBGALA* (**b**) and *LBMS* (**c**). Green: Alpha helix, lines and arrows: Beta strand. Brown: Helix. Red to blue representative confidence key from high to low.

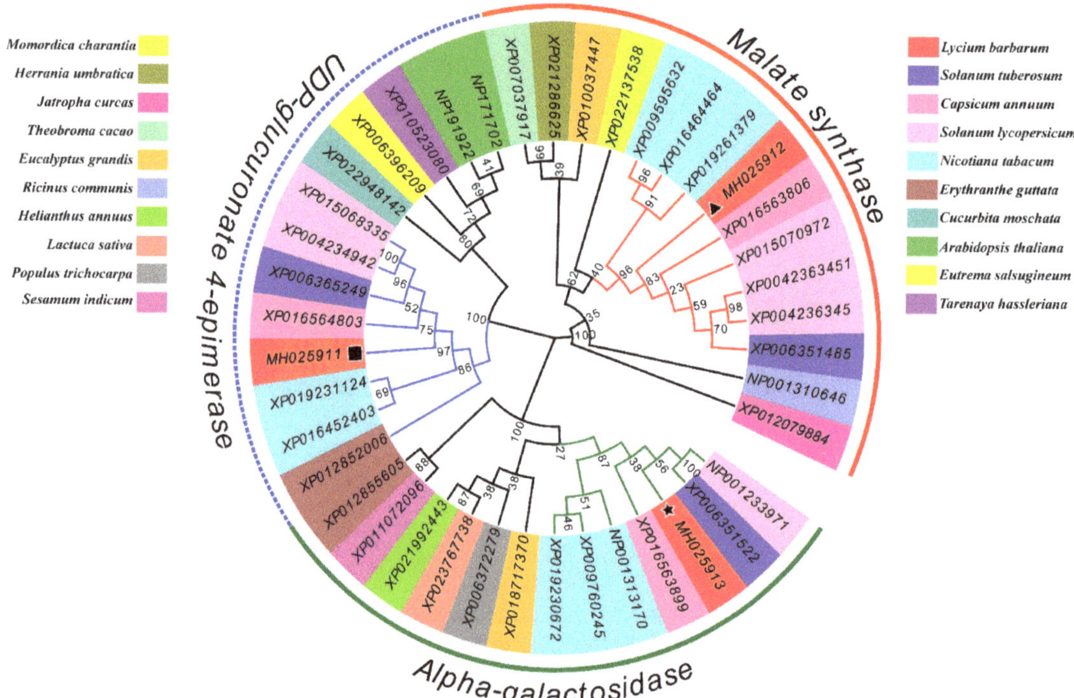

Figure 3. Phylogenetic tree based on the genes: *LBGAE*, *LBGALA* and *LBMS*, with another 20 species by neighbor-joining analysis. Numbers of the branches indicate relative bootstrap support (1000 replications). ★: *LBGALA* ▲: *LBMS* ■: *LBGAE*.

2.4. Regulation Pathways of the Genes LBGAE, LBGALA and LBMS

LBGAE, *LBGALA* and *LBMS* genes in different pathways and different stages of sugar conversion are detailed in Figure 4. These three genes are involved in the regulation of four sugar metabolism pathways: amino sugar and nucleotide sugar metabolism (ko00520), starch and sucrose metabolism (ko00500), pyruvate metabolism (ko00620), and galactose metabolism (ko00052) (Figure 4). The gene *LBGAE* (EC: 5.1.3.6, UDP-glucuronate 4-epimerase) is mainly responsible for the interconversion of UDP-D-Galacturonate and UDP-D-Glucuronate in starch and sucrose metabolism (ko00500), and UDP-GlcA and UDP-Gala in amino sugar and nucleotide sugar metabolism (ko00520) as represented in Figure 4.

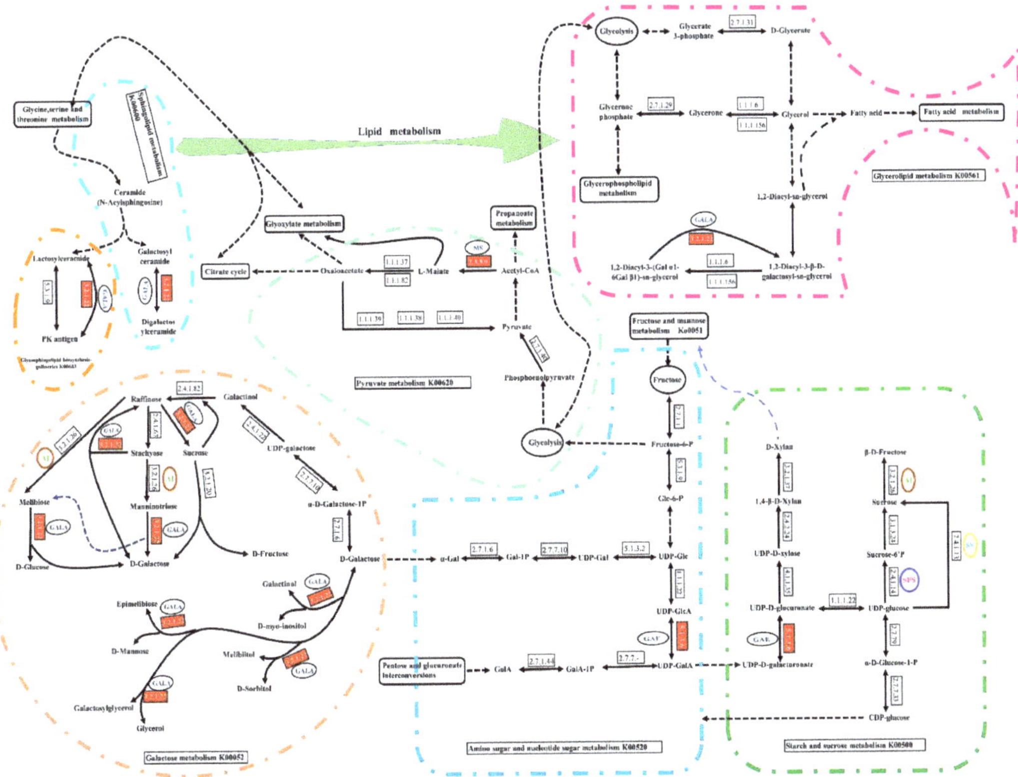

Figure 4. Schematic representation of involvement of genes *LBGAE* (*Lycium barbarum* UDP-glucuronate 4-epimerase), *LBGALA* (*Lycium barbarum* alpha-galactosidase), and *LBMS* (*Lycium barbarum* malate synthase) in the regulating of four sugar metabolism and lipid metabolism pathways. Red representative gene significantly upregulated.

In galactose metabolism (ko00052), *LBGALA* (EC: 3.2.1.22, alpha-galactosidase) acts at several places, including the conversion of stachyose, mannotriose, melibiose, epimelibiose and D-mannose, as well as glycerol and galactosyl-glycerol; it also converts raffinose into sucrose and stachyose. *LBGALA* is also involved in the regulation of lipid metabolism (K00561). Similarly, *LBMS* is involved in glyoxylate and dicarboxylate metabolism (K00630) and pyruvate metabolism (ko00620). The MS (EC: 2.3.3.9, malate synthase) is primarily responsible for the conversion of Acetyl-CoA to L(S)-Malate (Figure 4).

2.5. Expression Analysis of Genes LBGAE, LBGALA, and LBMS

Further evaluation of *LBGAE*, *LBGALA*, and *LBMS* to eCO$_2$ by qRT-PCR was not consistent among leaves, stems, and fruits. The expression levels of the three genes in stems and leaves measured at 90 and 120 days indicated that the expression of *LBGAE* and *LBGALA* was significantly higher under ambient conditions at 120 days in stems (Figure 5e). In fruits, three genes' expression levels were significantly higher under ambient conditions at 90 days (Figure 5c). However, the same genes showed higher expression under elevated CO$_2$ than ambient conditions at 120 days. Among them, the expression of *LBGAE* was significantly higher than *LBGALA* and *LBMS* ($p < 0.05$, $p < 0.01$) (Figure 5).

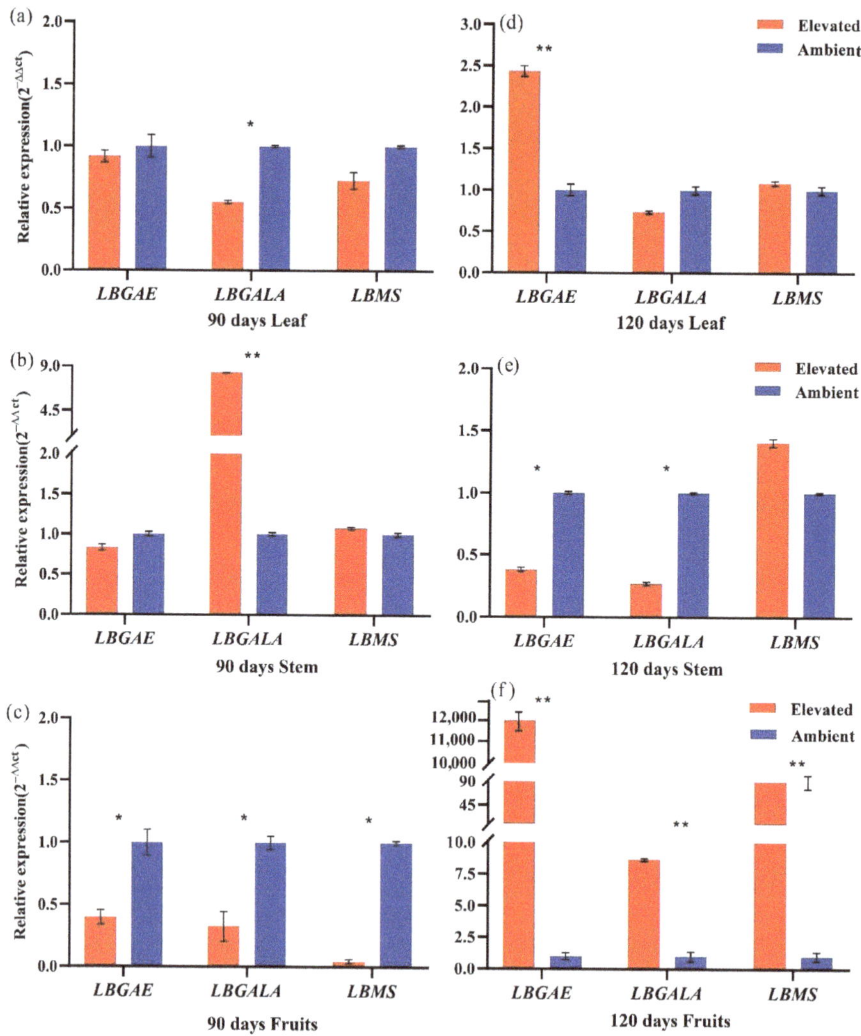

Figure 5. Relative expression of the genes *LBGAE*, *LGALA* and *LBMS* in leaves (**a,d**), stems (**b,e**), and fruits (**c,f**) for 90 days and 120 days under ambient and elevated CO$_2$ conditions. Red and blue bars represent elevated and ambient CO$_2$ concentration treatment, respectively. * $p < 0.05$, ** $p < 0.01$.

2.6. Sugar Content and Sucrose Metabolism-Related Enzymes Activities

Various sugar components and sucrose metabolism-related enzymes in goji berry fruits grown under ambient and elevated CO_2 for 90 and 120 days were determined to understand three genes' role in sugar metabolism (Figure 6). All four estimated sugars, including reducing sugar, starch, polysaccharide, and sucrose, followed a downward trend under elevated CO_2. The levels of reducing sugar and starch decreased significantly in fruit samples grown under eCO_2 at both 90 and 120 days (Figure 6a,b), while polysaccharide and sucrose declined only in 120-day samples (Figure 6c,d) ($p < 0.05$), compared to ambient CO_2 conditions.

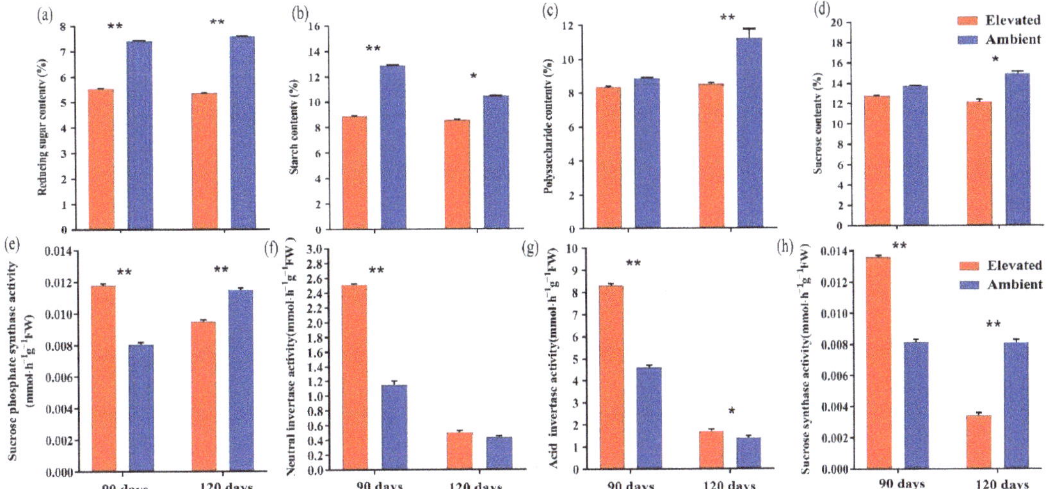

Figure 6. The content of reducing sugar (**a**), starch (**b**), polysaccharide (**c**), sucrose (**d**) and activity of sucrose metabolism-related enzymes sucrose phosphate synthase (**e**), neutral invertase (**f**), acid invertase (**g**), and sucrose synthase (**h**) of goji berry fruit samples grown under ambient and elevated CO_2 for 90 and 120 days. Red and blue bars represent elevated and ambient CO_2 concentration treatment, respectively. ** $p < 0.01$, * $p < 0.05$.

The activity of four sucrose metabolism-related enzymes, sucrose phosphate synthase, neutral invertase, acid invertase, and sucrose synthase were also determined in the samples collected at 90 and 120 days. Fruits grown under elevated CO_2 for 90 days showed a significant increase in the activity of four enzymes (Figure 6e–h) ($p < 0.01$). However, the activity of sucrose phosphate synthase and sucrose synthase decreased significantly in the samples collected at 120 days (Figure 6e,h) ($p < 0.01$) compared to ambient CO_2 levels.

3. Discussion

3.1. Cloning and Sequence Analysis of Sugar Metabolism Genes

The current study isolated three sugar metabolism genes *LBGAE, LBGALA,* and *LBMS* from goji berry fruit under elevated CO_2 and analyzed their sequence characteristics and potential functions. The gene *LBGAE* has 1236 bp ORF that encodes a 411-amino acid polypeptide, a putative protein containing SDR (short-chain dehydrogenases/reductases) superfamily conserved domain. It constitutes a large family of NAD(P)(H)-dependent oxidoreductases and a structurally diverse C-terminal region, with enzymes having critical roles in carbohydrate, lipid, and amino acid metabolism [16]. The gene *LBGALA* has 1233 bp ORF that encodes a 410 amino acid polypeptide. The *LBGALA* putative protein includes an Amy-Ac (Alpha-amylase catalytic) domain family and is the largest family of glycoside hydrolases (GH), usually acting on starch, glycogen, related oligosaccharides, and polysaccharides. The enzyme Amy-Ac can catalyze the conversion of alpha-1,4 and alpha-1,6-glucose chains [17]. Similarly, *LBMS* has 1716 bp ORF that encodes a 571 amino acid polypeptide, the putative protein with a malate synthase conserved domain. The malate synthase catalyzes the Claisen condensation of glyoxylate and acetyl-CoA to malyl-CoA, which hydrolyzes to malate and CoA. In addition, malate is also involved in the glyoxylate cycle that allows certain organisms, such as plants and fungi, to derive their carbon requirements from two carbon compounds by bypassing the two carboxylation steps of the citric acid cycle [18]. These three genes influence the sucrose metabolism-related enzymes sucrose phosphate synthase, neutral invertase, acid invertase, and sucrose synthase estimated in response to eCO_2 in this study (see the schematic representation of sugar metabolism in Figure 4) and strongly suggest the involvement of *LBGAE, LBGALA,* and *LBMS* in the regulation of sugar metabolism under elevated CO_2. Phylogenetic alignment of amino acid sequences revealed a very high sequence homology of putative *LBGAE, LBGALA,* and *LBMS* sequences with *Capsicum annuum, Solanum tuberosum, Solanum lycopersicum,* and *Nicotiana tabacum* of Solanaceae family proteins.

Overall, the sequence analysis of these three genes revealed that the genes' protein structures contain some functional domains, which provide initial evidence of their involvement in sugar metabolism, and this information can be used to explore sugar metabolism to eCO_2.

3.2. Regulation and Expression

The schematic representation of the transcriptome data of the three genes [9] in sugar metabolism pathways shows genes involved in four sugar metabolism pathways and their overexpression under elevated CO_2 in fruits (Figures 4 and 5). Along with sugar metabolism, *LBGALA* displayed an up-regulation in the other two lipid metabolism pathways (Figure 4). These genes amplified their expression levels to eCO_2 treatment compared to ambient conditions in fruits at 120 days (Figure 5). However, gene transcripts measured in leaf and stem were different from the results of the fruits. Mostly, the expression of these genes in eCO_2 treatment was significantly lower or not different than ambient CO_2 conditions both at 90 and 120 days in leaf and stem samples. Elevated atmospheric CO_2 concentration in the long-term is known to affect the photosynthesis and its components as a response to increased carbohydrate production and metabolism. Long-term exposure to elevated CO_2 can result in reduced levels of mRNAs encoding specific photosynthesis genes due to increased glucose or sucrose levels in the leaves [5,19,20]. The differential expression of the sugar metabolism genes, hexokinase and sucrose phosphate synthase, was found to play a role in exerting regulatory influence on sucrose biosynthesis to accommodate enhanced photosynthesis in *Jatropha* subjected to a CO_2-enriched environment [5]. In contrast, the variation in leaf soluble carbohydrate amount at elevated and ambient CO_2 concentrations reduced with crop development, whereas the difference in transcript levels increased [21]. In this study, the reduction in the levels of the three genes under eCO_2 over the ambient CO_2 conditions in leaves and stems might be related to increased sucrose and glucose levels in the leaves [19]. The increase in the levels of the genes expressed in the fruit to elevated CO_2 could be a feedback regulation to compensate for reduced sugar components. These results demonstrate that three genes' potential regulatory function in the sugar metabolism of goji berry influences enzyme activity and metabolism.

3.3. Response of Sugar and Sucrose Metabolism-Related Enzymes to Elevated CO_2

In previous studies with goji berry, CO_2 enrichment showed an influence on growth, photosynthesis, flavonoids, carotenoids, and sugars [9,22]. Enhanced CO_2, especially in C3 plants, causes increased photosynthesis, which generally leads to increased production of carbohydrates [23]. However, in the present study, the estimated four different kinds of sugars showed a reduction under eCO_2 in samples treated for 90 and 120 days compared to ambient conditions. The results agree with other studies that the elevated CO_2 reduces the sugar content in goji berry fruits [9,22,24]. Some studies demonstrated an increase in reducing sugars, total sugar, and starch content under short-term eCO_2 (850 ppm) compared to ambient conditions. However, long-term exposure reduces their levels, possibly due to photosynthesis acclimation [25]. Other studies observed higher carbon exchange rates to eCO_2 in the first few weeks of treatment but later decreased due to photosynthesis acclimation and low sink source or high sugar accumulation [8,11]. The possible explanation for the reduction in the four sugar components estimated is photosynthesis acclimation. However, no significant differences in the sugars were observed between 90 and 120 days, which suggests the occurrence of photosynthesis acclimation before 90 days. Further studies measuring photosynthesis along with sugar levels at different developmental stages are required to understand the process of photosynthesis acclimation in goji berry.

Sucrose, glucose, and fructose are the main soluble sugars in the fruit of the plant, and the contents of these sugars play a key role in the quality of fruit. Research in goji berry and other fruit crops indicated a close relationship between fruit sugar accumulation and sucrose-metabolizing enzymes, including sucrose synthase, sucrose phosphate synthase, acid invertase, and neutral invertase [12,26,27]. In contrast to the sugar components response, a significant variation from 90 days to 120 days in the activity levels of sucrose metabolism-related enzymes was observed (Figure 6e–h). The activity levels were higher in samples of eCO_2 than ambient CO_2 at 90 days. However, a significant drop at 120 days, which might be related to a reduction in the sugar levels was observed. These results agree with previous studies where sucrose metabolism-related enzymes were affected; sugar and secondary metabolite levels were reduced [10,28]. In a study with *Arabidopsis*, overexpression of sucrose phosphate synthase increased foliar sucrose/starch ratios and decreased foliar carbohydrate contents when plants were grown for long periods under enriched CO_2 conditions [29]. In rice, a high leaf CO_2 exchange rate was observed to enhanced CO_2 resulting in increased peduncle exudate sucrose levels but decreased levels in the developing grains. The whole process was linked to the poor responses to enhanced CO_2 in the activity of sucrose synthase, UDP-glucose pyrophosphorylase, ADP-glucose pyrophosphorylase, and starch synthases enzymes that are involved in the conversion of sucrose to starch [30]. Overall, this study demonstrates that the elevated CO_2 levels affected the enzyme involved in the sucrose metabolism and quality of goji berry fruit by reducing its sugar components.

4. Materials and Methods

4.1. Plant Material and Elevated CO_2 Treatment

The cultivar used in the present study was "Ningqi NO.1", and experiments were conducted with a one-year-old goji berry seedling obtained from the Ningxia Academy of Agriculture and Forestry Science, China. Experiments were conducted in open-top chambers (OTCs) [31,32] with two treatments ambient (400 $\mu mol \cdot mol^{-1}$ CO_2) and elevated CO_2 (700 $\mu mol \cdot mol^{-1}$ CO_2) for 120 days. The experimental site is located in Yongning county of yellow river alluvial plain in Central Ningxia of China (38°13′50.34″ N; 106°14′22.19″ E, 1116.86 m above sea level, located in the northwest inland). The experimental farm's climatic conditions are arid and middle temperate with a frost-free period of 140–160 days. The annual sunshine duration in the location was 3000 h, with an average annual temperature of 8.5 °C and precipitation of 180~300 mm. Matured fruit samples (red fruit after the color change period) were selected from the well-grown (treatment 120 days) and similar flowering period plants for the analysis. The collected samples were frozen immediately in liquid nitrogen and stored at −80 °C for further analysis.

4.2. Extraction of Total RNA and cDNA Synthesis

Total RNA was extracted from the fruits of goji berry by using RNAprep Pure Plant Kit (Tiangen, Beijing, China), according to the manufacturer's protocol. RNA concentration was measured using a NanoDrop 2000 spectrophotometer (Thermo Scientific, Waltham, MA, USA), and the quality was assessed by performing electrophoresis on 0.8% agarose gel. First-strand cDNA synthesis was synthesized using the PrimeScript™RT Reagent Kit with gDNA Eraser (Takara, Dalian, China), and the samples were stored at $-20\ ^\circ$C until sequenced.

4.3. Cloning Analyses of the LBGAE, LBGALA and LBMS

The sequences of *LBGAE*, *LBGALA*, and *LBMS* obtained from transcriptome sequencing were searched in the NCBI database. Based on the search, the sequences of *LBGAE* and *LBGALA* were incomplete, and the *LBMS* gene has a complete sequence. Rapid amplification of cDNA ends (RACE) was used to obtain the full-length cDNA sequence of *LBGAE* and *LBGALA* genes. The $3'$ and $5'$ RACE reactions were carried out using a SMARTer® RACE $5'/3'$ Kit (Clontech, Beijing, China) following the manufacturer's recommendations. The RACE primers were synthesized using SNAPGENE and primer 5.0. The specific primers $3'$-RACE (GSP1) and $5'$-RACE (GSP2) were designed to be specific to the target gene (Table S1).

A primer pair, F2 and R2 (Table S1), was designed by employing PrimeSTAR Max DNA (Takara, Dalian, China) to amplify the full length of *LBGAE* and *LBGALA*. The steps of PCR amplification include a pre-denaturation at 94 $^\circ$C for 5 min, followed by 33 cycles of 94 $^\circ$C for 40 s, 58 $^\circ$C for 30 s, 72 $^\circ$C for 1 min 20 s, and then a final extension at 72 $^\circ$C for 10 min and stored at 16 $^\circ$C. Products of PCR were visualized and verified using 0.8% agarose gel. The product was then extracted and purified utilizing a PCR Purification Kit (Tiangen, Beijing, China), and then ligated into the pMD18-T vector (Takara, Dalian, China) and sequenced.

4.4. Bioinformatics Analysis of Sequences

DNAMAN (V8.0) software was used to evaluate the molecular mass, base composition, and base distribution of the nucleic acid sequences. In addition, BioEdit (V7.0.5.3) software was used to perform sequence analysis of the enzyme digestion. Base-homology analysis was conducted utilizing BLAST of NCBI (http://www.ncbi.nlm.nih.gov/BLAST/) (accessed on 5 January 2021), and protein structural domain analysis of the protein sequence was studied employing SMART (Simple Modular Architecture Research Tool) (http://smart.embl-heidelberg.de/) (accessed on 5 January 2021) tool. The open reading frame was identified by ORF finder (https://www.ncbi.nlm.nih.gov/orffinder/) (accessed on 5 January 2021). Protein tertiary structure was modeled by the homology-modeling server SWISS-MODEL (http://swissmodel.expasy.org/) (accessed on 5 January 2021), and the models were further analyzed and refined using Swiss-Pdb Viewer v4.1.0.

The molecular weight and isoelectric point (pI) of protein were estimated by ProtParam (http://web.expasy.org/protparam/) (accessed on 5 January 2021). Multiple sequence alignment and phylogenetic analysis were performed using ClustalX 2.0, MEGA 7.0. Software tools PSort II (http://psort.hgc.jp/) (accessed on 5 January 2021), TMHMM (http://www.cbs.dtu.dk/services/TMHMM/) (accessed on 5 January 2021), and SignalP4.1 (http://www.cbs.dtu.dk/services/SignalP/) (accessed on 5 January 2021) were employed to predict the protein subcellular location, the single transmembrane region, and the signal peptide, respectively. Protein hydrophobicity and hydrophilicity were assessed through Protscale (http://web.expasy.org/protscale/) (accessed on 5 January 2021). A phylogenetic tree of proteins from different species was constructed using the neighbor-joining (NJ) method in the MEGA7.0 software (https://www.megasoftware.net) (accessed on 5 January 2021). The reliability of the tree was measured by bootstrap analysis with 1000 replications.

4.5. Gene Expression Analysis by qRT-PCR

The specific primers for qPCR of *LBGAE*, *LBGALA*, and *LBMS* and β-actin (internal reference) were designed based on the full cDNA sequence (Table S1), and the primers were designed using PrimerQuest Tool (http://sg.idtdna.com/Primerquest/Home/Index) (accessed on 5 January 2021). The qRT-PCR assay was conducted using a Quantitative Fluorescent assay that used the abm®EvaGreen qPCR MasterMix-ROX kit (ABM, Vancouver, BC, Canada).

The qRT-PCR was performed with a 20 μL of reaction mix containing 10 μL abm®EvaGreen qPCR MasterMix-No Dye (2×), 0.4 μL of each Primer (10 μM, 0.8 μL of PCR Reverse Primer), 10 μM, 1.0 μL of cDNA template, and 7.4 μL of sterilized ddH$_2$O. The cycle of qRT-PCR reaction program was denaturation at 95 °C for 10 min, followed by 45 cycles of 94 °C for 15 s, 60 °C for 1 min, and stored at 16 °C. The qPCR assays were performed by using ABI StepOne Plus (Applied Biosystems, Foster City, USA) with nine biological and three technical replicates. The relative expression levels of *LBGAE*, *LBGALA*, and *LBMS* were calculated using the $2^{-\Delta\Delta ct}$ method [33].

4.6. Polysaccharide, Sucrose, Reducing Sugar, Starch, and Sucrose Metabolism-Related Enzymes Estimation

Fruit samples were collected in nine replicates from ambient and elevated CO$_2$ OTC chambers on 90 and 120 days. The samples were stored at −80 °C until further the next stage. The content of polysaccharide, sucrose, reducing sugars, and starch was measured by following the method described by Ha et al. [10] using approximately 5.0 g sample. The enzyme activity of sucrose phosphate synthase, neutral invertase, acid invertase, and sucrose synthase was estimated as described by Liu et al. [28].

5. Conclusions

This study confirms the effect of eCO$_2$ on various sugar components and enzyme activities of goji berry fruit estimated at both 90 and 120 days. Furthermore, the three sugar metabolism genes that upregulated to eCO$_2$ were analyzed in detail to understand the molecular mechanism of the goji berry response to eCO$_2$. The full-length cDNAs of *LBGAE*, *LBGALA*, and *LBMS* were cloned and analyzed for their protein characteristics. The potential regulatory functions of the proteins were identified; in addition, significant upregulation transcriptome of these three genes in fruits to CO$_2$ was confirmed. The expression of these genes might have played an important role in the sucrose-metabolizing enzyme activity and reduced the content of four sugars estimated. However, further studies are required to understand the underlying regulatory mechanism of these genes to eCO$_2$. The results of this study provide theoretical references and lay the molecular basis to study metabolic regulation of sugar accumulation and the subsequent understanding of gene functions in goji berry to improve fruit quality in changing climate conditions. However, a detailed study on the role of three genes and their mechanism in regulating sugar metabolism by knock-out or knock-down mutants is needed.

Supplementary Materials: The following are available online at https://www.mdpi.com/2223-774 7/10/2/321/s1, Figure S1: Complete nucleotide and deduced amino acid sequence of cDNA of gene *LBGAE*, *LBGALA*, *LBMS*; Figure S2: Prediction of protein secondary structure of *LBGAE*, *LBGALA* and *LBMS* sequence; Figure S3: Multiple sequence alignment of *LBGAE*, *LBGALA* and *LBMS* amino acid sequences with other known species; Table S1: Primer sequences for ORF sequences cloning and qPCR.

Author Contributions: B.C. and L.S.; methodology, Y.M.; validation and data curation, Y.M., and M.J.D.; writing—original draft preparation, M.J.D.; writing—review and editing, Y.M.; visualization, Y.M.; supervision, B.C.; V.R.R. and H.G.; project administration, B.C.; funding acquisition, B.C. All authors have read and agreed to the published version of the manuscript.

Funding: This research was funded by the National Natural Science Foundation of China, grant numbers 31660199 and 31160172.

Institutional Review Board Statement: Not applicable.

Informed Consent Statement: Not applicable.

Data Availability Statement: Data are contained within the article or Supplementary Material.

Acknowledgments: The authors acknowledge the Ningxia University and USDA-ARS, Adaptive Cropping Systems Laboratory for providing the research facilities.

Conflicts of Interest: The authors declare no conflict of interest.

References

1. Amagase, H.; Farnsworth, N.R. A review of botanical characteristics, phytochemistry, clinical relevance in efficacy and safety of *Lycium barbarum* fruit (Goji). *Food. Res. Int.* **2011**, *44*, 1702–1717. [CrossRef]
2. Cheng, J.; Zhou, Z.; Sheng, H.; He, L.; Fan, X.; He, Z.; Sun, T.; Zhang, X.; Zhao, R.J.; Gu, L. An evidence-based update on the pharmacological activities and possible molecular targets of *Lycium barbarum* polysaccharides. *Drug Des. Dev. Ther.* **2015**, *9*, 33.
3. Potterat, O. Goji (*Lycium barbarum* and *L. chinense*): Phytochemistry, pharmacology and safety in the perspective of traditional uses and recent popularity. *Planta Med.* **2010**, *76*, 7–19. [CrossRef]
4. Jönsson, A.M.; Anderbrant, O.; Holmér, J.; Johansson, J.; Schurgers, G.; Svensson, G.P.; Smith, H.G. Enhanced science–stakeholder communication to improve ecosystem model performances for climate change impact assessments. *Ambio* **2015**, *44*, 249–255. [CrossRef] [PubMed]
5. Kumar, S.; Sreeharsha, R.V.; Mudalkar, S.; Sarashetti, P.M.; Reddy, A.R. Molecular insights into photosynthesis and carbohydrate metabolism in Jatropha curcas grown under elevated CO_2 using transcriptome sequencing and assembly. *Sci. Rep.* **2017**, *7*, 11066. [CrossRef] [PubMed]
6. Mahdu, O. The Impacts of Climate Change on Rice Production and Small Farmers' Adaptation: A Case of Guyana. Ph.D. Thesis, Virginia Tech, Blacksburg, VA, USA, 2019.
7. Parry, M.L. *Climate Change and World Agriculture*; Routledge: New York, NY, USA, 2019; pp. 35–78.
8. Yelle, S.; Beeson, R.C.; Trudel, M.J.; Gosselin, A. Acclimation of two tomato species to high atmospheric CO_2: I. Sugar and starch concentrations. *Plant Physiol.* **1989**, *90*, 1465–1472. [CrossRef]
9. Ma, Y.; Reddy, V.R.; Devi, M.J.; Song, L.; Cao, B. *De novo* characterization of the Goji berry (*Lycium barbarium* L.) fruit transcriptome and analysis of candidate genes involved in sugar metabolism under different CO_2 concentrations. *Tree Physiol.* **2019**, *39*, 1032–1045.
10. Ha, R.; Ma, Y.; Cao, B.; Guo, F.; Song, L. Effects of simulated elevated CO_2 concentration on vegetative growth and fruit quality in *Lycium barbarum*. *Sci. Silvae Sin.* **2019**, *55*, 28–36.
11. Ainsworth, E.A.; Rogers, A.; Nelson, R.; Long, S.P. Testing the "source–sink" hypothesis of down-regulation of photosynthesis in elevated CO_2 in the field with single gene substitutions in *Glycine max*. *Agric. For. Meteorol.* **2004**, *122*, 85–94. [CrossRef]
12. Hubbard, N.L.; Pharr, D.M.; Huber, S.C. Sucrose phosphate synthase and other sucrose metabolizing enzymes in fruits of various species. *Physiol. Plant.* **1991**, *82*, 191–196. [CrossRef]
13. Ruan, Y. Sucrose metabolism: Gateway to diverse carbon use and sugar signaling. *Annu. Rev. Plant Biol.* **2014**, *65*, 33–67. [CrossRef] [PubMed]
14. Schaffer, A.A.; Aloni, B.; Fogelman, E. Sucrose metabolism and accumulation in developing fruit of *Cucumis*. *Phytochemistry* **1987**, *26*, 1883–1887. [CrossRef]
15. Winter, H.; Huber, S.C. Regulation of sucrose metabolism in higher plants: Localization and regulation of activity of key enzymes. *Crit. Rev. Plant. Sci.* **2000**, *19*, 31–67. [CrossRef]
16. Kavanagha, K.; Jçrnvallb, H.; Perssonc, B.; Oppermanna, U. The SDR superfamily: Functional and structural diversity within a family of metabolic and regulatory enzymes. *Cell. Mol. Life Sci* **2008**, *65*, 3895–3906. [CrossRef]
17. Kuriki, T.; Takata, H.; Yanase, M.; Ohdan, K.; Fujii, K.; Terada, Y.; Takaha, T.; Hondoh, H.; Matsuura, Y.; Imanaka, T. The concept of the α-amylase family: A rational tool for interconverting glucanohydrolases/glucanotransferases, and their specificities. *J. Appl. Glycosci.* **2006**, *53*, 155–161. [CrossRef]
18. Erb, T.J.; Frerichs-Revermann, L.; Fuchs, G.; Alber, B.E. The apparent malate synthase activity of Rhodobacter sphaeroides is due to two paralogous enzymes, (3S)-malyl-coenzyme A (CoA)/β-methylmalyl-CoA lyase and (3S)-malyl-CoA thioesterase. *J. Bacteriol.* **2010**, *192*, 1249–1258. [CrossRef]
19. Krapp, A.; Hofmann, B.; Schäfer, C.; Stitt, M. Regulation of the expression of rbcS and other photosynthetic genes by carbohydrates: A mechanism for the 'sink regulation'of photosynthesis? *Plant J.* **1993**, *3*, 817–828. [CrossRef]
20. Moore, B.; Cheng, S.; Sims, D.; Seemann, J. The biochemical and molecular basis for photosynthetic acclimation to elevated atmospheric CO_2. *Plant Cell Environ.* **1999**, *22*, 567–582. [CrossRef]
21. Nie, G.; Hendrix, D.L.; Webber, A.N.; Kimball, B.A.; Long, S.P. Increased accumulation of carbohydrates and decreased photosynthetic gene transcript levels in wheat grown at an elevated CO_2 concentration in the field. *Plant Physiol.* **1995**, *108*, 975–983. [CrossRef]
22. Cao, B.; Hou, J.; Pan, J.; Song, L.; Kang, J. Effects of doubled atmospheric CO_2 concentration on sugar accumulation of different organs in *Lycium barbarum*. *J. Northwest For. Univ.* **2014**, *3*, 13–17.
23. Thompson, V.; Dunstone, N.J.; Scaife, A.A.; Smith, D.M.; Slingo, J.M.; Brown, S.; Belcher, S.E. High risk of unprecedented UK rainfall in the current climate. *Nat. Commun.* **2017**, *8*, 107. [CrossRef]
24. Hou, J.; Cao, B. Effect of elevated CO_2 concentration on carbon and nitrogen allocation in *Lycium barbarum*. *J. Northeast For. Univ.* **2011**, *39*, 75–77.

25. Yadav, S.; Mishra, A.; Jha, B. Elevated CO_2 leads to carbon sequestration by modulating C4 photosynthesis pathway enzyme (PPDK) in *Suaeda monoica* and *S. fruticosa*. *J. Photochem. Photobiol. B* **2018**, *178*, 310–315. [CrossRef]
26. Wang, T.; Wright, D.; Xu, H.; Yang, Y.; Zheng, R.; Shi, J.; Chen, R.; Wang, L. Expression patterns, activities and sugar metabolism regulation of sucrose phosphate synthase, sucrose synthase, neutral invertase and soluble acid invertase in different Goji cultivars during fruit development. *Russ. J. Plant Physiol.* **2019**, *66*, 29–40. [CrossRef]
27. Zhang, X.; Liu, S.; Du, L.; Yao, Y.; Wu, J. Activities, transcript levels, and subcellular localizations of sucrose phosphate synthase, sucrose synthase, and neutral invertase and change in sucrose content during fruit development in pineapple (*Ananas comosus*). *J. Hortic. Sci. Biotechnol.* **2019**, *94*, 573–579. [CrossRef]
28. Liu, Y.; Zhang, Y.; Cao, B. Effects of hight atmospheric CO_2 concentrations on activities of sucrose metabolism-related enzymes in *Lycium barbarum* fruit. *J. Northwest For. Univ.* **2016**, *31*, 44–47.
29. Signora, L.; Galtier, N.; Skøt, L.; Lucas, H.; Foyer, C.H. Over-expression of sucrose phosphate synthase in *Arabidopsis thaliana* results in increased foliar sucrose/starch ratios and favours decreased foliar carbohydrate accumulation in plants after prolonged growth with CO_2 enrichment. *J. Exp. Bot.* **1998**, *49*, 669–680. [CrossRef]
30. Chen, C.; Li, C.; Sung, J. Carbohydrate metabolism enzymes in CO_2-enriched developing rice grains of cultivars varying in grain size. *Physiol. Plant.* **1994**, *90*, 79–85. [CrossRef]
31. Ma, Y.; Cao, B.; Song, L.; Ha, R.; Jia, H.; Liu, K. Gas Saving Type Control System of Top-Open Gas Simulation Chamber and Its Control Method. Invention Patent ZL201711003932.1 27 October 2020.
32. Ma, Y.; Wang, N.; Jia, H.; Cao, B. Evaluation of a modified open-top chamber simulation system on the study of elevated CO_2 concentration effects. *J. Earth Environ.* **2019**, *10*, 307–315.
33. Adnan, M.; Morton, G.; Hadi, S. Analysis of rpoS and bolA gene expression under various stress-induced environments in planktonic and biofilm phase using $2^{-\Delta\Delta CT}$ method. *Mol. Cell. Biochem.* **2011**, *357*, 275–282. [CrossRef]

Review

Crop Adaptation: Weedy and Crop Wild Relatives as an Untapped Resource to Utilize Recent Increases in Atmospheric CO$_2$

Lewis H. Ziska

Mailman School of Public Health, Columbia University, New York, NY 10032, USA; lhz2103@cumc.columbia.edu

Abstract: Adaptation measures are necessary to ensure the stability and performance of the food supply relative to anthropogenic climate change. Although a wide range of measures have been proposed (e.g., planting dates, crop choices, drought resistance), there may be a ubiquitous means to increase productivity relatively quickly. Numerous studies have shown that the projected increase in atmospheric CO$_2$ can stimulate crop growth and seed yield with noted intra-specific differences within crop cultivars, suggesting potential differences to CO$_2$ that could be exploited to enhance seed yield in the future. However, it is worth emphasizing that atmospheric CO$_2$ has already risen substantially ($\approx$27% since 1970) and that, at present, no active effort by breeders has been made to select for the CO$_2$ increase that has already occurred. In contrast, for weedy or crop wild relatives (CWR), there are indications of evolutionary adaptation to these recent increases. While additional steps are needed, the identification and introgression of these CO$_2$-sensitive traits into modern crop cultivars may be a simple and direct means to increase crop growth and seed yield.

Keywords: adaptation; breeding; CO$_2$; CWR; seed yield

Citation: Ziska, L.H. Crop Adaptation: Weedy and Crop Wild Relatives as an Untapped Resource to Utilize Recent Increases in Atmospheric CO$_2$. *Plants* **2021**, *10*, 88. https://doi.org/10.3390/plants10010088

Received: 20 December 2020
Accepted: 31 December 2020
Published: 4 January 2021

Publisher's Note: MDPI stays neutral with regard to jurisdictional claims in published maps and institutional affiliations.

1. Introduction

Maintaining food security is a seminal objective for the remainder of the century. While there are a number of recognized obstacles to achieve this objective, environmental limitations associated with unprecedented anthropogenic change threaten core aspects, including production, access, and quality. Specifically, climate change can alter abiotic environments, such as water availability (too much or too little), irregular temperatures, and extreme climatic events (flash droughts, derechos, etc.) [1,2]. Climatic-induced changes in biotic competition from agronomic pests (insects, disease, weeds) pose another significant constraint to global food production [3,4].

Such vulnerabilities within the agronomic food chain necessitate an immediate need to begin adapting crops to empirical threats associated with climatic change. Adaptation per se represents a wide range of approaches, but one facet of obvious consequence is genetics. In that regard, there are a number of exemplary and ongoing efforts to select for crop lines that can respond to climatic extremes, such as drought or extreme temperature [5–7].

There is another genetic approach that is becoming recognized as a potential adaptation tool: the selection of intra-specific variation in seed yield among C$_3$ crop cultivars in response to projected increases in atmospheric CO$_2$. Such an approach is being used to determine CO$_2$ sensitivity in conjunction with economic yield for cassava [8], rice [9], soybean [10], wheat [11], inter alia. At present, there is ample evidence that considerable variation exists within crop cultivars for anticipated increases in CO$_2$ and that selection for such variation holds promise as an adaptive means to increase crop yields.

Yet, the anticipated increase in atmospheric CO$_2$ is, in the short term, slow, 2–3 ppm per year, and selection for cultivars for projected CO$_2$ levels 30–50 years into the future does not address the current need to adapt crop systems.

On the other hand, atmospheric CO_2 has already increased substantially from $\approx$325 to 412 ppm since 1970, which is an increase of $\approx$27%. Has this recent increase been exploited through ongoing artificial selection to choose current crop lines that are CO_2 sensitive? Has the increase in CO_2 been sufficient to begin evolutionary selection for increased growth and seed yield for weedy or crop wild relatives (CWR)? The objective of the current review is to compare and contrast selection efforts from breeders and nature with respect to CO_2 sensitivity and potential seed yield for these two groups and to provide insight into metrics that could be of immediate (and future) benefit for utilizing CO_2 to increase crop growth and seed yield.

2. Breeder Efforts to Select for CO_2 Responsiveness in Crops

While intra-specific variation in response to future CO_2 is being evidenced experimentally, there is no verification of any directed attempts by breeders to select for seed yield sensitivity to the increase in CO_2 that has already occurred. It has been proposed that rapid-cycle breeding and new cultivar introductions could, over time, incrementally improve crop lines in adapting to new climates [12]. If true, then breeders could be already selecting (albeit passively) the most CO_2-responsive cultivars over time. As such, there would be little need to initiate any active CO_2 breeding programs to exploit recent (or projected) changes in CO_2 with respect to seed yield.

Has such an approach worked to date? If modern, recently introduced lines are adapted to current CO_2 levels ($\approx$412 ppm), then they should demonstrate a greater CO_2 response to recent CO_2 increases relative to cultivars that were developed during the early twentieth century.

There are a number of studies that have explicitly tested this question. Sakai et al. [13] examined five japonica rice lines: 'Aikoku' (released in 1882), 'Norin 8' (1934), 'Koshihikari' (1956), 'Akihikari' (1976), and 'Akidawara' (2009). No differences in seed yield were noted between the oldest line (Aikoku, 1882) and the newest line (Akidawara, 2009) to increased CO_2 (19 and 19.3%, increase at 600 ppm CO_2, relative to ambient, respectively). For wheat, yield responses to rising CO_2 actually declined with the release of newer cultivars [14,15]. For oat, Ziska and Blumenthal [16] examined the growth and vegetative characteristics of cultivated oat (*Avena sativa* L.) from seven geographical locations to CO_2 concentration increases that corresponded roughly to the CO_2 from the 1920s (300 ppm), as well as current (400 ppm) and mid-21st century projected levels (500 ppm). Newer lines were less responsive than older lines to rising CO_2 in terms of both leaf area and tiller number. Overall, there is little evidence that cultivars in current use are those best adapted to maximize productivity in response to increasing atmospheric CO_2 (but see [17] for barley).

Interestingly, for the oat study [16], significant age $\times$ CO_2 interactions were observed with greater phenotypic variation noted for the older cultivars (Table 1). Diminished variance and increasing genetic uniformity is not unexpected, given increased farm size and greater mechanization during the 20th century [18]. However, uniformity can also limit responses to environmental factors, including rising CO_2 [19], suggesting that the ability to respond to rising CO_2 or other environmental perturbations may be constrained through modern breeding efforts. Overall, the narrow genetic base of modern cultivars may constitute a major bottleneck for crop improvement efforts [20–22].

Table 1. Significance of equality of variance for "old" and "new" oat cultivars for select vegetative and growth characteristics averaged over all CO_2 concentrations. An asterisk indicates a greater degree of phenotypic variation. Note the greater degree of phenotypic variation for oat cultivars released in the 1920s. Data are from 16.

Variable	"Old"	"New"
Leaf area	*	
Leaf wt.	**	
Tiller wt.	**	
Weight tiller^{-1}	***	
Tiller No.	*	
Root wt.	*	
Total wt.	***	
RGR	**	

* $p < 0.05$; ** $p < 0.01$; *** $p < 0.001$. n.s. = not significant.

3. Weedy and Wild Crop Relatives

Crop wild relatives (CWR) are those undomesticated "cousins" of cultivated crop lines that represent a potential untapped genetic resource that could be used as a means to adapt to new pests or abiotic changes, including, at least theoretically, increased levels of CO_2. Of course, there are no directed efforts to adapt CWR to the recent increase in CO_2; however, it is worth determining if any evolutionary adaptive changes that have allowed CWRs to adapt to recent CO_2 increases have occurred.

It can be argued that such evolutionary changes are unlikely given the recent increase in CO_2. However, the traditional paradigm of weed evolution as a very slow process is incomplete, and there are a number of examples demonstrating that rapid evolutionary change (years or decades) can occur within weed biology, (e.g., *Microstegium vimineum*, [23]; *Lythrum salicaria*, [24]; *Brassica rapa*, [25]; *Avena fatua*, [26]). In turn, such changes could include evolution in response to anthropogenic climate change [27–29].

At present, there is initial evidence indicating that recent increases in CO_2 may have already altered the adaptive response of some annual weeds. For example, Bunce [30] examined recent increases in CO_2 on the growth response of four annual weeds over a narrow CO_2 range (90 ppm below and above ambient) and demonstrated that the efficiency by which these weeds utilized CO_2 declined at concentrations above ambient, indicating that these weeds had adapted to recent CO_2 increases.

With respect to CWR, comparisons of relative fitness to cultivated lines suggest differential adaptation to recent CO_2 increases. Comparisons of six cultivated and six wild or weedy biotypes of rice indicated a greater overall growth response among wild relative to cultivated rice to recent (300–400 ppm) increases in CO_2 [31] (Figure 1), suggesting that the rapid evolution of weedy biotypes may have increased their fitness relative to the crop. Greater seed yields were also recorded for Stuttgart, a CWR to rice, compared to Clearfield, a cultivated rice line, for the same recent CO_2 increase [32]. Similarly, using a resurrection approach [33], seeds of two temporally distinct populations of CWR wild oat (*Avena fatua* L.) from the same location, one from the 1960s and one from 2014, (a relative CO_2 increase of 80 ppm, or 25% from 1960) demonstrated different competitive abilities against a cultivated oat (*A. sativa*) line, with the more recent (2014) *A. fatua* population having greater growth and competitive ability at current CO_2 levels [34].

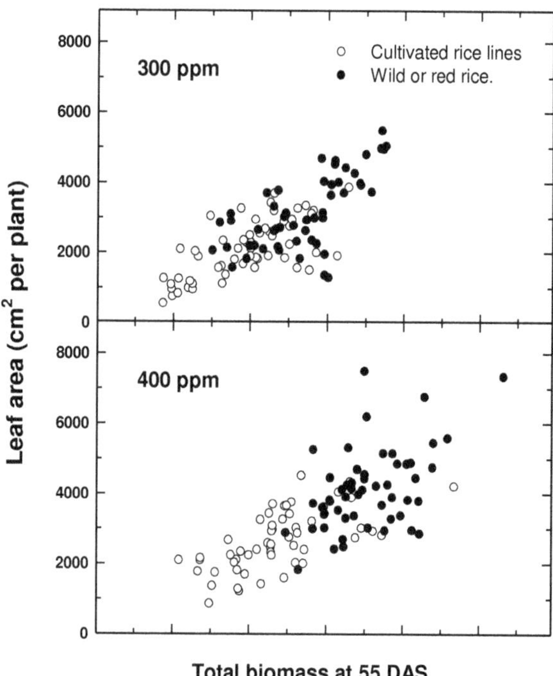

Figure 1. Response of six red rice (RR, filled circles) and six cultivated rice (CR, open circles) varieties to the recent change in atmospheric CO_2 concentrations (300–400 ppm) for leaf area as a function of total above-ground biomass 55 days after sowing. Note the increase in growth for RR for the recent CO_2 increase. Data are adapted from Ziska and McClung [31].

4. Differences in Selection

Additional data to confirm and expand upon these results is obviously needed. However, it is important to recognize differences within the selection process for cultivated and CWR biotypes. Weeds and wild crop relatives have been subject to natural environmental perturbations for millennia and have as a consequence maintained a much higher level of genetic diversity [21]. Both are characterized by rapid growth, high seed production, environmental plasticity, and genetic variability, and they are considered highly adaptable [35,36]. Such plant species, with short generation times and high seed production, generally show more rapid rates of molecular evolution [37].

Conversely, it is generally recognized that advances in plant breeding are associated with recurrent selection, usually in field environments. As a consequence, selection for say, pest resistance, should be occurring concurrently with rising CO_2, and, as a result, reflect CO_2 adaptation. However, plant breeding is a long-term process that can extend over decades, and indirect selection for yield under field conditions is likely to be inefficient because yield is related to a number of abiotic and biotic factors. The unintended consequences of recurring selection are that genetic variation and associated phenotypes can be reduced relative to the available hereditary potential.

Genetic exchange among breeders has also shifted over time. As documented by Atlin [12], prior to the 1990s, breeders typically exchanged varieties. However, as Genetically modified organism (GMO) lines were introduced and breeding became more commercialized, Intellectual Property (IP) protection increased, resulting in the current U.S. practice of issuing utility patents that prevent proprietary lines from being used as parents by other breeders. Unfortunately, simultaneous with these changes, a number of countries recognized that their own indigenous crops were unique genetic resources and restricted their usage in public breeding programs. Consequently, obtaining elite or novel varieties internationally has become restrictive. Overall, such actions have reduced reliance on publically available seed sources, reducing genetic diversity [38].

Given the need for the mechanization of large land holdings and economic consistency in response to water and fertilizer so as to achieve high crop productivity, dedomestication and genetic uniformity are necessary, even if such practices lead to "Domestication Bottlenecks" [39]. However, such uniformity in management may also limit the extent of genetic variation in response to environmental changes, such as CO_2 [40].

Overall, opportunities for increasing production may be missed if we assume that current breeding efforts have resulted in crop plants that are adapted to the recent increase in CO_2 concentration. Rather, it suggests that CWR and natural selection may serve as a starting point for active intervention to enhance genetic diversity to meet new environmental uncertainty and optimize crop yields to ongoing CO_2 increases. In addition, land races, populations of a cultivated plant that are distinct to a given geographic and environmental locale, are worth additional evaluation and may also provide a useful genetic resource in that regard.

5. Challenges and Next Steps

If CWR represent an opportunity to exploit the recent increase in CO_2, it is an opportunity that includes a number of pragmatic challenges. What follows is by no means inclusive but rather representative of next steps.

5.1. Phenotypic and Genotypic CO_2 Sensitivity Traits

Greater effort is necessary to document and determine those traits that may have already contributed to a greater sensitivity to recent CO_2 increases (e.g., red rice). A consensus is needed to identify those phenological, morphological, and/or physiological characteristics that are associated with CO_2 responsiveness. Initial studies have suggested different organismal levels associated with CO_2 sensitivity including genetic (e.g., carbohydrate regulation of RNA, [41], biochemical (e.g., Rubisco activase) [42], leaf (e.g., stomatal density [43] or photosynthesis [44], whole plant (relative growth rate) [45], management (e.g., planting density) [46] and canopy (e.g., nitrogen applications) [47], but specific organismal characteristics consistently associated with CO_2 responsiveness and crop yield for CWRs have not been identified. A definitive set of these parameters is necessary for breeders to select CO_2-sensitive crop archetypes.

5.2. Introgression of CWR Traits

While CWRs can make an effective contribution to broadening the genetic diversity of crops, their direct use in breeding has primarily focused on introgressing loci for disease resistance, not abiotic stress [48,49]. This is due, in part, to the presence of objectionable traits in CWRs (e.g., transfer of undesirable QTLs) as well as breeding barriers with the crop.

Initiatives have commenced that aim to adapt agricultural sustainability to climate change through the use of CWRs to broaden and improve the cultivated gene pool as part of the International Treaty on Plant Genetic Resources for Food and Agriculture (ITPGRFA, [50]). This effort is designed to collect, preserve, and prepare CWRs for evaluation and potential adaptation of crops to climate change. Other initiatives, such as Diversity Seek (DivSeek) are underway to begin evaluating the potential of crop and wild relative diversity present within gene banks [51].

To overcome difficulties in facilitating gene introgression, Prohens et al. [52] have suggested a novel approach, 'introgressiomics', a mass-scale expansion of plant materials and populations that confer genetic introgressions from CWRs into crops. Their goal is to generate chromosome substitution lines (CSLs), introgression lines (ILs), and multi-parent advanced inter-cross (MAGIC) populations through the use of marker-assisted selection as a means to characterize genetic traits present in CWRs, but to also develop genetically relevant elite materials that can be incorporated into breeding programs as a means to adapt crops to climatic change.

5.3. Additional Climatic Variables

The climatic consequences of increasing CO_2 are obvious and include increases in surface temperature, changes in precipitation and extreme events that will have negative consequence for crop productivity. Temperatures, especially during anthesis, may be critical in maintaining yield performance [53]. In addition, there are numerous studies indicating that rising temperature per se may negate any stimulatory effect for projected CO_2 increases [54–56].

At present, evaluations of CWR to both recent increases in CO_2 and temperature are unavailable; however, it is of interest to note one seminal study that has examined Indica, Indica-like, Japonica, and CWR of rice to projected CO_2 concentrations four temperature treatments [57]. They reported an increased yield sensitivity to high temperature stress at higher (600 ppm) CO_2, but their results also showed that CWR for rice demonstrated superior CO_2 x temperature interactions with respect to yield, supporting the idea of using wild or unadapted gene pools in rice to enhance breeding efforts for climate change adaptation.

In any case, a fundamental challenge in CO_2 selection will be to consider multiple environmental interactions with a merited focus on temperature and moisture conditions to assess possible negative interactions with respect to yield. By necessity, such selections will include evaluations of multiple-gene responses.

5.4. Nutritional Considerations

For projected CO_2 increases, there is considerable evidence indicating that stimulatory effects of increased CO_2 may be accompanied by declines in nutritional quality, including but not limited to protein and minerals [58–61]. The basis for the CO_2-induced changes in crop quality are still being elucidated, in part because there are a number of biological and physical processes that are influenced by increasing CO_2 [62]. However, any efforts to adapt CWRs and cultivars to recent CO_2 increases must include concurrent selection efforts or the co-development of suitable management practices that will maintain the desired quality and nutritional characteristics necessary for human health.

5.5. Scaling Up to the Field Level

Genomic and molecular traits that are assimilated into new cultivars may increase performance at the whole plant or leaf level; however, as emphasized by Sinclair et al. [63], such improvements need to scale up to field responses. As such, the identification of CO_2-induced increases in the photosynthetic rate or seed yield of single leaves or whole plants in the laboratory require complementary management approaches that will illicit similar responses in the field. In this regard, it may be worthwhile to examine historical trends by cultivar. For example, comparisons of yield performance from 1950s cultivars, relative to the same cultivars under today's CO_2 concentration using similar metrics (row spacing, soil types, pesticide usage) could help identify cultivar x CO_2 sensitivity and, potentially, elucidate management practices that could maximize CO_2 yield responses in situ.

6. Conclusions

The challenge of adapting to an uncertain climate is paramount to maintaining global food security. High yielding crop varieties with tolerance to biotic and abiotic stresses associated with climatic change are needed to meet this challenge. Currently, there is recognition that the current genetic base of modern cultivars may be too narrow for crop improvement efforts in that regard. In turn, this may reflect a lack of genetic ideotypes that encompass traits associated with climatic change or uncertainty associated with the rate of change. As such, the use of CWR (or landraces) may be a means of enhancing genetic diversity of cultivated crops as a means to adapt to these changes.

Among adaptation efforts, there is an opportunity to exploit the recent increase in atmospheric CO_2 as a means to stimulate plant growth and yield. Currently, comparisons of cultivated and CWR indicate differential selection in regard to this change, with initial evidence suggesting evolutionary adaptation for some CWR, e.g., red rice. Introgression of these traits into current cultivars, and appropriate management practices, may provide a means to utilize the increase in CO_2 that has already occurred ($\approx 27\%$ increase since 1970) to stimulate seed yield.

At present, there is no empirical evidence that breeders are selecting for CO_2 responsiveness. This may be due, in part, to current economic and agronomic practices associated with modern agriculture. Yet, it seems surprising that CO_2 as a potential adaptation strategy to maintain global food production is not being utilized. One can ask if other abiotic resources such as sunlight, water, or nutrients had increased to a similar extent in recent decades whether incentives to optimize that increase would be underway.

Adaptation to recent CO_2 increases will not be a complete solution to the complications associated with climate change. However, adaptation may represent one of the simplest research strategies to help maintain global food security relative to the anthropogenic stresses associated with climatic change. As such, it is hoped that this review can serve as a sounding board and starting point for additional efforts.

Funding: This research received no external funding.

Institutional Review Board Statement: Not applicable.

Informed Consent Statement: Not applicable.

Data Availability Statement: The article is a review synopsis and does not contain original data.

Acknowledgments: The author thanks James Bunce for his invitation and for his years of mentorship.

Conflicts of Interest: The author declares no conflict of interest.

References

1. Harrison, M.T.; Cullen, B.R.; Rawnsley, R.P. Modelling the sensitivity of agricultural systems to climate change and extreme climatic events. *Agric. Syst.* **2016**, *148*, 135–148. [CrossRef]
2. Petrie, M.D.; Bradford, J.B.; Lauenroth, W.K.; Schlaepfer, D.R.; Andrews, C.M.; Bell, D.M. Non-analog increases to air, surface, and belowground temperature extreme events due to climate change. *Clim. Chang.* **2020**, *158*, 1–24.
3. Deutsch, C.A.; Tewksbury, J.J.; Tigchelaar, M.; Battisti, D.S.; Merrill, S.C.; Huey, R.B.; Naylor, R.L. Increase in crop losses to insect pests in a warming climate. *Science* **2018**, *361*, 916–919. [CrossRef] [PubMed]
4. Ziska, L.H.; Bradley, B.A.; Wallace, R.D.; Bargeron, C.T.; LaForest, J.H.; Choudhury, R.A.; Garrett, K.A.; Vega, F.E. Climate change, carbon dioxide, and pest biology, managing the future: Coffee as a case study. *Agronomy* **2018**, *8*, 152. [CrossRef]
5. Singh, R.; Singh, G.S. Traditional agriculture: A climate-smart approach for sustainable food production. *Energy Ecol. Environ.* **2017**, *2*, 296–316. [CrossRef]
6. Henry, R.J. Innovations in plant genetics adapting agriculture to climate change. *Curr. Opin. Plant Biol.* **2020**, *56*, 168–173. [CrossRef]
7. Javadinejad, S.; Dara, R.; Jafary, F. Analysis and prioritization the effective factors on increasing farmers resilience under climate change and drought. *Agric. Res.* **2020**, *9*, 1–17. [CrossRef]
8. Ruiz-Vera, U.M.; De Souza, A.P.; Ament, M.R.; Gleadow, R.M.; Ort, D.R. High sink strength prevents photosynthetic down-regulation in cassava grown at elevated CO_2 concentration. *J. Exp. Bot.* **2020**. [CrossRef]
9. Hasegawa, T.; Sakai, H.; Tokida, T.; Nakamura, H.; Zhu, C.; Usui, Y.; Yoshimoto, M.; Fukuoka, M.; Wakatsuki, H.; Katayanagi, N.; et al. Rice cultivar responses to elevated CO_2 at two free-air CO_2 enrichment (FACE) sites in Japan. *Funct. Plant Biol.* **2012**, *40*, 148–159. [CrossRef]
10. Bishop, K.A.; Betzelberger, A.M.; Long, S.P.; Ainsworth, E.A. Is there potential to adapt soybean (G lycine max M err.) to future [CO_2]? An analysis of the yield response of 18 genotypes in free-air CO_2 enrichment. *Plant Cell Environ.* **2015**, *38*, 1765–1774. [CrossRef]
11. Bunce, J. Using FACE systems to screen wheat cultivars for yield increases at elevated CO_2. *Agronomy* **2017**, *7*, 20. [CrossRef]
12. Atlin, G.N.; Cairns, J.E.; Das, B. Rapid breeding and varietal replacement are critical to adaptation of cropping systems in the developing world to climate change. *Glob. Food Sec.* **2017**, *12*, 31–37. [CrossRef] [PubMed]

13. Sakai, H.; Tokida, T.; Usui, Y.; Nakamura, H.; Hasegawa, T. Yield responses to elevated CO_2 concentration among Japanese rice cultivars released since 1882. *Plant Prod. Sci.* **2019**, *22*, 352–366. [CrossRef]

14. Manderscheid, R.; Weigel, H.J. Photosynthetic and growth responses of old and modern spring wheat cultivars to atmospheric CO_2 enrichment. *Ag. Ecosys. Environ.* **1997**, *64*, 65–73. [CrossRef]

15. Ziska, L.H.; Morris, C.F.; Goins, E.W. Quantitative and qualitative evaluation of selected wheat varieties released since 1903 to increasing atmospheric carbon dioxide: Can yield sensitivity to carbon dioxide be a factor in wheat performance? *Glob. Chang. Biol.* **2004**, *10*, 1810–1819. [CrossRef]

16. Ziska, L.H.; Blumenthal, D.M. Empirical selection of cultivated oat in response to rising atmospheric carbon dioxide. *Crop Sci.* **2007**, *47*, 1547–1552. [CrossRef]

17. Schmid, I.; Franzaring, J.; Müller, M.; Brohon, N.; Calvo, O.C.; Högy, P.; Fangmeier, A. Effects of CO_2 enrichment and drought on photosynthesis, growth and yield of an old and a modern barley cultivar. *J. Agron. Crop Sci.* **2016**, *202*, 81–95. [CrossRef]

18. Doyle, J. *Altered Harvest: Agriculture, Genetics and the Fate of the World's Food Supply*; Viking Press: New York, NY, USA, 1985.

19. Reich, P.B.; Knops, J.; Tilman, D.; Craine, J.; Ellsworth, D.; Tjoelker, M.; Lee, T.; Wedin, D.; Naeem, S.; Bahauddin, D.; et al. Plant diversity enhances ecosystem responses to elevated CO_2 and nitrogen deposition. *Nature* **2001**, *410*, 809–810. [CrossRef]

20. Mammadov, J.; Buyyarapu, R.; Guttikonda, S.K.; Parliament, K.; Abdurakhmonov, I.Y.; Kumpatla, S.P. Wild relatives of maize, rice, cotton, and soybean: Treasure troves for tolerance to biotic and abiotic stresses. *Front. Plant Sci.* **2018**, *9*, 886. [CrossRef]

21. Zhang, H.; Mittal, N.; Leamy, L.J.; Barazani, O.; Song, B.H. Back into the wild—Apply untapped genetic diversity of wild relatives for crop improvement. *Evol. App.* **2017**, *10*, 5–24. [CrossRef]

22. Zhang, F.; Batley, J. Exploring the application of wild species for crop improvement in a changing climate. *Curr. Opin. Plant Biol.* **2020**, *56*, 218–222. [CrossRef] [PubMed]

23. Novy, A.; Flory, S.L.; Hartman, J.M. Evidence for rapid evolution of phenology in an invasive grass. *J. Evol. Biol.* **2013**, *26*, 443–450. [CrossRef] [PubMed]

24. Colautti, R.I.; Barrett, S.C. Rapid adaptation to climate facilitates range expansion of an invasive plant. *Science* **2013**, *342*, 364–366. [CrossRef] [PubMed]

25. Franks, S.J.; Sim, S.; Weis, A.E. Rapid evolution of flowering time by annual plant in response to a climate fluctuation. *Proc. Natl. Acad. Sci. USA* **2007**, *104*, 1278–1282. [CrossRef] [PubMed]

26. Naylor, J.M.; Jana, S. Genetic adaptation for seed dormancy in Avena fatua. *Can. J. Bot.* **1976**, *54*, 306–312. [CrossRef]

27. Clements, D.R.; DiTommaso, A. Climate change and weed adaptation: Can evolution of invasive plants lead to greater range expansion than forecasted? *Weed Res.* **2011**, *51*, 227–240. [CrossRef]

28. Ravet, K.; Patterson, E.L.; Krähmer, H.; Hamouzová, K.; Fan, L.; Jasieniuk, M.; Lawton-Rauh, A.; Malone, J.M.; McElroy, J.S.; Merotto, A., Jr.; et al. The power and potential of genomics in weed biology and management. *Pest Manag. Sci.* **2018**, *74*, 2216–2225. [CrossRef] [PubMed]

29. Ziska, L.H.; Blumenthal, D.M.; Franks, S.J. Understanding the nexus of rising CO_2, climate change, and evolution in weed biology. *Invasive Plant Sci. Manag.* **2019**, *12*, 79–88. [CrossRef]

30. Bunce, J.A. Are annual plants adapted to the current atmospheric concentration of carbon dioxide? *Int. J. Plant Sci.* **2001**, *162*, 1261–1266. [CrossRef]

31. Ziska, L.H.; McClung, A. Differential response of cultivated and weedy (red) rice to recent and projected increases in atmospheric carbon dioxide. *Agron. J.* **2008**, *100*, 1259–1263. [CrossRef]

32. Ziska, L.H.; Tomecek, M.B.; Gealy, D.R. Competitive interactions between cultivated and red rice as a function of recent and projected increases in atmospheric carbon dioxide. *Agron. J.* **2010**, *102*, 118–123. [CrossRef]

33. Franks, S.J.; Hamann, E.; Weis, A.E. Using the resurrection approach to understand contemporary evolution in changing environments. *Evol. App.* **2018**, *11*, 17–28. [CrossRef] [PubMed]

34. Ziska, L.H. Could recent increases in atmospheric CO_2 have acted as a selection factor in Avena fatua populations? A case study of cultivated and wild oat competition. *Weed Res.* **2017**, *57*, 399–405. [CrossRef]

35. Baker, H.G. The evolution of weeds. *Ann. Rev. Ecol. Sys.* **1974**, *5*, 1–24. [CrossRef]

36. Radosevich, S.R.; Holt, J.S.; Ghersa, C. *Weed Ecology: Implications for Management*; John Wiley & Sons: New York, NY, USA, 1997.

37. Smith, S.A.; Donoghue, M.J. Rates of molecular evolution are linked to life history in flowering plants. *Science* **2008**, *322*, 86–89. [CrossRef]

38. Jacobsen, S.E.; Sørensen, M.; Pedersen, S.M.; Weiner, J. Feeding the world: Genetically modified crops versus agricultural biodiversity. *Agron. Sust. Develop.* **2013**, *33*, 651–662. [CrossRef]

39. Tanksley, S.D.; McCouch, S.R. Seed banks and molecular maps: Unlocking genetic potential from the wild. *Science* **1997**, *277*, 1063–1066. [CrossRef]

40. Treharne, K. The implications of the 'greenhouse effect' for fertilizers and agrochemicals. In *The Greenhouse Effect and UK Agriculture*; Bennet, R.C., Ed.; Ministry of Agriculture, Fisheries and Food: London, UK, 1989; pp. 67–78.

41. Sheen, J. Feedback control of gene expression. *Photosyn. Res.* **1994**, *39*, 427–438. [CrossRef]

42. Sage, R.F.; Coleman, J.R. Effects of low atmospheric CO_2 on plants: More than a thing of the past. *Trends Plant Sci.* **2001**, *6*, 18–24. [CrossRef]

43. Ward, J.K.; Kelly, J.K. Scaling up evolutionary responses to elevated CO_2: Lessons from Arabidopsis. *Ecol. Lett.* **2004**, *7*, 427–440. [CrossRef]

44. Bunce, J.A. Contrasting responses of seed yield to elevated carbon dioxide under field conditions within Phaseolus vulgaris. *Agric. Ecosyst. Environ.* **2008**, *128*, 219–234. [CrossRef]

45. Poorter, H. Do slow-growing species and nutrientstressed plants respond relatively strongly to elevated CO_2? *Glob. Change Biol.* **1998**, *4*, 693–697. [CrossRef]

46. Shimono, H. Rice genotypes that respond strongly to elevated CO_2, also respond strongly to low planting density. *Agric. Ecosyst. Environ.* **2011**, *141*, 240–243. [CrossRef]

47. Ziska, L.H.; Weerakoon, W.; Namuco, O.S.; Pamplona, R. The influence of nitrogen on the elevated CO_2 response in field-grown rice. *Austral. J. Plant Physiol.* **1996**, *23*, 45–52. [CrossRef]

48. Hajjar, R.; Hodgkin, T. The use of wild relatives in crop improvement: A survey of developments over the last 20 years. *Euphytica* **2007**, *156*, 1–13. [CrossRef]

49. Maxted, N.; Kell, S.P. Establishment of a global network for the in situ conservation of crop wild relatives: Status and needs. *FAO Comm. Genet. Resour. Food Agric. Rome Italy* **2009**, *266*, 3–94.

50. Dempewolf, H.; Eastwood, R.J.; Guarino, L.; Khoury, C.K.; Müller, J.V.; Toll, J. Adapting agriculture to climate change: A global initiative to collect, conserve, and use crop wild relatives. *Agroecol. Sustain. Food Syst.* **2014**, *38*, 369–377. [CrossRef]

51. Meyer, R.S. Encouraging metadata curation in the Diversity Seek initiative. *Nat. Plants* **2015**, *1*, 1–2. [CrossRef]

52. Prohens, J.; Gramazio, P.; Plazas, M.; Dempewolf, H.; Kilian, B.; Díez, M.J.; Fita, A.; Herraiz, F.J.; Rodríguez-Burruezo, A.; Soler, S.; et al. Introgressiomics: A new approach for using crop wild relatives in breeding for adaptation to climate change. *Euphytica* **2017**, *213*, 158–166. [CrossRef]

53. Schlenker, W.; Roberts, M.J. Nonlinear temperature effects indicate severe damages to US crop yields under climate change. *Proc. Natl. Acad. Sci. USA* **2009**, *106*, 15594–15598. [CrossRef]

54. Chavan, S.G.; Duursma, R.A.; Tausz, M.; Ghannoum, O. Elevated CO_2 alleviates the negative impact of heat stress on wheat physiology but not on grain yield. *J. Exp. Bot.* **2019**, *70*, 6447–6459. [CrossRef] [PubMed]

55. Wang, W.; Cai, C.; He, J.; Gu, J.; Zhu, G.; Zhang, W.; Zhu, J.; Liu, G. Yield, dry matter distribution and photosynthetic characteristics of rice under elevated CO_2 and increased temperature conditions. *Field Crops Res.* **2020**, *248*, 107605. [CrossRef]

56. Wang, J.; Wang, C.; Chen, N.; Xiong, Z.; Wolfe, D.; Zou, J. Response of rice production to elevated [CO_2] and its interaction with rising temperature or nitrogen supply: A meta-analysis. *Clim. Chang.* **2015**, *130*, 529–543. [CrossRef]

57. Wang, D.R.; Bunce, J.A.; Tomecek, M.B.; Gealy, D.; McClung, A.; McCouch, S.R.; Ziska, L.H. Evidence for divergence of response in Indica, Japonica, and wild rice to high CO $2\times$ temperature interaction. *Glob. Chang. Biol.* **2016**, *22*, 2620–2632. [CrossRef]

58. Cotrufo, M.F.; Ineson, P.; Scott, A. Elevated CO_2 reduces the nitrogen concentration of plant tissues. *Glob. Chang. Biol.* **1998**, *4*, 43–54. [CrossRef]

59. Loladze, I. Hidden shift of the ionome of plants exposed to elevated CO_2 depletes minerals at the base of human nutrition. *Elife* **2014**, *3*, e02245. [CrossRef]

60. Loladze, I. Rising atmospheric CO_2 and human nutrition: Toward globally imbalanced plant stoichiometry? *Trends Ecol. Evol.* **2002**, *17*, 457–461. [CrossRef]

61. Myers, S.S.; Zanobetti, A.; Kloog, I.; Huybers, P.; Leakey, A.D.; Bloom, A.J.; Carlisle, E.; Dietterich, L.H.; Fitzgerald, G.; Hasegawa, T.; et al. Increasing CO_2 threatens human nutrition. *Nature* **2014**, *510*, 139–142. [CrossRef]

62. McGrath, J.M.; Lobell, D.B. Reduction of transpiration and altered nutrient allocation contribute to nutrient decline of crops grown in elevated CO_2 concentrations. *Plant Cell Environ.* **2013**, *36*, 697–705. [CrossRef]

63. Sinclair, T.R.; Purcell, L.C.; Sneller, C.H. Crop transformation and the challenge to increase yield potential. *Trends Plant Sci.* **2004**, *9*, 70–75. [CrossRef]

MDPI
St. Alban-Anlage 66
4052 Basel
Switzerland
Tel. +41 61 683 77 34
Fax +41 61 302 89 18
www.mdpi.com

Plants Editorial Office
E-mail: plants@mdpi.com
www.mdpi.com/journal/plants